# COMPUTER FUNDAMENTALS FOR CHEMISTS

COMPUTERS IN
CHEMISTRY AND INSTRUMENTATION

*edited by*

JAMES S. MATTSON    HARRY B. MARK, JR.    HUBERT C. MACDONALD, JR.

*Volume 1:* **Computer Fundamentals for Chemists**

*Volume 2:* **Electrochemistry: Calculations, Simulation, and Instrumentation**

*Volume 3:* **Spectroscopy and Kinetics**

*Volume 4:* **Computer Assisted Instruction in Chemistry: Introduction and Fundamentals** (IN PREPARATION)

*Volume 5:* **Computer Assisted Instruction in Chemistry: Applications** (IN PREPARATION)

# COMPUTER FUNDAMENTALS FOR CHEMISTS

*Edited by*

*James S. Mattson*
DIVISION OF CHEMICAL OCEANOGRAPHY
ROSENSTIEL SCHOOL OF MARINE
AND ATMOSPHERIC SCIENCES
UNIVERSITY OF MIAMI
MIAMI, FLORIDA

*Harry B. Mark, Jr.*
DEPARTMENT OF CHEMISTRY
UNIVERSITY OF CINCINNATI
CINCINNATI, OHIO

*Hubert C. MacDonald, Jr.*
KOPPERS COMPANY, INC.
MONROEVILLE, PENNSYLVANIA

MARCEL DEKKER, INC. New York

MARCEL DEKKER, INC.
270 Madison Avenue, New York, New York 10016

LIBRARY OF CONGRESS CATALOG CARD NUMBER: 72-91432
ISBN: 0-8247-1432-6

Current printing (last digit):

10 9 8 9 7 6 5 4 3 2

Printed in the United States of America

## INTRODUCTION TO THE SERIES

In the past decade, computer technology and design (both analog and digital) and the development of low cost linear and digital "integrated circuitry" have advanced at an almost unbelievable rate. Thus, computers and quantitative electronic circuitry are now readily available to chemists, physicists, and other scientific groups interested in instrument design. To quote a recent statement of a colleague, "the computer and integrated circuitry are revolutionizing measurement and instrumentation in science." In general, the chemist is just beginning to realize and understand the potential of computer applications to chemical research and quantitative measurement. The basic applications are in the areas of data acquisition and reduction, simulation, and instrumentation (on-line data processing and experimental control in and/or optimization in real time).

At present, a serious time lag exists between the development of electronic computer technology and the practice or application in the physical sciences. Thus, this series aims to bridge this communication gap by presenting comprehensive and instructive chapters on various aspects of the field written by outstanding researchers. By this means, the experience and expertise of these scientists is made available for study and discussion.

It is intended that these volumes will contain articles covering a wide variety of topics written for the nonspecialist but still retaining a scholarly level of treatment. As the series was conceived it was hoped that each volume (with the exception of Volume 1 which is an introductory discussion of basic principles and applications) would be devoted to one subject; for example, electrochemistry, spectroscopy, on-line analytical service systems. This format will be followed wherever possible. It soon became evident, however, that to delay publication of completed manuscripts while waiting to obtain a volume dealing with a single subject would be unfair to not only the authors but, more important, the intended audience. Thus, priority has been given to speed of publication lest the material become dated while awaiting publication. Therefore, some volumes will contain mixed topics.

The editors have also decided that submitted as well as the usual invited contributions will be published in the series. Thus, scientists who have recent developments and advances of potential interest should submit detailed outlines of their proposed contribution to one of the editors for consideration concerning suitability for publication. The articles should be imaginative, critical, and comprehensive survey topics in the field and/or other fields and which are written on a high level, that is, satisfying to specialists and nonspecialists alike. Parts of programs can be used in the text to illustrate special procedures and concepts, but, in general, we do not plan to reproduce complete programs themselves, as much of this material is either routine or represents a particular personality of either the author or his computer.

# PREFACE

This volume of the series is intended to present, in a simple and concise manner, the basic general principles and theories that are necessary for the chemist to read, digest, and apply the more specialized material discussed in the volumes on specific applications to computation, data reduction, simulation, and instrumentation.

Chapter 1 is a general introduction to the subject of computer applications to chemistry and instrumentation. Chapter 2 and Chapter 3 present the basic principles of analog and digital logic circuitry, respectively. Chapter 4 discusses the various computer languages and their applicability. Chapter 5 illustrates the application of the computer to the simulation of physical systems. The application of the digital computer to simulate analog logic is given in Chapter 6. Chapter 7 is a brief discussion of the basic principles of hybrid (analog-digital) computer systems as applied to on-line experimental operations (a very comprehensive chapter on this subject will appear in Volume 4). Chapter 8 discusses the important application of the computer as a learning machine.

The editors wish to acknowledge with thanks the efforts of Bonnie Koran, who produced most of the line drawings for the figures in this volume. We also acknowledge the help of many of our colleagues who have contributed helpful comments concerning this volume.

## LIST OF CONTRIBUTORS

T. L. ISENHOUR, Department of Chemistry, University of North Carolina, Chapel Hill, North Carolina

P. C. JURS, Department of Chemistry, The Pennsylvania State University, University Park, Pennsylvania

JOHN J. KOZAK, Kings College, Wilkes-Barre, Pennsylvania

VINCENT A. LoDATO, The RAND Corporation, Santa Monica, California

HARRY B. MARK, Jr., Department of Chemistry, University of Cincinnati, Cincinnati, Ohio

D. K. MEANS, Research and Development Center for Electronics, Reliance Electric Company, Ann Arbor, Michigan

RICHARD D. SACKS, Department of Chemistry, University of Michigan, Ann Arbor, Michigan

J. G. SELLERS, Imperial Chemical Industries Limited, Corporate Laboratory, The Heath, Runcorn, Cheshire, England

CLARENCE H. THOMAS, Department of Chemistry, University of Cincinnati, Cincinnati, Ohio

CONTENTS

# COMPUTER FUNDAMENTALS FOR CHEMISTS

# Chapter 1

# INTRODUCTION TO COMPUTERS

John J. Kozak

Kings College
Wilkes-Barre, Pennsylvania



## I. INTRODUCTION

The electronic computer is one of the most useful calculation tools available to chemists today. The simplification of data handling and the ability to perform very complicated calculations with relative ease make the computer almost a necessity in modern chemical practice. In fact, many undergraduate and graduate chemistry programs are substituting the knowledge of a computer language for the traditional foreign language requirement.

The computer language is the medium through which the chemist communicates with the computer. The computer does not have the ability to think or to create ideas. It can only perform mathematical and logical operations which it has been programmed to perform by an operator. It can, however, perform these operations with amazing speed. Before being able to use the computer successfully the chemist must be able to write his commands to the computer in a language that can be "understood" by the computer or that can be translated into such a language. A discussion of some of the languages available to be used for this purpose will be given later.

There are two types of electronic computers available for use, analog and digital computers. An analog computer can represent continuous functions by measuring an amplitude, for example, a length, a voltage, or an intensity. A slide rule would be an example of an analog computer. It can be used to mechanically perform operations on a continuous set of numbers. This introduction, however, is not primarily concerned with analog computers (the basics of which are discussed in Chapter 2 of this volume) but with the use of digital computers.

The digital computer can represent only a finite number of discrete characters. These characters may be in the form of numbers such as 12 or 1.2 or alphabetic information such as the letter A or the word STOP. In addition to these the digital computer can represent special characters for mathematical operations such as multiplication (*), division (/), addition (+), and subtraction (-), among others. The abacus would be an example of a mechanical digital computer. It can be used to perform operations on a finite set of numbers represented by the possible arrangements of a set of beads.

## II. THE COMPONENTS OF A DIGITAL COMPUTER

The basic components of a modern digital computer are represented in Fig. 1. The control unit and the arithmetic-logical unit together are

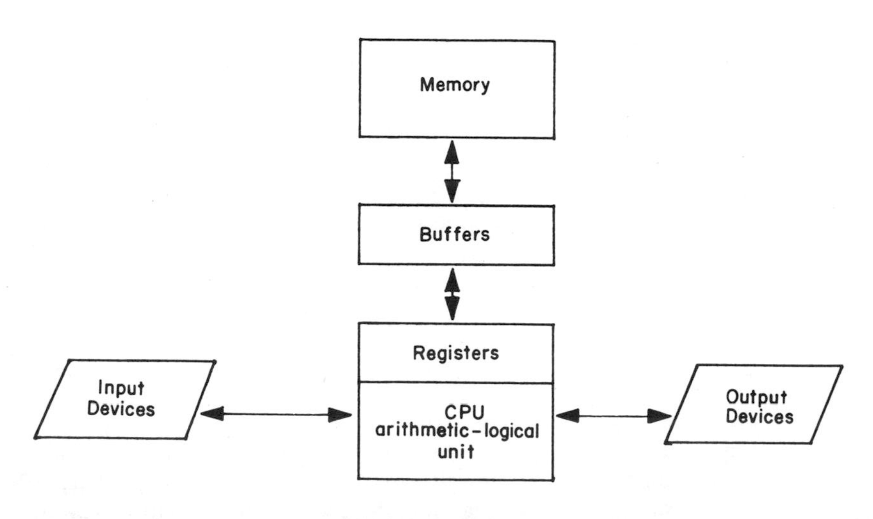


Fig. 1

referred to as the central processing unit or CPU. The CPU is the basic control device of the modern digital computer. The CPU directs and coordinates the operations of all the other components. It accesses and interprets instructions from the input devices along with data, directs the operation of arithmetic and logical functions on the data as required by the instructions, and transfers the results to some output device. The CPU can also direct data and instructions to and from the storage area.

The storage or memory core of the computer is a device capable of receiving information from an external input device or the CPU, holding that information, and making it available for retrieval by the CPU or for transference to some output device. The internal storage of the computer consists of a three-dimensional array of magnetic cores in the shape of a toroid. The magnetic core element is made of ferromagnetic materials such as iron oxide and arranged near electrical conductors in such a way that they can be magnetically polarized in one of two states as depicted in Fig. 2. Current flowing in one direction near the core element sets up a

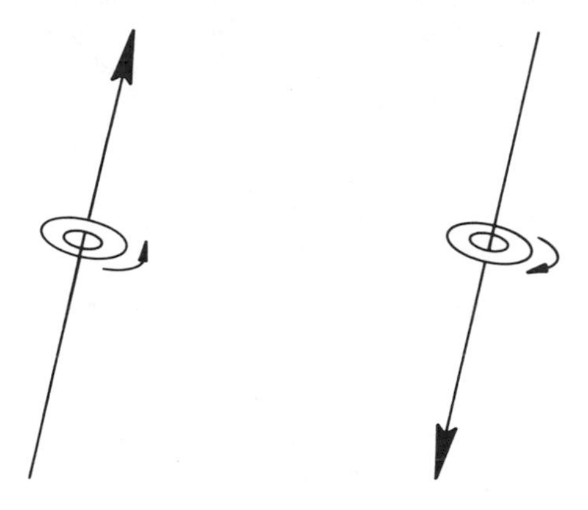

Fig. 2

clockwise magnetic field in the toroid and current flowing in the opposite direction will set up a counterclockwise magnetic field. Since these elements can exist in one of only two possible states they are called bistable. Consequently, data and instructions are most conveniently represented internally by the computer in the binary number system. Clockwise polarization will represent the binary digit 1 and counterclockwise polarization will represent the binary digit 0. Each individual magnetic core element is called a bit and combination of 1 and 0 bits can be used to represent various units such as a numerical digit, a letter, or a word. The number of bits required to represent one character is called a byte and varies with each computer. A series of bytes arranged in sequence to represent a piece of data or an instruction is called a word. The word is the basic unit of storage in the computer and the various sizes of computers usually refers to the number of words it can accommodate in storage.

The operations carried out by the CPU are usually accomplished in an extremely short period of time. The multiplication of two numbers carried out internally within the computer may require only a microsecond or less in some computers. It requires a much longer period of time, called access time, to transfer information to and from storage making it available for the CPU. Because of the great differences in times required for these operations, the CPU is nonoperational for part of the time as it waits for information to be transferred. In order to minimize this delay caused by access time temporary storage areas called buffers are used. Blocks of information are transferred from storage to the buffers making them readily available to the CPU for processing in sequence.

In addition to these buffers, there are short-term storage units called registers that are utilized in the operations of the CPU. Words or groups of words are transferred from the buffer units to the register so that the information contained in the words may be processed. A register may contain an instruction which will be executed by the computer. Then, the next instruction in sequence will be located in memory storage, placed in the register thereby erasing the first instruction, and the designated operation will be performed. The registers can also contain numbers or other data to be acted upon. Numbers in one register can be added to, subtracted from, multiplied, or divided by numbers in other registers. Numbers in a register can also be interacted with numbers in core storage in the same way. The result of using temporary and short-term storage areas in the computer operations is an increase in efficiency by making information more readily available to the CPU thereby lowering the access time of the computer.

The input and output devices are used to transmit information to and from the computer. One very common input device is the card reader. Before utilizing this device, information that is to be read into the computer is coded onto a punched card. The punched card contains 80 columns each of which can contain a character of information. Each character is represented by a combination of punched holes in that column. The card reader is a device that detects the punched holes in each card and converts each symbol into a form that can be placed in storage.

Magnetic tape is another frequently used input medium which can be read by a magnetic tape-drive device. Information on magnetic tape can be read into the computer more quickly than with punched cards, thereby increasing the efficiency of operation. In order to maximize the efficiency in this respect, however, magnetic tapes are usually prepared by translating information from a set of punched cards onto the magnetic tape using a second, smaller computer as a preprocessor to perform this function.

Another common input device is the electric typewriter which is connected to the computer via electric lines. The operator, sitting at the electric typewriter, can input his instructions to the computer at which time they will be translated into the machine language and operated upon as defined. The typewriter terminal is one end of a two-way communication path to the computer so that the output data requested by the operator can be typed directly onto the terminal by the computer as it processes the instructions. In most cases the operator can interrupt the processing and make corrections to his instructions, if necessary. This "editing" ability is a great advantage in the use of a typewriter input device but a major disadvantage to its use is the fact that it is a much slower type of input device than either the punched-card reader or magnetic tape drive. It is most applicable, therefore, to the solution of small problems and problems which demand a great deal of interaction between the operator and the computer.

Another less frequently used input medium is punched paper tape that can be used with a paper tape reader, a similar process to the punched card reader.

All of the above devices can also be used as output devices with the computer. That is, the results of the computer operations can be punched onto a set of cards using a card-punch or read onto a magnetic tape with a tape-drive device or printed on a typewriter terminal, as indicated, or punched onto paper tape. The most commonly used output device, however, is the high speed printer which is capable of printing many lines of information per minute. There are two types of printers that are usually encountered in a computer system. One type is an <u>on-line printer</u> which is connected directly to the computer. In this type of system printed results are generated directly from internal memory. The high speed printer can generate 600 lines per minute of printed information containing both alphabetic and numerical characters or up to 1285 lines per minute containing numbers only. An <u>off-line</u> printer processes punched cards or magnetic tapes or punched cards which have been produced by the computer and transported to the printer.

## III. PROGRAMMING LANGUAGES

A computer has the electronic circuitry which makes it capable of storing information as a series of electromagnetic signals. It can also operate on these signals, for example, by adding two series each of which represents a number to produce a third resultant series which represents the sum of the two numbers. Before this operation can be accomplished, however, a human operator must have given the computer an instruction or set of instructions telling it to add two numbers, defining the location of these two numbers in storage and defining the location for the storage of the resulting sum. An important fact illustrated by the preceding statement is that the computer must be instructed in detail on every phase of an operation. When directing a human operator one can usually assume that he or she has a certain amount of background knowledge necessary to solve a problem. If the sum of all integers between 1 and 10 is to be found, you need not say:

1. Add the number 1 to the number 2.
2. Add the sum of the first step to the number 3.
3. Add the sum of the second step to the number 4.

and so on. But, it is exactly this type of precision and detail that is required when instructing a computer.

The series of instructions to the computer written to solve a problem or produce a desired result is called a computer program. The person who writes the program is usually called a programmer. The program must be written in a coded form that can be recognized and interpreted by the computer. A computer language is a system of symbols that can be interpreted by the computer and used to communicate with it. Some languages are designed specifically for a particular computer while others are in general use and computer independent. Computer languages can be divided into three general types: (1) machine languages, (2) symbolic languages, and (3) problem-oriented languages. The three categories actually represent three different levels of programming. The machine languages are oriented completely to the needs of the computer. Symbolic languages represent a system partially machine-oriented and partially operator-oriented. Problem-oriented languages are designed with the human operator in mind and have characteristics of conversional and mathematical statements.

## A. Machine Languages

Machine languages are designed with a specific computer in mind. Because of the bistable nature of the storage unit of computers, all instructions are represented internally in the computer in binary form. Different series of 0 and 1 bits represent different instructions or characters of information. Ultimately, all instructions in any other symbolic form must be reduced to binary notation before they can be executed by the computer. This function is accomplished by processing programs. These are previously written machine language programs which can translate instructions written in a symbolic or problem-oriented form into binary notation.

The binary code, then, is the fundamental language of the computer. However, in order to simplify operator requirements other digital notations such as octal or hexadecimal may be used as machine languages. These notations are also reduced to binary form by a processor prior to execution.

Computer programs written in any language other than machine language are referred to as source programs. The programs written in these languages by a programmer activate a processor to translate the instructions into machine language form. The machine language version is called the object program and is executed by the computer.

A major difficulty encountered in machine-language programming is keeping track of all the instructions and data contained in the program. Each storage location in the memory of the computer that can accommodate a word has a fixed reference number called an address associated with it. The instruction and data that are part of a computer program are stored in the memory of the computer and can only be accessed by reference to the address or storage location of the instruction or data. Except for the simplest programs it becomes a monumental task to keep track of all of the addresses containing information that are to be utilized. Fortunately, the processors that have been developed to translate symbolic and problem-oriented languages usually have the capability to automatically keep track of these addresses where it stores data and instructions.

## B. Symbolic Languages

Symbolic languages consist of a collection of mnemonic symbols that replace the binary or digital code of the machine language. For example, addition in machine language code may be represented by the binary notation 1100. In a symbolic language addition may be represented by the symbol A or ADD. A symbolic code is more convenient to the operator thereby facilitating the writing of complex programs.

The program that processes instructions written in a symbolic language and converts them into machine-compatible form is called an Assembler. Programs written in a symbolic language are usually referred to as assembly language programs.

In general, an instruction in an assembly language program must specify three things:

(1) the operation to be performed, sometimes called the operation code or "op code,"
(2) the operands, that is, the registers or storage locations or input/output devices that are to be used in the operation,
(3) the location of the next instruction to be executed.

The following illustrates symbolic language coding to find the product of two numbers, X and Y:

```
L    2, X
M    2, Y.
```

The first statement says to load the number with the symbol X into register 2. The second statement says to multiply the number with the symbol Y by the contents of register 2. Unless otherwise specified, the result will be automatically stored in register 2 erasing the value that was there previously.

Instructions are normally stored in sequence in memory. This means that the next instruction to be located and executed is understood to be the next in sequence unless there is a specific instruction directing otherwise.

## C. Problem-Oriented Languages

Problem-oriented languages are, for the most part, machine independent. They are designed with the operator in mind and usually utilize key words in English as codes for the different operations. The same program in a problem-oriented language can often be used on different computer models with little or no alterations.

The processing program that translates the source program written in a problem-oriented language into machine language form is called a compiler. The entire translating process is called compilation. The compiler decodes the key-words in the program converting them to symbolic or binary form and keeps track of where it stores all the data and instructions. A very useful built-in feature of most compilers is that it checks the program to find any mistakes that might have been made violating the rules of the language and gives a listing of these mistakes.

Two of the most common problem-oriented languages are FORTRAN and ALGOL. Both of these languages are mathematically oriented and suitable for the solving of scientific problems. They utilize symbols and statements that are algebralike in nature and easily learned and recognized by the potential programmer. COBOL is another common problem-oriented language which is business-oriented and ideal for business-type file processing. PL/1, which stands for programming language, version 1, is a relatively new language that incorporates many of the features of FORTRAN, ALGOL, and COBOL.

In addition to the above-mentioned languages there are many other problem-oriented languages designed for very specific applications. The FORTRAN language, however, is most extensively used for scientific applications and a detailed study of FORTRAN and FORTRAN program will be undertaken in Section IV. A detailed discussion of computer languages is given in Chapter 4 of this volume.

## Flow-Charting

The importance of using unambiguous instructions to the computer has been emphasized in the preceding section. The program must be presented to the computer as a series of related instructions. The best way to organize these instructions for a complex problem is through the use of flow charts. A flow chart is a schematic representation of the logical steps in a problem. Standard symbols are used to represent various processes leading to the solution. The steps outlined in the flow chart are then transcribed into statements which can be entered into the computer. The basic symbols that can be used to construct most flow charts are pictured in Fig. 3.

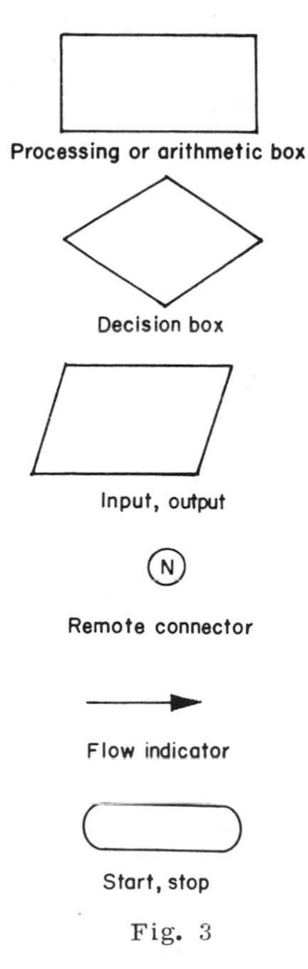


Fig. 3

Arithmetic calculations are indicated by the rectangular box. The diamond-shaped box indicates a decision. Input or output are indicated by a parallelogram. The remote connector allows transferring to another part of the flow chart without using lengthy flow indicators. The start or stop symbol indicates the first and last step in a program. Normally there is more than one path exiting from a decision box. The correct exit path depending upon the result of the decision is indicated on the arrows outlining the direction of flow.

As an example of flow-charting, consider the solution of a quadratic equation by use of the quadratic formula. For an equation $ax^2 + bx + c = 0$,

$$x = \frac{-b \pm \sqrt{b^2 - 4ac}}{2a}$$

There are two roots to the equation to be calculated. However, when the quantity $b^2 - 4ac$ is negative, the roots are complex and the program should be terminated. Only real roots will be calculated. A flow chart outlining the method of solution is given in Fig. 4.

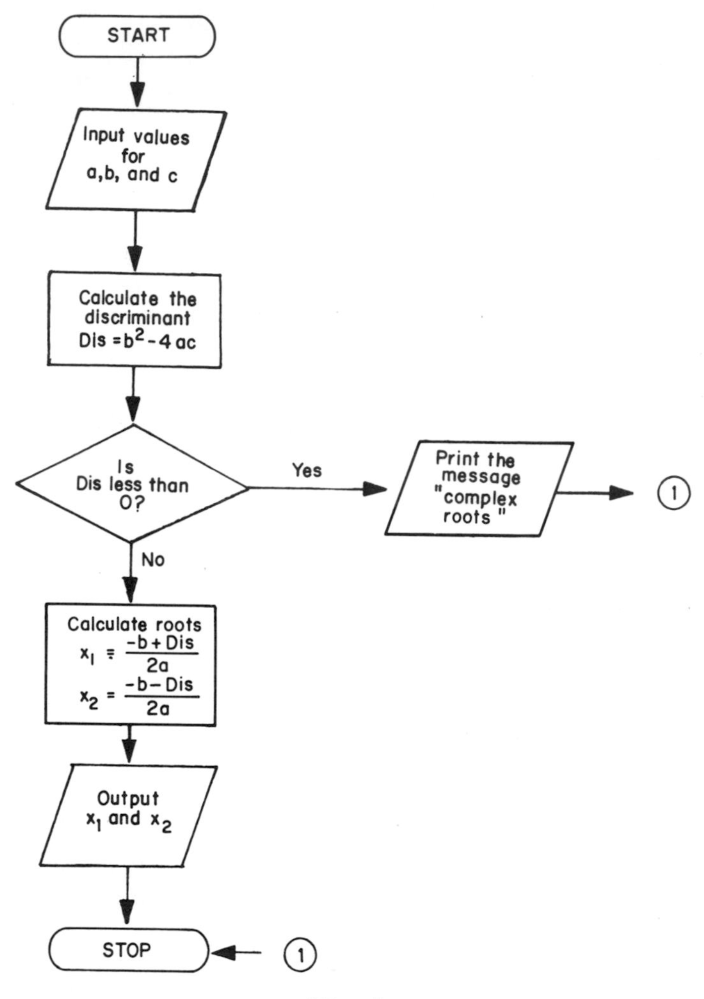


Fig. 4

The first step is to read in values for a, b, and c. The discriminant is calculated in step two. If the discriminant is negative, the message "complex roots" is printed and the program is terminated. If the discriminant is greater than or equal to zero, the roots are calculated and printed before the program is terminated.

Once this flow chart has been developed, the steps in the program can be translated into the computer language form.

## IV. FORTRAN PROGRAMMING

### A. Fundamentals of FORTRAN

FORTRAN, an acronym for FORmula TRANslation, is a problem-oriented language designed for use in science and mathematics. There are many different levels or versions of FORTRAN, the most commonly used versions being FORTRAN II and FORTRAN IV. In general, each higher version of FORTRAN incorporates all of the features of the lower versions with some additional features or refinements. There are general rules that govern the use of FORTRAN programs making it possible to execute the same FORTRAN programs on different computers with little or no alteration. This is not strictly true, however, and some specific rules for writing FORTRAN programs will vary from computer to computer depending on the nature of the FORTRAN compiler available with any given computer. There are usually syntax rules or there will be differences in the features incorporated in various FORTRAN compilers. This work will deal with the version of FORTRAN known as FORTRAN IV since this is the most widely used version today. General rules for writing programs in the FORTRAN IV language will be given and discussed. Keep in mind, however, that there will be slight variations in these rules from computer to computer and that these system-oriented rules must be learned before any programming is attempted.

### B. FORTRAN Symbols

There are 47 characters or symbols available for use in the FORTRAN language. These include the 26 capital letters and the ten digits from 0 to 9. These 36 symbols are referred to as alphanumeric or more simply alphameric characters. In addition to these characters, there are 11 special characters available. The list of special characters is given in Table 1.

In some versions of FORTRAN the dollar sign ($) is not included in the character set. Most of the special characters have the same meaning that they have in ordinary mathematical or verbal usage. There are some differences, however. The plus sign represents addition and the minus sign represents subtraction while the equals sign indicates equality as in the example, 2 + 2 - 1 = 3. The asterisk is the symbol for multiplication and the slash is the symbol for division as in the example, 4 * 3/2 = 6. The left and right parentheses are used to enclose expressions meant to be separated from other symbols. The comma and period are punctuation characters that are used to clarify expressions. The period is also the decimal point in the number system. The blank which will be symbolized by a lower case b is a legal character and is obtained by depressing the space bar on a typewriter or key punch. When a FORTRAN program is compiled, blanks are usually ignored. The number of blanks used in a FORTRAN program, then, has no

TABLE 1

Special Characters

| Name of character | Symbol |
|---|---|
| Plus sign | + |
| Minus sign | - |
| Asterisk | * |
| Slash | / |
| Equals sign | = |
| Left parenthesis | ( |
| Right parenthesis | ) |
| Comma | , |
| Period | . |
| Blank | |
| Dollar sign | $ |

effect on the results. They are used to make an expression more readable by separating words and symbols. There are some instances, though, that will be discussed later where the blank is required in some FORTRAN expressions and it does have meaning. The dollar sign is a special character used in most higher versions of FORTRAN. It usually signifies that the command that follows it is a special command to be executed.

## C. Coding FORTRAN Instructions

In order for the computer to accomplish a given objective it must follow a set of coded instructions. Each instruction or sentence is called a FORTRAN statement. The FORTRAN statement may be written on a punched card or typed on a typewriter terminal, but in either case it must follow a definite form. The card used for punching a FORTRAN statement is seen in Fig. 5. It is 80 columns long and each column can contain only one character. The same limitation is present when typing FORTRAN statements on a terminal. That is, only the first 80 columns may be used. The columns are grouped into various sections designed for specific functions. Columns 1-5 may contain an identifying number for each FORTRAN statement. The computer will follow each statement in order. However, we can specify that it should skip forward or backward to another instruction, if needed. In this case, that instruction must have an identifying number that can be referenced. The statement number can be anything from 1 to 99999, and the statement numbers do not have to increase in the same order as their position in the program. The statement with a statement number of 10 can come before

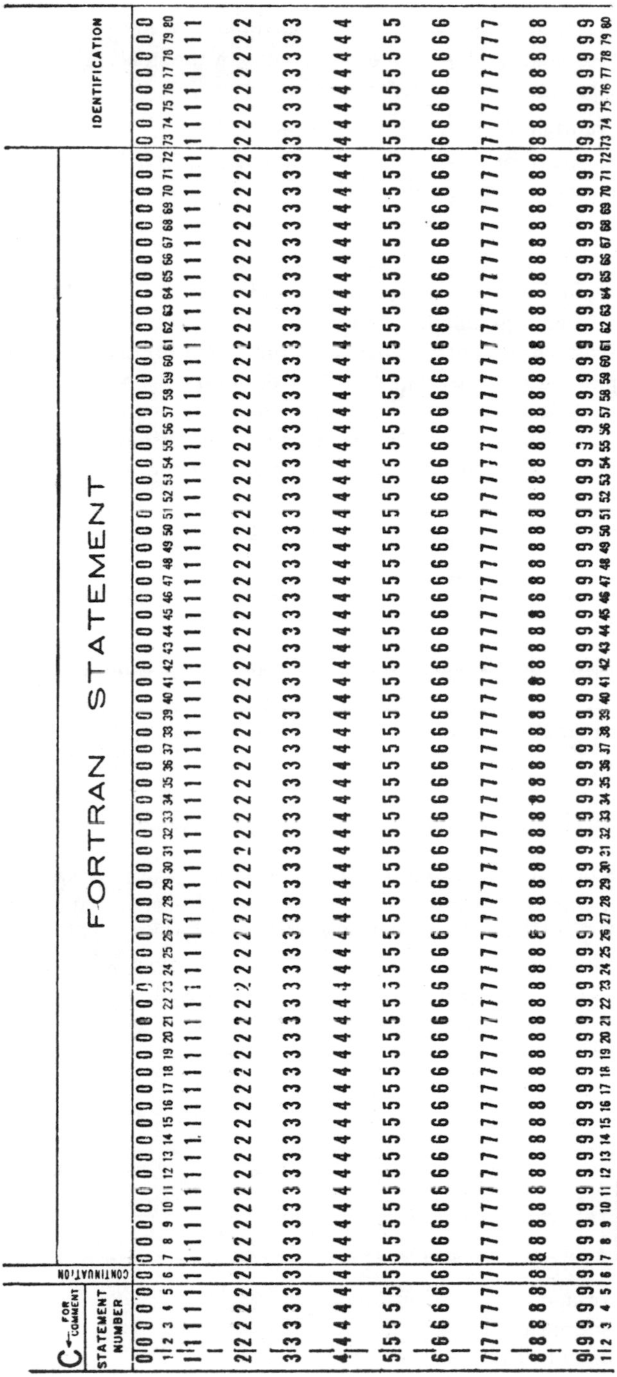

C ← FOR COMMENT
STATEMENT NUMBER
CONTINUATION
FORTRAN STATEMENT
IDENTIFICATION
00000 0 000000000000000000000000000000000000000000000000000000000000000000 00000000
1 2 3 4 5 6 7 8 9 10 11 12 13 14 15 16 17 18 19 20 21 22 23 24 25 26 27 28 29 30 31 32 33 34 35 36 37 38 39 40 41 42 43 44 45 46 47 48 49 50 51 52 53 54 55 56 57 58 59 60 61 62 63 64 65 66 67 68 69 70 71 72 73 74 75 76 77 78 79 80
11111 1 111111111111111111111111111111111111111111111111111111111111111111 11111111
22222 2 222222222222222222222222222222222222222222222222222222222222222222 22222222
33333 3 333333333333333333333333333333333333333333333333333333333333333333 33333333
44444 4 444444444444444444444444444444444444444444444444444444444444444444 44444444
55555 5 555555555555555555555555555555555555555555555555555555555555555555 55555555
66666 6 666666666666666666666666666666666666666666666666666666666666666666 66666666
77777 7 777777777777777777777777777777777777777777777777777777777777777777 77777777
88888 8 888888888888888888888888888888888888888888888888888888888888888888 88888888
99999 9 999999999999999999999999999999999999999999999999999999999999999999 99999999
1 2 3 4 5 6 7 8 9 10 11 12 13 14 15 16 17 18 19 20 21 22 23 24 25 26 27 28 29 30 31 32 33 34 35 36 37 38 39 40 41 42 43 44 45 46 47 48 49 50 51 52 53 54 55 56 57 58 59 60 61 62 63 64 65 66 67 68 69 70 71 72 73 74 75 76 77 78 79 80
IBM 888157

Fig. 5

the one with a statement number of 9 since these are only names used to refer to the statements. They have no effect, in themselves, on the order in which the list of statements is executed.

Column 1 is also used to signify a comment card. Any card with a C in column 1 is not considered a source statement and is not compiled. Comment cards have no effect on the program. They are included in the printed listing of the program and can be used to explain part or all of the program to anyone who uses it.

Column 6 is used to contain a continuation code if a FORTRAN statement is too long to fit onto one card. We may continue a statement on the following card or cards but we must indicate to the compiler that is still the same statement by placing a digit from 1 to 9 in column 6. The number of continuation cards that can be used for any one statement varies from compiler to compiler. The continuation cards cannot have a statement number since any statement must be referenced at the beginning of the statement.

Columns 73-80 are actually ignored by the compiler and may be used for any purpose by the program. Statements may be numbered in order or the name of the programmer may be entered here for identification purposes or they may be left blank.

The FORTRAN statement can be written only in columns 7 to 72 on any given card. Any part of the statement outside of this range will either be ignored by the compiler or treated as something else and lead to an error in either case.

## D. Numbers and Variables in FORTRAN

There are two kinds of numbers used in FORTRAN: "fixed-point" or integer numbers and "floating-point" or real numbers. An integer number or integer constant is a series of digits without a decimal point. It is called a fixed-point constant because the decimal point is always implied to the right of the last digit but never written. Integer constants may have a + or - sign but no other special characters or letters are allowed. Some correct integer constants are

1
+37
-265
0
312

Some incorrect integer constants are

3, 2 (comma)
12.75 (decimal point)
4 *37 (special character)

The length of an integer constant allowed depends on the level of FORTRAN being used and the type of computer. The IBM 360, for example, has a maximum number of five digits in an integer constant. Integers are used whenever only whole numbers are needed. If some quantity is to be incremented by a whole number in the program an integer constant would be used.

When performing arithmetic operations using integers you must keep in mind the maximum allowable size for integers and the fact that they can only be whole numbers. For example 200 * 500 = 100000. But if the maximum number of digits allowed is five, only the five right-most digits will be stored as an answer of 00000. Also, 10/4 = 2.5. However, the numbers to the right of the decimal point will be dropped and 2 will be stored as the number.

Real numbers or "floating-point" numbers have a decimal point. They are used in calculations that require numbers with a fraction. As with integer constants, they may or may not have a sign, but no other special characters are allowed. Some examples are

1.
1.0
-17.6
0.015
37.321

The maximum length of real constants is also limited by the level of FORTRAN and type of computer in use.

Real numbers can also be written with exponents in a manner analogous to scientific notation. The number 35 can be written as $3.5 \times 10^1$. Numbers in this form can be written in FORTRAN using the letter E to represent a number multiplied by 10 raised to a certain power. This type of number consists of a string of digits with a decimal point in any desired position followed by the letter E followed by a one or two digit number. The first part of the number is the number to be multiplied by 10 raised to a certain power. This part of the number is called the precision part. The one or two digit number following the E is the power to which 10 is raised. Both the precision part and the power of 10 may or may not have a plus or minus sign. The maximum length of exponential numbers is also determined by the particular system in use.

Some examples of real numbers and their exponential equivalents are

357.0 = 3.57E2 or 3.57E+2 or 35.7E+1
0.0031 = 3.1E-3 or 31.0E-4
0.000000000001 = 1.0E-12

The decimal point may be placed anywhere in the precision part with the proper exponent. Unless otherwise directed, however, the computer will print out a number in exponential form with the decimal point to the left of the first digit.

Complex numbers can also be expressed in FORTRAN by writing the real and imaginary parts to the number enclosed in parentheses and separated by a comma. For example, the number 6.2 - 3.1 i would be written in FORTRAN as (6.2, -3.1). The real and imaginary parts may be signed or unsigned and the i is understood for the imaginary part on the right. Any real or floating-point numbers may be used as the parts of a complex number, including those in exponential form.

Quantities that take on various values as a program is executed are called variables. As in any mathematical expression, variables are given variable names to distinguish one from another. For example, X and Y may represent two variables in an algebraic expression. In FORTRAN, variable names are formed from any of the alphameric characters. No special characters are allowed. A variable name can contain any of the alphameric characters but it must begin with an alphabetic character. Also, variable names in FORTRAN are limited to a maximum length of six characters. Examples of some correct variable names are

| | |
|---|---|
| X | A37 |
| BOY | ABCDEF |
| INK | CD14AB |
| TILL | JOHN |

Some incorrect variable names are

| | |
|---|---|
| 3AB | (First character must be alphabetic.) |
| AB*C | (special character) |
| SOMETHING | (more than 6 characters) |

In general, there are two types of variables, integer or fixed-point and real or floating-point. Integer variables can represent only integer constants. The name of an integer variable must begin with one of the six letters I, J, K, L, M, or N. The other letters in the variable name may be any of the available alphameric characters. With most compilers, this limitation can be overruled by the programer by specifying any variable name as an integer variable.

Real variable names, then, must begin with any letter other than I, J, K, L, M, or N. This rule can also usually be altered by the programmer. Real variables can represent only real or floating-point numbers.

Complex numbers can also be represented by complex variables in FORTRAN. Names for complex variables follow the same rules as those for integer and real variables, but there is no restriction on the first letter of the variable name.

Two other types of variables are frequently encountered in FORTRAN programs. As explained above, there is a limit to the number of digits allowed in integer or real numbers. This limit varies with different computers. When the need arises for using more significant digits than is allowed by a system, the programmer may override this limitation by specifying a variable to be a double precision variable. In this way, a number can contain up to sixteen digits in the mantissa if this accuracy is needed.

Double precision numbers are written in exponential form with a D replacing the E. Therefore, if $1.76 \times 10^3$ were a double precision constant it could be expressed in FORTRAN as 1.76D+3.

It is also sometimes necessary to make logical comparisons between data as well as to perform arithmetic operations upon them. There would be only two logical constants .TRUE. and .FALSE.. A logical variable would be capable of representing one or the other value at any point in the program.

## E. FORTRAN Statements

A FORTRAN program consists of a series of FORTRAN statements each of which specifies some instruction to be executed. These statements can be categorized into general groups depending upon the nature of the instruction.

A TYPE or declaration statement allows the programmer to determine the nature of a variable in a program. For example, a variable name such as INIT would be an integer variable because it begins with the letter I. The programmer can override this specification by declaring that INIT should be a real variable by using the proper TYPE statement. The general form of the TYPE statement would be

TYPE $\quad v_1, v_2, \ldots, v_n$

where TYPE is a word that specifies the nature of the variable and $v_1, v_2, \ldots, v_n$ is a list of the variables to be specified. The keywords to be used in the TYPE statement are INTEGER, REAL, COMPLEX, LOGICAL, and DOUBLE PRECISION. An example of a TYPE statement is

```
REAL    INIT, JOB, K
```

where the variable names INIT, JOB, and K are now real variables and can represent only real or floating-point numbers. Another example is

```
LOGICAL    MORE, LESS
```

The variables MORE and LESS are declared to be logical variables and can have only a value of .TRUE. or .FALSE..

An arithmetic statement in FORTRAN is a list of variables and/or constants separated by operators that define the operations to be performed. The list of FORTRAN operators and the operations they signify is as follows:

| Operator | Operation |
|---|---|
| ** | exponentiation |
| + | addition |
| - | subtraction |
| * | multiplication |
| / | division |

The general form of an arithmetic statement would be

A = expression

where A is a variable and expression is an arithmetic expression, variable, or constant. Some examples are

```
A = 3.0
B = A**2
```

In the first statement, A is assigned the value of 3.0. In the second statement, B is assigned the value of A raised to the second power. The variable B, therefore, has the value 9.0.

FORTRAN arithmetic statements are written in an analogous manner to algebraic statements except that none of the operators can be omitted. In algebra, the expression AB would represent A multiplied by B. But, in FORTRAN, AB would be a real variable name of two letters. If multiplication is intended, the expression must be written as A*B.

Parentheses are used in arithmetic statements for clarification or to indicate the order in which operations are to be performed. If the sum of A and B were to be divided by the sum of C and D and this result set equal to X, the expression would be

```
X = (A+B)/(C+D)
```

In the absence of parentheses, the order of operations within an expression is the same as in algebra. Exponentiation is followed by multiplication and division from left to right, followed by addition and subtraction from left to right. In the expression

```
A = B*C**2+D/E+3.0
```

C would be raised to the second power. Then, this value would be multiplied by B. Next, D divided by E would be calculated. This value would be added to the previous result and the final answer would be added to 3.0 and this value assigned to A. This process is also demonstrated by the following two numerical expressions:

2+2/2+2=5

but,

(2+2)/(2+2)=1

Within an arithmetic statement no two operators can be used in sequence.

```
A/-B
```

is wrong, but A divided by the negative of B can correctly be expressed as

```
A/(-B)
```

Parentheses are not operators and are used for clarification.

In the FORTRAN language, all variables and constants in an arithmetic expression must be of the same mode; that is, they must all be either real or integer. The expression

```
A=2.0+1
```

is invalid because 2.0 is a real constant and 1 is an integer constant.

Other examples of invalid arithmetic statements would be

```
I=3+4-1.0E+2
X=INT+Y
```

However, it is permissible for an expression to be evaluated using all real variables or all integer variables and storing this result in a variable of opposite mode. For example,

```
I=(3.0+2.0)/2.0
```

The expression to the right of the equals sign consists of only real constants. In floating-point arithmetic the result would be 2.5. However, this answer is to be stored with the integer variable name I. The final result is that all numbers to the right of the decimal point are dropped and I has the value of 2.

Another point about the structure of arithmetic statements is that real numbers can be raised to real or integer powers. Therefore, both of the following are valid expressions for setting A equal to the square of B.

```
A=B**2.0
A=B**2
```

However, integers cannot be used when the exponent is a fraction.

```
A=B**(1/2)
```

is not valid because 1/2 in integer arithmetic is 0. Fractional exponents must be expressed as real numbers.

```
A=B**(1./2.)
```

Any variables which appear on the right-hand side of an arithmetic statement must have been assigned numerical values before that statement is encountered in the program. Consider the following three statements as being the first three statements in a FORTRAN program.

```
A=1.0
B=2.0
D=A+B+C
```

The third statement would be invalid because no value had been assigned to C. The program would not be executed because there would be no way of determining the sum of A + B + C without knowing the value of C.

When arithmetic statements are executed, the procedure followed is that the right-hand side of the expression is evaluated and this result is assigned to the variable name on the left. In the expression

```
A=B+C
```

the sum of B + C is computed from their numerical values and this result is stored with the variable name A. If A had a numerical value before this operation, the old value is destroyed or erased and replaced with the new result. The values of B and C, individually, are not altered. A statement such as

```
A=B
```

results in A and B both having the same numerical value.

Since the result of the arithmetic expression is assigned to the variable on the left when the operation is executed, statements can be written which have the same variable on both sides of the equal sign. For example, in the following two statements

```
A=2.0
A=A+1.0
```

A is originally assigned the value of 2.0. In the second statement, this original value of A is added to 1.0 and this new result is then stored with the variable name A. The final result is that A = 3.0. The statement A = A + 1.0 would be illogical in algebra, but the equals sign does not have the same meaning in a FORTRAN statement as in an algebraic statement. The equals sign in the FORTRAN statement indicates that the variable name on the left is to be given the value of the expression on the right.

Arithmetic statements then can be used to assign a numerical value to a variable as in A = 3.0, or to evaluate a mathematical expression as in A = 3.0 + 2.0 - 1.0, or to increment the value of a variable as in I – I + 1, or to convert integer mode into real mode, or vice versa as in I = 5.0/2.0.

## 1. Logical Statements

FORTRAN IV allows for the manipulation of logical quantities. Logical variables may be assigned values as in

```
L=.TRUE.
WHY=.FALSE.
```

Keep in mind that the variables L and WHY must have been declared logical variables before being used in these expressions. In addition to assigning values to logical variables, logical expressions may be formed using six relational operators. These operators are:

| Relational operator | Meaning |
|---|---|
| .GT. | Greater than |
| .GE. | Greater than or equal to |
| .LT. | Less than |
| .LE. | Less than or equal to |
| .EQ. | Equal to |
| .NE. | Not equal to |

The relational operators can be used in combination with arithmetic expressions, constants, or variables to form logical expressions. For example, consider the following expressions:

```
1.NE.2
X+3.0.EQ.Y
```

The first expression has the logical value of .TRUE. The second expression will be .TRUE. or .FALSE. depending on the values of X and Y when the statement is executed.

There are also three logical operators available in FORTRAN IV:

| Logical operator | Example | Meaning |
|---|---|---|
| .NOT. | .NOT.X | If X is .TRUE., then .NOT. X is .FALSE. If X is .FALSE., then .NOT.X is .TRUE. |
| .AND. | X.AND.Y | The expression X.AND.Y is .TRUE. only if both X and Y are .TRUE. |
| .OR. | X.OR.Y | The expression X.OR.Y is .TRUE. if either X or Y is .TRUE. |

Logical operators can be used with any logical expressions, variables, or constants to form larger logical expressions.

```
(I.NE.3).OR.(J.GE.5)
```

The above expression has the value .TRUE. if either I does not equal 3 or J is greater than or equal to 5. The expression is .FALSE. only if both conditions are .FALSE..

## 2. DATA Statement

Many variables within a computer program will require an initial value before the program is executed. This initialization may be accomplished by using an arithmetic statement such as

A=3.0

where A would be assigned the value of 3.0. In cases where a number of variables are to be initialized FORTRAN IV allows for such initialization in one statement. The general form of the DATA statement is

DATA D1, D2, D3/V1, V2, V3/

where DATA is the keyword that indicates initialization, D1, D2, D3 is the list of variables to be assigned initial values, and V1, V2, V3 enclosed within the slashes is the list of values for the listed variables. The first variable, D1, would be assigned the first value, V1, and so on.

## 3. Control Statements

When a program is executed it may be desirable to change the sequence of operations from the order in which they were written depending upon conditions existing at the time of execution. For example, an intermediate result may be calculated and one of several possible operations performed depending on the sign and magnitude of the result. When calculating the roots of a quadratic equation using the quadratic formula, it would be necessary to terminate the program and indicate that the roots are imaginary if the expression $(b^2-4ac)$ is negative. But, if it is positive the program should continue and calculate the real roots.

Control statements in a FORTRAN IV program allow you to specify the order of execution of the statements.

GO TO Statement. The GO TO statement, sometimes called the unconditional GO TO, allows branching to any executable statement in the program when it is encountered. It has the general form

GO TO n

where n is the number of another executable statement within the program. In the following example of part of a FORTRAN IV program

```
      .
      .
      .
10    X=Y+1.0
      GO TO 12
11    X=Y-1.0
12    Z=X**2
      .
      .
      .
```

statement number 10 is executed and then statement number 12. Statement number 11 is skipped and can only be executed by branching to it directly from some other part of the program.

Computed GO TO Statement. The computed GO TO statement allows branching to one of several points in one step. It has the general form

GO TO ($n_1$, $n_2$, $n_3$, . . . , $n_m$), I

where $n_1$, $n_2$ and so on are a list of executable statement numbers and I is an integer variable which may have any value from 1 to m. Then, if I = 1, control is transferred to the first statement number $n_1$ ; if I = 2, the program branches to the second statement number $n_2$ and so on. If the integer variable is less than 0 or greater than the number of statements in the list, the next statement in sequence is executed. For example,

```
        .
        .
        .
     GO TO (3, 13, 23), INT
33   X=3.0
        .
        .
        .
```

If INT has the value 1, the program branches to statement number 3. If INT has the value 2, the program branches to statement number 13. If INT has the value 3, the program branches to statement number 23. If INT has the value 4, no branching occurs and the next statement in sequence, number 33, is executed.

Arithmetic IF Statement. The arithmetic IF statement allows branching within the program depending upon conditions at the time of execution. The general form is

IF(expression)n1, n2, n3

where an arithmetic expression is contained within the parentheses and n1, n2, n3 are three statement numbers within the program. The IF statement transfers control to one of the three specified statements depending on the value of the expression within the parentheses. If the value of the expression is less than 0, the program branches to statement number n1; if the value of the expression equals 0, the program branches to n2; if the value of the expression is greater than 0, the program branches to n3. In the statement

IF(X)10, 20, 30

control is transferred to statement number 10 if X is negative, to number 20 if X equals 0, and to number 30 if X is positive. An example of an IF statement with an expression to be evaluated is

IF (X-2.) 10,20,30

The same rules hold concerning the value of the expression. Thus, if X is less than 2 (the expression is negative), the program branches to statement number 10; if X = 2, to number 20; and, if X is greater than 2, to number 30.

Logical IF Statement. The logical IF statement evaluates a logical rather than an arithmetic expression and executes a statement if the value of the logical expression is .TRUE.. The general form is

```
IF(Logexpression)S
```

where Logexpression represents any logical expression and S is any executable statement except another IF statement or a DO statement, to be explained later. For example,

```
IF (I .EQ. 0)Y = X**2
```

The expression Y=X**2 will be executed only if the expression I.EQ.0 is .TRUE.. If the logical expression is .FALSE. the next statement in sequence is executed. Another example of a logical IF statement is

```
IF (I .GT. 0 .OR. J .GT. 0) GO TO 10
```

Here, the program would branch to statement number 10 if either I .GT. 0 or J .GT. 0 is .TRUE..

DO Statement. When a series of statements are to be executed within a program a number of times, it is not efficient to write them into the program each time. The DO statement causes a program to execute a series of statements called a DO loop a specified number of times. The general form is

```
DO n i=m1, m2, m3
```

where n is a statement number that follows the DO statement by one or more statements, where i is an integer variable used to count the number of times the DO loop is executed, and m1, m2, and m3 are integer variables or integer constants. If m3 is not listed, a value of 1 is assumed. The DO statement causes all the statements following it to be executed up to and including the statement number n a specified number of times. The first time through the DO loop, I is assigned the value of m1. The second time through the loop, I is incremented by m3. I is incremented by m3 each time through the loop until the value of I exceeds m2. Then, control is transferred to the next statement following statement number n. In the following example all of the integer numbers between 1 and 10 are added using a DO loop.

```
      SUM=0.0
      DO 20 I=1, 10
      X=I
20    SUM=SUM+X
      GO TO 30
```

In the first statement, the value of the variable SUM is initialized at 0. The DO statement causes the lines following it up to and including statement number 20 to be executed 10 times. The integer variable I acts as a counter for the number of times the program passes through the DO loop. I is initially set equal to 1 and is incremented by 1 each time the DO loop is executed. The I assumes the values 1, 2, 3 and so on until it exceeds 10 at which time control is transferred to the statement immediately following the DO loop, that is, the statement GO TO 30. Since I assumes all of the integer values between 1 and 10, it is also used to find the sum of these numbers. I is converted to a floating-point number in the third statement, and this number is added to the total sum previously calculated. So, the result in sum is equal to $0.0 + 1.0 + 2.0 + \cdots + 10.0$ or 55.0.

There are important rules to be observed in the use of DO statements.

1. The DO parameters m1, m2, and m3 cannot be changed by any statement within the DO loop.
2. The final statement within the DO loop cannot be a control statement.
3. Control cannot be transferred into a DO loop from outside the loop since the indexing parameter m1 would not have been initialized. You can, however, transfer control out of a DO loop at anytime except within the final statement.
4. DO loops can be placed within other DO loops. This is called nesting of DO loops. But, the inner DO loop must be contained entirely within the outer DO loop.

## 4. DIMENSION Statements

Part or all of a FORTRAN program may be used to repeat some calculation a number of times with new data. It would be much more advantageous to assign the same variable name to each of these data than to assign individual variable names for all of them. FORTRAN has this capability of using the same variable name for several data by allowing the use of subscripted variables. If, for example, a program was written to calculate the square of any number and there were 5 numbers to be used, the first could be referred to as $X_1$, the second as $X_2$ and so on. That is, these 5 numbers would form an array and each member of the array could be accessed by reference to the same variable name with the proper subscripts. In FORTRAN, the subscript is enclosed within parentheses immediately following the variable name. Thus, $X_1$ in FORTRAN is X(1). The subscripts may be any positive integer, integer variable, or integer expression. Zero and negative subscripts are not allowed. Any variable in FORTRAN can have one, two, or three subscripts corresponding to one-, two-, and three-dimensional arrays. Thus, $X_{ij}$ in a two-dimensional array would become X(I, J) in FORTRAN. Note that when there are more than one subscript they must be separated by commas.

Before a subscripted variable can be used in a FORTRAN program the maximum size of the array must have been previously defined. This is accomplished with a DIMENSION statement. The general form is

DIMENSION X(n), Y(n, n), Z(n, n, n), ...

where X, Y, and Z are variables that will be subscripted in the program, and the n values indicate the maximum number to be used as a subscript in the array. Subscripted variables can be thought of as numbers of 1-, 2-, or 3-dimensional matrices. The maximum number for each subscript, n, indicates the amount of storage to be allocated for the array. In the example

DIMENSION X(5), Y(3, 3), Z(4, 3)

five storage locations are allocated for the X array, nine for the Y array, and twelve for the Z array. All of the subscripts need not be used in a program even though storage area has been reserved for them. Therefore, when the exact number of variables to be accommodated in an array is not known, the n values in the DIMENSION statement should be large enough to accommodate all possible variables. However, once a variable has been defined in a DIMENSION statement, it cannot be used in the program without a subscript.

## 5. INPUT/OUTPUT Statements

The FORTRAN statements outlined thus far have dealt with routines to be performed once the program has begun execution. Input and output statements provide for the transmission of data to the computer through some input device and the outputting of results on an output device.

In reading data from a punched card or magnetic tape or some other input device and in writing output on an output device, the computer must be told exactly what type of information is to be read and where it is located on the input medium or what type of output to print and where it is to be located on the output medium. It must be told whether the data is in the form of numbers or letters or a combination of both, whether these numbers are real or integer or complex, and the length of the numbers or series of letters. If a punched card is used as an input medium, for example, the computer must be told in which columns the data appear.

The input statement in FORTRAN IV has the general form

READ(i, n)list

where i is an integer number that indicates the input device to be used, n is a statement number of the FORMAT statement to be followed, and list refers to the series of variables whose values are to be inputted.

An example of a READ statement is

READ(5, 20)X, Y, Z

The statement says to read values for the variables X, Y, and Z from the input device with the code 5 according to the form outlined in statement number 20. The particular i value, sometimes called the logical unit number, for each input medium varies with different computers. These should be ascertained before the program is written. Logical unit 5 most often refers to a card reader, however, and this is the meaning used here. The above READ statement then says to read the values for X, Y, and Z from a punched card according to the form outlined in statement nunber 20.

The general form of the output statement is

```
WRITE(i, n)list
```

where i refers to the logical unit number of the output device to be used. Logical unit 6 usually refers to the high-speed printer and logical unit 7 usually refers to the card punch.

```
WRITE(6, 20)X, Y, Z
```

This statement says to print the values of X, Y, and Z on the printer using the FORMAT in statement number 20.

```
WRITE(7, 20)X, Y, Z
```

This statement says to punch the values of X, Y, and Z on a punched card using the FORMAT in statement number 20.

Note that in both the READ and WRITE statements, the logical unit number and the FORMAT statement number are enclosed within parentheses. There is no comma between the parentheses and list of variables but each variable must be separated by a comma.

## V. CONCLUSION

We have seen a brief overview of the makeup of computers and how one uses them to speed up calculation tasks. The FORTRAN language has also been introduced and basic programming philosophy given.

Since this chapter serves only as a brief introduction to this enormous field, mastery of this chapter will not make one a master of computer programming. It should, however, serve as a launch point for delving deeper into this science. Another source is the information published by your particular computer installation. Here you find the capabilities and conventions which are used at this installation and the basic information on how to access your machine. You will also learn of any idiosyncrasy your installation possesses, which obviously cannot be covered in general discussions. A further source of information should be the manufacturers of the equipment. Thus, IBM, Digital Equipment Corporation, CDC, and others are willing to furnish you with detailed information about their systems. They will also discuss the programming of their systems and how this programming is implemented on their hardware. Finally, there are many excellent books in the area of computers and programming of computers. These are useful because they give you tips on how to approach problems and show you how to solve similar problems, making the job of programming easier. Most important, they give us the detailed basics we need to prepare a proper program. Some of these books are listed in the bibliography.

Computers aid us by freeing us from the drudgery of repetitious calculations: they are extremely valuable tools. It is becoming increasingly important for us to understand this technology and to gain the ability to apply it. It is the purpose of this chapter to provide a base on which the reader can develop his expertise in this area. The following chapters, and other volumes in the series, will allow the reader to explore what is available and give him a working ability to add to his present knowledge.

## BIBLIOGRAPHY

B. W. Arden, An Introduction to Digital Computing, Addison-Wesley, Reading, Mass., 1963.

G. B. Davis, An Introduction to the IBM System/360 Computer, McGraw-Hill, New York, 1965.

C. B. Germain, Programming the IBM 360, Prentice-Hall, Englewood Cliffs, N.J., 1967.

T. L. Isenhour and P. C. Jurs, Introduction to Computer Programming for Chemists, Allyn and Bacon, Boston, 1972.

D. E. Knuth, The Art of Computer Programming, Addison-Wesley, Reading, Mass., 1968.

D. D. McCracken and W. S. Dorn, Numerical Methods and FORTRAN Programming with Applications in Engineering and Science, Wiley, New York, 1964.

R. W. Southworth and S. L. Deleeuw, Digital Computation and Numerical Methods, McGraw-Hill, New York, 1965.

D. D. Spencer, Game Playing With Computers, Spartan Books, New York, 1968.

Chapter 2

# BASIC PRINCIPLES OF THE ELECTRONIC ANALOG COMPUTER

Harry B. Mark, Jr.

Department of Chemistry
University of Cincinnati
Cincinnati, Ohio



## I. INTRODUCTION

The purpose of this chapter is to introduce the reader to the basic electronic principles of the analog computer and how to employ it for both basic computation and analog simulation of physical and chemical systems. A working knowledge of this fundamental material will be useful, if not necessary, in order to fully understand the chapters dealing with the more complex and sophisticated specific applications of analog computers (and techniques) to be contained in subsequent volumes of this series. The topics to be discussed will not only deal with computation and simulation with respect to chemical problems (such as kinetics), but will deal extensively with quantitative instrumentation based on operational amplifier electronics and hybrid computers (mixed analog and digital circuitry).

It can be said that an analog computer represents the parameters, coefficients, and variables of a specific physical or chemical system and/or problem by corresponding electronic quantities, such as continuously variable voltages, values of resistance and capacitance in a network, etc. These so-called machine variables are made to obey the exact mathematical relations analogous to those that describe the original system or problem. The desired analogous relations are established by the appropriate interconnection of individual computing elements or modules (operational amplifier circuits) to constrain the over-all electronic circuit to a true representation. Thus, certain conveniently measured machine variables (usually voltages) at appropriate points in the circuit are recorded or measured and their values or behavior represent the solution(s) of the given system or problem. In general, but not always, the machine variables are a function of time as time is generally an independent variable in any system described by differential equations. The basic computing element or module is the operational amplifier with resistive and capacitive passive networks as the input and feedback circuits.

In order to understand the application of the analog computer to calculation and instrumentation, this chapter will describe the basic principles and use of Laplace circuit analysis of passive networks, the principles, characteristics, and computation applications of operational amplifiers, and finally the solution of a few simple example problems.

## II. THE LAPLACE TRANSFORM METHOD OF ANALYSIS OF PASSIVE NETWORKS

The analysis of complex passive networks, which contain capacitances and/or inductances as well as resistances and in which voltages may vary with time, is a relatively complex and difficult problem. There are three basic approaches to such circuit analysis: (i) a complex number method, (ii) an integrodifferential equation method, and (iii) a Laplace transform method. The complex number method is applicable only for sinusoidal varying voltages and combines Kirchhoff's rules [1, 2] with a special form of vector representation to yield a solution as a function of time. However, the response obtained is only the steady-state part of the solution. Method (ii) used Kirchhoff's rules to establish integrodifferential loop or point equations which, when solved, yield a general solution containing both the steady-state and the transient responses. Although this method is rather general, solution of the differential equations by ordinary methods can be quite difficult, especially for discontinuous excitations which would be common in physical and chemical problems. Method (iii), equivalent in principle to (ii) and reducible to (i), is a considerably simpler approach to circuit analysis, and without loss of generality. The method, based on the use of Laplace transforms, leads more readily to a specific solution containing both steady-state and transient responses — especially in the case of discontinuous excitations. This method is used also in all electronic handbooks and tables to describe transfer functions and impedances in analog computer circuits.

## A. The Laplace Transform

A major advantage of the Laplace transform is that it reduces a first-degree differential function of the type which describes typical electronic circuits or networks to an algebraic function. This allows the linear differential equation with constant coefficients, of the form

$$\sum_n a_n \frac{d^n y}{dt^n} = F(t) \qquad (n = 0, 1, 2, \ldots)$$

(where the $a_n$ terms are constants), to be solved by algebraic manipulations. The differential equations are solved by first taking the Laplace transform of the equation, then performing algebraic manipulations, and then taking the inverse Laplace transform [3-7].

Another important advantage of the Laplace method is that the initial conditions of the system are inserted into a transformed differential equation before the inverse transform is taken. The resulting solution is, therefore, a specific solution for these specific initial conditions. This procedure circumvents the necessity of first obtaining a general solution which in many cases may be a difficult if not impossible task for the average chemist.

The Laplace transform F(s) of a function F(t) (which exists only at positive values of time t) is defined by [3]

$$\bar{F}(s) = \int_0^\infty e^{-st} F(t)\, dt \qquad (1)$$

where the Laplace parameter s (which may be real, imaginary, or complex) must have a value such that the integral is finite. Because the term $e^{-st}$ approaches zero as the value t approaches infinity, then $e^{-st}F(t)$ will also vanish at infinite t even though F(t) may not [provided that F(t) does not increase as fast as $e^{-st}$]. Hence the area under $e^{-st}F(t)$ may be finite and will be a function of s rather than t. For a more complete explanation of the nature of the Laplace transform, the treatment in refs. [3, 4] are recommended.

In practice it is usually not necessary to perform the integration indicated by Eq. (1) because Laplace operations and transforms are tabulated for many common functions as well as common circuit networks. Such tables appear in the more recent editions of pertinent handbooks [5, 8, 10]. For the convenience of the reader, Table 1, which includes most of the transforms and operations used in this chapter, is given.

TABLE 1

Laplace Operations and Transforms

| | Function | Laplace transform |
|---|---|---|
| 1. | $F(t)$ | $\int_0^{\infty} e^{-st}F(t)\,dt = \bar{F}(s)$ |
| 2. | $(d/dt)F(t)$ | $s\bar{F}(s) - F(+0)$ |
| 3. | $(d^2/dt^2)F(t)$ | $s^2\bar{F}(s) - sF(+0) - (dF/dt)(+0)$ |
| 4. | $\int_0^t F(\gamma)\,d\gamma$ | $(1/s)\bar{F}(s)$ |
| 5. | $F(t - a)(= 0 \text{ at } t < a)$ | $e^{-as}\bar{F}(s)$ |
| 6. | $F(t) = F(t - a)$ | $\dfrac{\int_0^a e^{-st}F(t)\,dt}{1 - e^{-st}}$ |
| 7. | Unit impulse | 1 |
| 8. | $t$ | $1/s^2$ |
| 9. | $e^{at}$ | $1/(s - a)$ |
| 10. | $\sin at$ | $a/(s^2 + a^2)$ |
| 11. | $\cos at$ | $s/(s^2 + a^2)$ |

## B. Linear Circuit Elements

The common linear elements of passive networks and of interest in analog computation are the resistor R, the capacitor C, and the inductor L. (A common nonlinear component used in sucn applications is the ordinary diode.) The electronic properties of these linear components are described by the manner in which they conduct current i when a voltage E is placed across them. The current-voltage relationships for each of these components are listed in Table 2. The terms R, C, and L are, of course, constants with respect to time.

TABLE 2

Properties of Linear Circuit Elements

| Element name | Symbol | Descriptive equation | Transfer impedance $\bar{Z}(s) = \bar{E}/\bar{i}$ |
|---|---|---|---|
| Resistor | | $R = \frac{E}{i}$ | R |
| Capacitor | | $C = \frac{i}{dE/dt}$ | $\frac{1}{sC}$ |
| Inductor | | $L = \frac{E}{di/dt}$ | sL |

## C. Transfer Impedance

A useful function in circuit analysis is the ratio of the input voltage transform $\bar{E}_i$ to the output current transform $\bar{i}_o$ of a network when the output is short-circuited to ground (see diagrams of Fig. 1, for example).

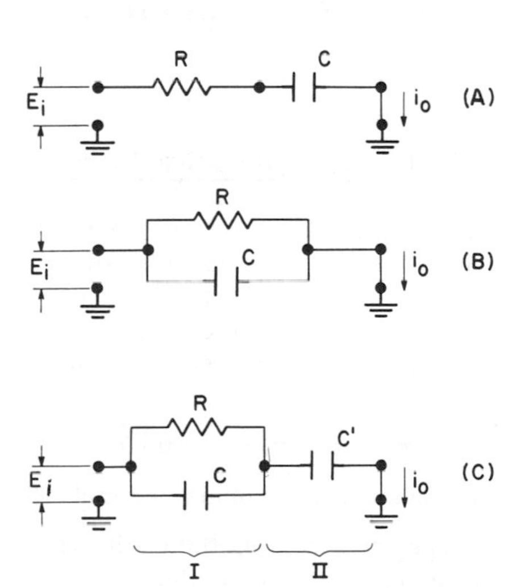


Fig. 1. Simple passive networks used in deriving example transfer impedances: (A) series connection; (B) parallel connection; (C) combination connection.

This ratio is termed transfer impedance $\bar{Z}(s)$. The transfer impedances of the linear circuit elements are obtained simply by taking the Laplace transform of the current-voltage equation, for a single element or a complex network.

For a simple resistive element, E = Ri; therefore, the transfer impedance is given by

$$\bar{Z}(s) = \frac{\bar{E}_i}{\bar{i}_o} = R. \tag{2}$$

For a simple capacitor, i = CdE/dt; and in the Laplace domain (see Table 1, transform 2),

$$\bar{i} = C[s\bar{E} - E(+0)]. \tag{3}$$

[With respect to using the Laplace transform only for circuit analysis as discussed in this chapter, it will be assumed that all experiments and/or measurement start at time t equal zero. In this section on passive networks the initial voltages across capacitors E(+0) and the initial currents through inductors i(+0) are considered also to be zero for simplicity of the discussion. But, as will be discussed later in Section IV, this simple initial condition is not always general and appropriate and initial conditions must be considered.] Thus, the simple transfer impedance of a capacitor is

$$\bar{Z}(s) = \frac{\bar{E}}{\bar{i}} = \frac{1}{sC}. \tag{4}$$

Similarly, for an inductor, E = Ldi/dt; and the transfer impedance is

$$\bar{Z}(s) = \frac{\bar{E}}{\bar{i}} = sL. \tag{5}$$

## D. Excitation-Response Principle

An important property of the Laplace transform is embodied in the definition of transfer impedance $\bar{Z}$. [From this point on any function symbol with a superscript will denote the Laplace transition of that function and that it is a function of (s) will be understood.] It is generally true that when a linear system is excited by an external source, the resulting response will be such that

$$\text{response transform} = \text{excitation transform} \times \text{system transform} \tag{6}$$

where the excitation and system transforms are mutually independent [5, 11].

In electronic networks and circuits, the excitation may be a varying current and the response may be voltage. Hence,

$$\bar{E} = \bar{i} \times \bar{Z} \tag{7}$$

where the transfer impedance $\bar{Z}$ is the system transform, which depends only on the properties of the network. Conversely, if the excitation is a varying voltage applied to a network and the current measured as the response function, then

$$\bar{i} = \bar{E} \times \bar{Y} \tag{8}$$

where the transfer admittance $\bar{Y} = 1/\bar{Z}$ is now the appropriate system transform.

This principle offers the means whereby the response to an excitation of a passive network can be readily predicted with a minimum of cumbersome mathematics as all operations are carried out in the Laplace plane. Once the manipulations are carried out and the response transform obtained, the real (time) plane form of the response is obtained by carrying out the inverse Laplace transformation. Hence, it is of interest to examine some common excitation functions and passive networks as examples which will introduce the analysis of operational amplifier networks in Section III.

### 1. Excitation Functions and Transforms

Several common excitation function waveforms and their Laplace transforms are given in Table 3 (see Table 1). In all cases there is no excitation before $t = 0$.

Notice that in general the Laplace transform of any periodic function is equal to the Laplace transform of the first wave or cycle divided by $1 - e^{-sT}$ where T is the period of the wave (see Table 1, number 6 [3, 11]).

The excitation form most commonly encountered in electronics is the sinusoid. Its transform is readily obtained by considering the exponential forms of the circular functions,

$$\sin \omega t = \frac{1}{2j}(e^{j\omega t} - e^{-j\omega t}) \tag{9}$$

$$\cos \omega t = \frac{1}{2}(e^{j\omega t} + e^{-j\omega t}) \tag{10}$$

where $j = \sqrt{-1}$, $\omega = 2\pi f$, and f is the frequency of the sinusoid.

### 2. Network or System Functions and Transforms

The principles of the method employed to obtain an appropriate system transform which describes a passive network are illustrated by a few detailed examples which are given below.

Generally, linear electronic networks are composed of various combinations of the circuit elements in series or parallel connections as shown in the examples given in Fig. 1.

TABLE 3

Excitation Transforms

| Type | Waveform | Laplace transform |
|---|---|---|
| 1. Impulse | 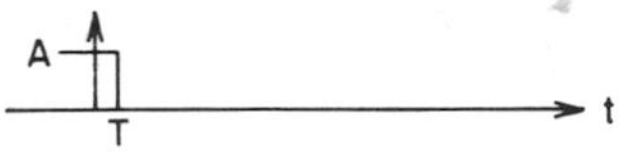 | $AT$ |
| 2. Step | 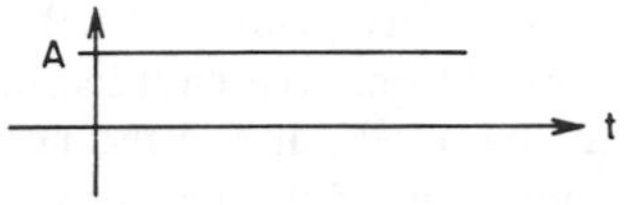 | $\frac{A}{s}$ |
| 3. Ramp | 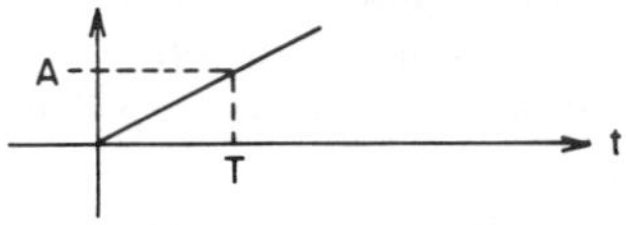 | $\frac{A}{s^2T}$ |
| 4. General periodic wave | 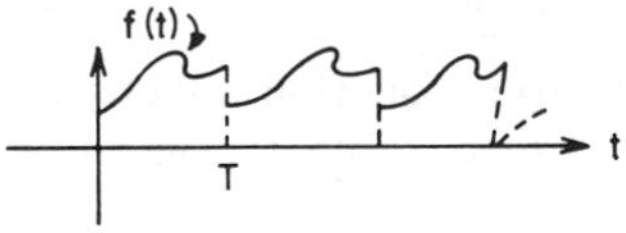 | $\frac{\int_0^T e^{-st} f(t)\,dt}{1 - e^{-s}}$ [a] |
| 5. Square wave | 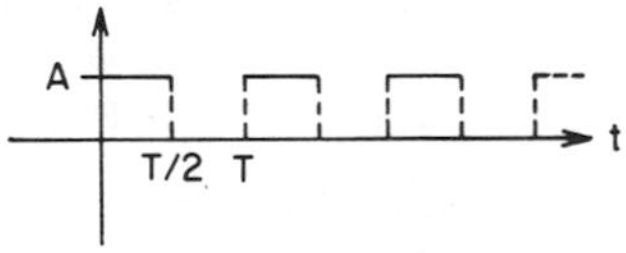 | $\frac{A(1 - e^{-sT/2})}{s(1 - e^{-sT})}$ |
| 6. Sawtooth | 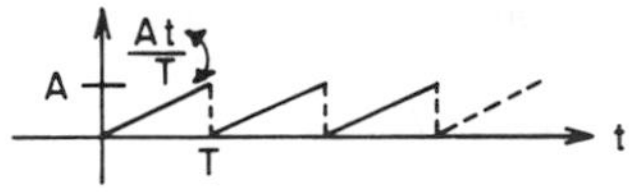 | $\frac{A[1 - (1 + sT)e^{-sT}]}{s^2T(1 - e^{-sT})}$ |
| 7. Triangular wave | 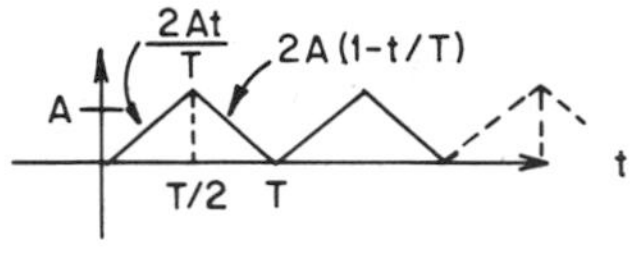 | $\frac{2A(1 + e^{-sT} - 2e^{-sT/2})}{s^2T(1 - e^{-sT})}$ |
| 8. Sinusoidal wave | 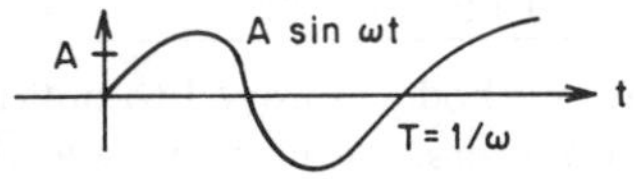 | $\frac{A\omega}{s^2 + \omega^2}$ |
| | 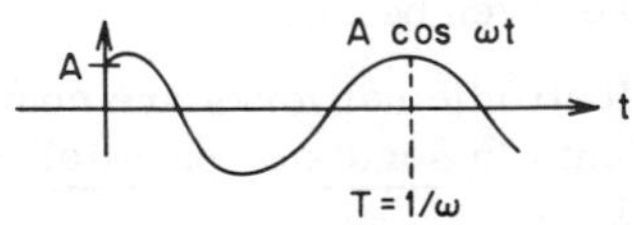 | $\frac{As}{s^2 + \omega^2}$ |



[a] Zero-shift theorems [3, 11]; see also Table 1, transform 5.

In general, for series connection the transfer impedance of the network is the sum of the transfer impedances of the individual parts. Thus, the transfer impedance of the series resistor-capacitor combination of Fig. 1(A) is given by

$$\bar{Z} = R + \frac{1}{sC}. \tag{11}$$

Again, in general, for a parallel connection, the reciprocal of the transfer impedance is the sum of the reciprocals of the transfer impedances of the individual parts. Thus, the transfer impedance of the parallel resistor-capacitor combination [Fig. 1(B)] is given by

$$\frac{1}{\bar{Z}} = \frac{1}{R} + sC \tag{12}$$

or

$$\bar{Z} = \frac{R}{1 + sRC} \tag{13}$$

Once the transfer impedance of a particular network is known, this network can be used as a part of a more complicated network. For example, in Fig. 1(C), a parallel resistor-capacitor network (part I) is connected in series with a capacitor (part II) to form the total network. Thus, the transfer impedance of the total network is

$$\bar{Z} = \bar{Z}_I + \bar{Z}_{II} = \frac{R}{1 + sRC} + \frac{1}{sC}. \tag{14}$$

These examples indicate that the transfer impedance of a very complicated network often can be readily obtained by considering the series and parallel combinations of successive subdivisions. The transfer impedances of a large number of typical networks are given in tables in many electronics texts [5-7].

## 3. Transfer Function

Another system transform commonly used in describing a passive network is the ratio of the output voltage transform $\bar{E}_o$ at some point of a network to the input voltage transform $\bar{E}_i$. This ratio which is more useful in considering operational amplifier circuits is called the transfer function, and can be obtained by considering the transfer impedance of appropriate sections of a network. For example, in network A of Fig. 2,

$$\frac{\bar{E}_i}{\bar{i}_g} = R + \frac{1}{sC} \tag{15}$$

$$\frac{\bar{E}_o}{\bar{i}_g} = \frac{1}{sC} \tag{16}$$

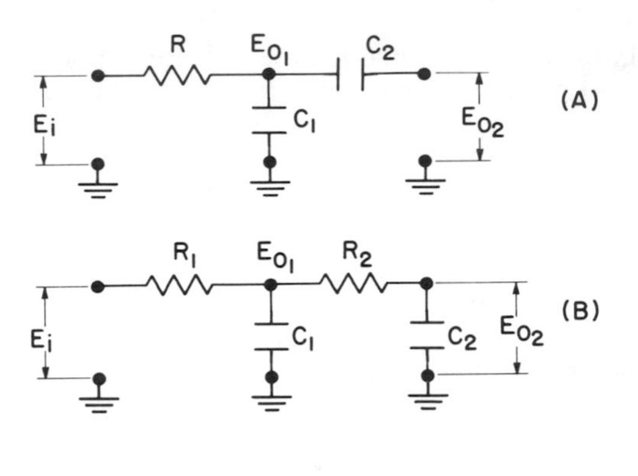


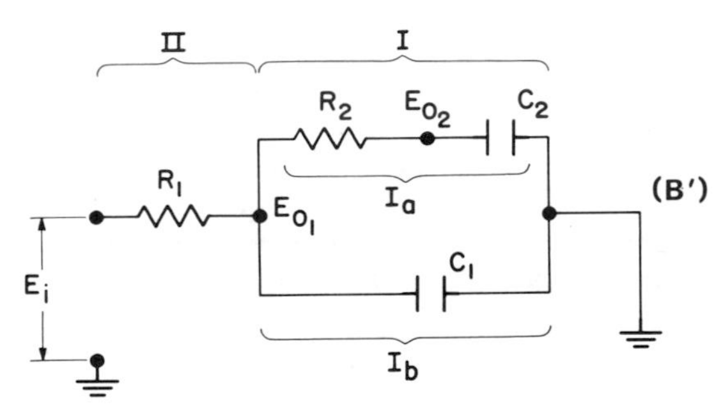


Fig. 2. Networks for the derivation of example transfer functions.

where $i_g$ is the current passing to ground through the capacitance C. Then, dividing Eq. (16) by Eq. (15), the transfer function is obtained:

$$\frac{\bar{E}_o^1}{\bar{E}_i} = \frac{1}{1 + sRC_1} \,. \tag{17}$$

Notice that since there can be no current through a "dangling" component such as the capacitance $C_2$, $E_o^1 = E_o^2$. Thus, transfer function (with respect to a different output point of the circuit) for this network is

$$\frac{\bar{E}_o^2}{\bar{E}_i} = \frac{1}{1 + sRC_1} \tag{18}$$

It is important to point out here that often a transfer function can be more easily derived by redrawing the configuration of a network. This operation is a matter of experience. For example, the transfer function, $\bar{E}_o^1/\bar{E}_i$, of network (B) of Fig. 2 can be obtained from the transfer impedances indicated in network (B′) [note that network (B′) is the equivalent of (B)]:

$$\bar{Z}_{Ia} = R_2 + \frac{1}{sC_2}$$

$$\frac{1}{\bar{Z}_I} = \frac{1}{\bar{Z}_{Ia}} + \frac{1}{\bar{Z}_{1b}} = \frac{1}{R_2 + 1/sC_2} + sC_1 \tag{19}$$

$$\bar{Z} = \bar{Z}_I + \bar{Z}_{II} = \frac{1 + sR_2C_2}{sC_2 + sC_1(1 + sR_2C_2)} + R_1.$$

The desired transfer function is therefore

$$\frac{\bar{E}_o^1}{\bar{E}_i} = \frac{\bar{Z}_I\bar{i}_g}{\bar{Z}\bar{i}_g} = \frac{1 + sR_2C_2}{sR_1C_1(1 + sR_2C_2) + sR_1C_2 + sR_2C_2 + 1}. \tag{20}$$

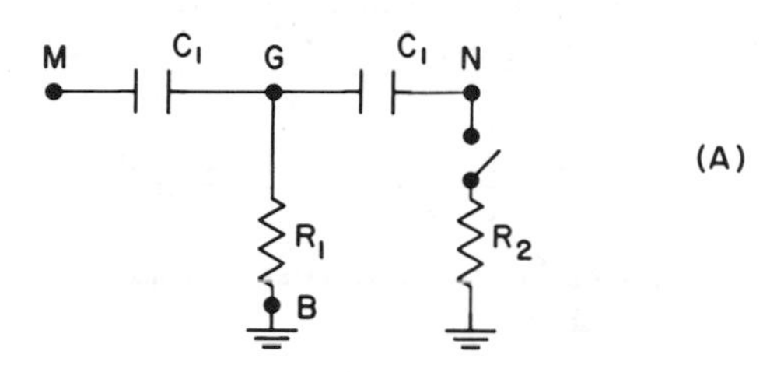


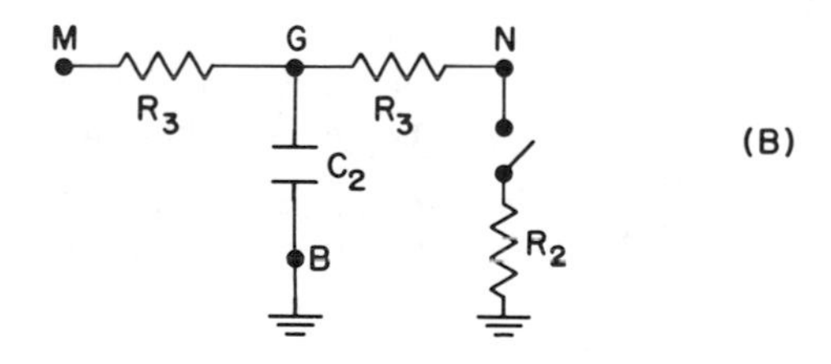


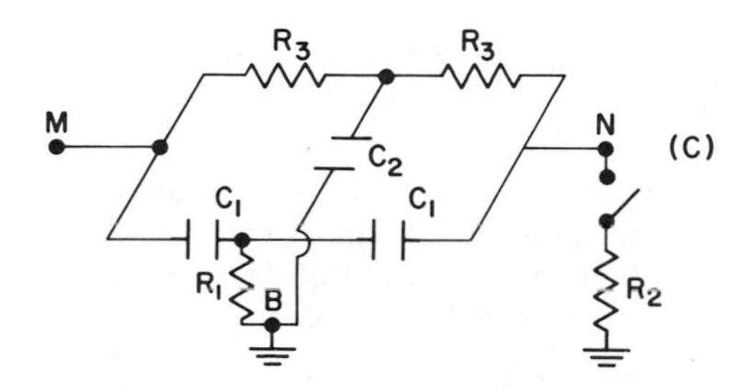


Fig. 3. Network combinations

Thus, although such system functions can always be obtained from simultaneous equations set up by the direct application of Kirchhoff's rules, a little practice and experience will demonstrate that a considerable amount of time and effort is saved by using a little intuition and ingenuity to determine a more expedient approach. Figure 3 shows a good example of this type of approach. Note that network (C) is a parallel combination of networks (A) and (B).

## 4. Response Transform

As was discussed above, the response transform is obtained by multiplying the excitation transform by the system transform. The response function (in the real plane) is then obtained by taking the inverse Laplace transform — as mentioned, also by consulting a Laplace transform table if possible. In order to make this operation clear, a detailed example calculation of a response function is given below. The objective is to calculate the response $E_o$ which occurs when network A in Fig. 2 is excited by an impulse voltage, waveform $E_1$, of area equal AT (see Table 3). The response transform is given by

$$\bar{E}_o = \left(\frac{1}{1 + sRC}\right)\bar{E}_i = \left(\frac{1}{1 + sRC}\right)AT = \frac{AT/RC}{s + (1/RC)}. \tag{21}$$

By consulting Table 1, transform 9, the response after making the inverse transformation is seen to be

$$E_o = \frac{AT}{RC} e^{-t/RC}. \tag{22}$$

As another example calculation using the same network, let the excitation voltage $E_1$ be a step function of height A. Then the response transform is

$$\bar{E}_o = \left(\frac{1}{1 + sRC}\right)\frac{A}{s}. \tag{23}$$

Table 1 does not list a transform of this form. However, by combining transforms 4 and 9, the response is found to be

$$E_o = \int_0^t \frac{A}{RC} e^{-\gamma/RC}\, d\gamma = A(1 - e^{-t/RC}). \tag{24}$$

Often, when a particular function cannot be found in transform tables, it is helpful to rearrange the transform by the method of partial fractions. Reconsider the case just described in Eq. (23). Suppose that there are arbitrary constants $\alpha$ and $\beta$ such that

$$\frac{1}{s(1 + sRC)} \equiv \frac{\alpha}{s} + \frac{\beta}{1 + sRC}. \tag{25}$$

To evaluate these constants, combine the fractions on the right side of Eq. (25) and cancel the denominators,

$$1 = \alpha(1 + sRC) + s\beta = \alpha + s(\alpha RC + \beta). \tag{26}$$

Then, by equating coefficients of s,

$$\alpha \equiv 1$$

and

$$\alpha RC + \beta = 0 \tag{27}$$

so that

$$\beta = -RC.$$

By putting these values into Eq. (25), one obtains

$$\frac{1}{s(1 + sRC)} = \frac{1}{s} - \frac{RC}{1 + sRC}. \tag{28}$$

Thus, Eq. (23) can be written

$$\bar{E}_o = \frac{A}{s} - \frac{A}{s + 1/RC}. \tag{29}$$

By looking up the individual transforms of each of these two terms on the right-hand side of Eq. (29), one obtains the response function

$$E_o = A(1 - e^{-t/RC}) \tag{30}$$

which is the same result obtained in Eq. (24).

Because sinusoidal waveforms are very commonly used as excitation functions, in electronic instruments, etc., it is of interest to examine a special procedure for obtaining the "steady-state" portion of the response to a sinusoidal excitation of a linear network or system. Under such conditions, the steady-state response must itself be a sinusoid with a frequency exactly equal to the frequency of the excitation. The amplitude and phase of the steady-state response sinusoid can be found from the appropriate system function by replacing the Laplace parameter s with $j\omega$ where $j = \sqrt{-1}$ and $\omega = 2\pi f$ and is the angular frequency of the excitation sinusoid. When this substitution is made, the system transform function becomes a complex number which can be put onto the usual form $a + jb$. This complex number is a representation (in complex vector notation) of the relative amplitude and phase of the response sinusoid.

As an example of the above procedure, consider a sinusoidal voltage $E_1 = H_i \sin \omega t$, applied to the passive network shown in Fig. 1(A). The transfer admittance of this network is given by [reciprocal of Eq. (11)]

$$\bar{Y}(s) = \frac{\bar{i}_o}{\bar{E}_i} = \frac{sC}{1 + sRC}. \tag{31}$$

Upon replacing of s with $j\omega$,

$$\bar{Y}(j\omega) = \frac{j\omega C}{1 + j\omega RC}. \tag{32}$$

This equation can be put into complex number form by multiplying the numerator and denominator by the complex conjugate of the denominator, $1 - j\omega RC$, thus yielding

$$\bar{Y}(j\omega) = \frac{\omega^2 R^2 C^2}{1 + \omega^2 R^2 C^2} + j\frac{\omega C}{1 + \omega^2 R^2 C^2}. \tag{33}$$

As shown in Fig. 4, the real and imaginary terms of Eq. (33) are the components of a complex vector. This type of complex vector can be described also by its magnitude $H_o$ and its angle $\theta$ ($\tan \theta = b/a$). The magnitude and angle of this complex vector are equal to the relative amplitude and phase, respectively, of the steady-state response. Hence, from Eq. (33),

$$H_o = \frac{\omega C}{(1 + \omega^2 R^2 C^2)^{1/2}}. \tag{34}$$

Then the steady-state current is

$$i_{0(ss)} = H_o H_i \sin(\omega t + \theta). \tag{35}$$

The above result can be obtained in another manner from Eq. (31) by substituting $\bar{E}_i = H_i \omega/(s^2 + \omega^2)$, taking the inverse Laplace transform of $\bar{i}_o$, then neglecting terms that approach zero as t becomes large.

## 5. General Rules for Verification of a Solution of a Network

After obtaining the solution to a particular problem, this solution should be carefully examined, of course, to determine if it is correct. There are several quick tests and/or "rules of the thumb" that can be employed in this check. One quick test is for dimensional consistency. Another test is

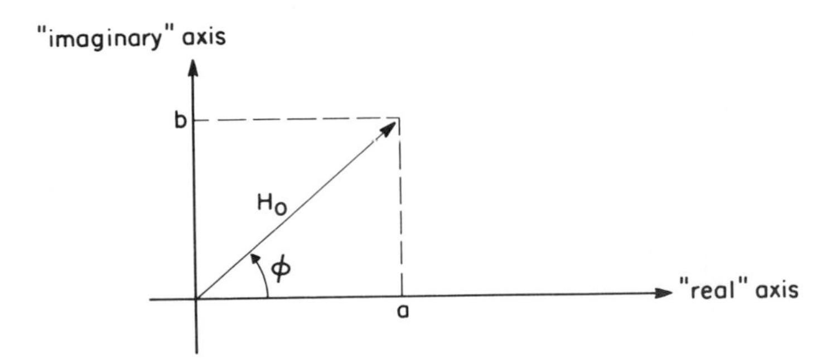


Fig. 4. Complex vector diagram [Eq. (33)].

achieved by examining the solution when certain specific or special conditions are inserted. When a solution fails the first test, a mathematical error in arriving at the solution is usually the source of error. If the failure occurs with the second test, this indicates either a mathematical error or perhaps a more serious error, the use of an incorrect starting equation.

In examining an equation for dimensional consistency, the following principles can be applied:

(i) Each term of a summation must have the same dimensions.
(ii) A trigonometric function and its argument, a logarithm and its argument, or an exponent must be dimensionless.
(iii) The dimensions of both sides of an equation must be the same.
(iv) The quantities $1/\omega$, $1/f$, $1/s$, RC, L/R, and $\sqrt{LC}$ all have the dimensions of time.

The examination of a solution under certain special or specific conditions of operation not only serves as a test for plausibility but also results in an intuitive feeling for the properties of the specific network. The solution generally will simplify to some predictable form when these special cases or conditions are introduced. The specific conditions most readily employed are the following.

(i) Specific times: Choose t equal to zero, infinity, or perhaps some other specific value which may be suggested by the form of the equation or the arrangement of the network.
(ii) Specific frequency: Choose f equal to zero, infinity, or some other value which may be suggested by the form of the equation or the arrangement of the network. For example, at high frequencies a capacitor acts like a short circuit while an inductor acts like an open circuit. At low frequencies a capacitor acts like an open circuit while an inductor acts like a short circuit.
(iii) Specific component values: Choose the resistance, capacitance, or inductance of a particular component equal to zero, infinity, or some other value suggested by the form of equation or the arrangement of the network. High resistance, low capacitance, and high inductance act like open circuits; low resistance, high capacitance, and low inductance act like short circuits.

As an example application of the above tests, consider the network shown in Fig. 1(A) and its transfer admittance, Eq. (31). When the capacitance is large (short circuit) the network reduces to just a resistor and the transfer admittance is then 1/R. When the capacitance is small or when the resistance is large (open circuit), the network is an open circuit, and the transfer admittance is zero. When the resistance is small (short circuit), the network is effectively just a capacitor, and the transfer admittance becomes sC. When s is small (i.e., when the slope of the input voltage is small), Eq. (31) becomes $\bar{Y} = sC$; hence, the current is approximately the derivative of the voltage. Thus, under conditions of high and low capacitance and resistance, Eq. (31) behaves properly. The equation also predicts that under certain conditions the network can be used as an approximate differentiator.

## III. THE OPERATIONAL AMPLIFIER AND ITS APPLICATION AS A COMPUTING ELEMENT

A simple electronic amplifier, which is a device that is used to make a small signal larger, can be considered to be performing a mathematical operation analogous to multiplication of the input signal by a constant factor which is the gain factor of the amplifier. Amplifiers of appropriate design (discussed below) called operational amplifiers can also be used for performing this operation as well as a large variety of other mathematical operations such as summation, subtraction, and integration and differentiation with respect to time, etc.

### A. General Electronic Characteristics

An operational amplifier is a wide-band amplifier [dc to MHz (in special cases) band pass] which has a very high open-loop gain ($10^4$ to $10^9$ depending on quality) and a high input impedance ($10^6$ to $10^{14}$ Ω). Also the operational amplifier is especially designed to be stable and free from drift. Also it is what is called an inverting amplifier, i.e., it has a negative gain, to provide for stabilization when feedback is employed as discussed below. The block diagram in Fig. 5 is a general schematic circuit which illustrates the basic principle of the feedback network of this amplifier and how it is connected to

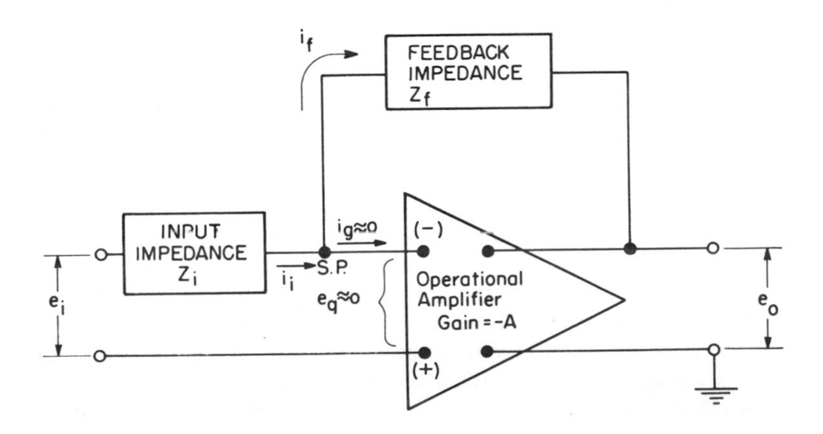


Fig. 5. Schematic diagram of an operational amplifier circuit.

perform a generalized mathematical operation. An input impedance network is connected from a signal input to the direct input of the operational amplifier. This direct input point is called the summing point (S. P.), and a feedback impedance network is connected from the amplifier output to the S. P. also. The most important characteristics of an operational amplifier, as explained below in detail, are that it draws only a negligible current $i_g$ and that the potential of the S. P., $e_g$, is practically constant and near ground potential, i.e., $e_g \approx 0$ is called a virtual ground [12-15]. Thus, in operation the current in the input impedance $i_i$ and in feedback impedance $i_f$ must, of course, be identical since the current into the operational amplifier is for all practical purposes zero:

$$i_i = i_f. \tag{36}$$

From Ohm's law, Eq. (36) may be written in terms of the operational voltages and impedances (see Fig. 5) as

$$\frac{e_i - e_g}{Z_i} = \frac{e_o - e_g}{Z_f}. \tag{37}$$

The expression for the gain of the operational amplifier is

$$e_o = -Ae_g. \tag{38}$$

Thus, by substituting for $e_g$ in Eq. (37) with Eq. (38) and solving for $e_o$, the following relationship is obtained:

$$e_o = \frac{-(Z_f/Z_i)e_i}{1 + (1 + Z_f/Z_i)/A}. \tag{39}$$

When the value of the gain A is very large compared to $1 + Z_f/Z_i$, as in the case for any well-designed operational amplifiers used with the input and feedback impedance values, then

$$e_o \approx \frac{Z_f}{Z_i} e_i. \tag{40}$$

Because of the high gain, the value of $e_g$ is virtually zero, hence, the reason for the S. P. to be termed a virtual ground. From Eq. (40) it is obvious that the mathematical operation performed on the input signal $e_i$ is $Z_f/Z_i$ and is thus determined only by the nature of the function $Z_f/Z_i$. The most general approach is to write Eq. (40) in so-called operational form: output voltage transform = network transform × input excitation transform or

$$\bar{e}_o(s) = -[\bar{Z}_f(s)/\bar{Z}_i(s)]\bar{e}_i(s) \tag{41}$$

where s corresponds to the Laplace parameter as discussed in the first

TABLE 4

Transfer Impedance Z(s) for Some Common Passive Networks

| Network | Transfer impedance Z(s) |
|---|---|
|  R | $R$ |
| 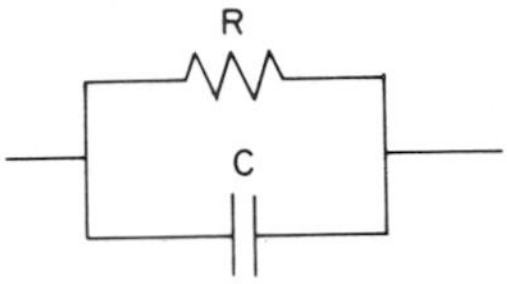 R, C | $\frac{R}{1 + sRC}$ |
| 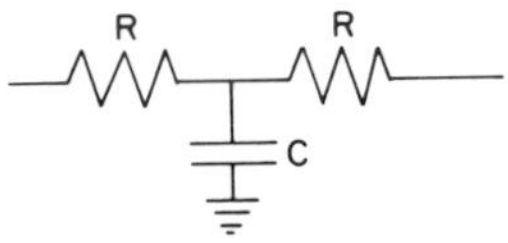 R, R, C | $2R\left(1 + s\frac{RC}{2}\right)$ |
|  C | $\frac{1}{sC}$ |
| 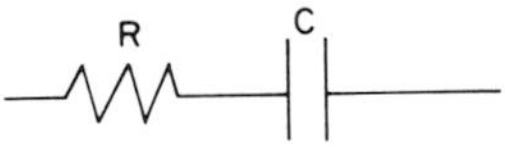 R, C | $R + \frac{1}{sC} = \frac{1}{sC}(1 + sRC)$ |
| 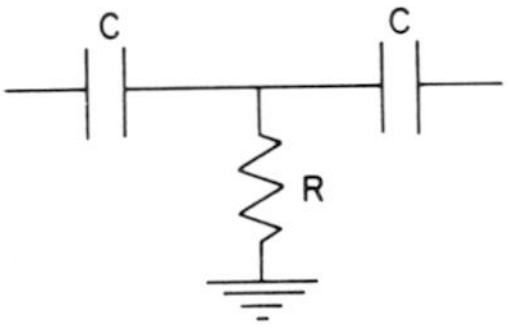 C, C, R | $\frac{1}{sC/2}\left(\frac{1 + s(2RC)}{s(2RC)}\right)$ |

section of this chapter. Table 4 lists the transfer impedances $\bar{Z}(s)$ for the more common networks employed in chemical computation and instrumentation. (For further tables of transfer impedances, see refs. [5-7].)

## B. Application as a Computing Element

A few typical and useful mathematical operation circuits will be described here in considerable detail so that the reader can fully understand the action of the amplifier itself and the way in which the operational (or network) transforms are obtained, how the operation is performed on the signal, and the principle of negative feedback. There are, of course, hundreds of operational configurations that are possible. Most of these, plus explanations, can be found in the manufacturers' literature [12-15] and handbooks [5].

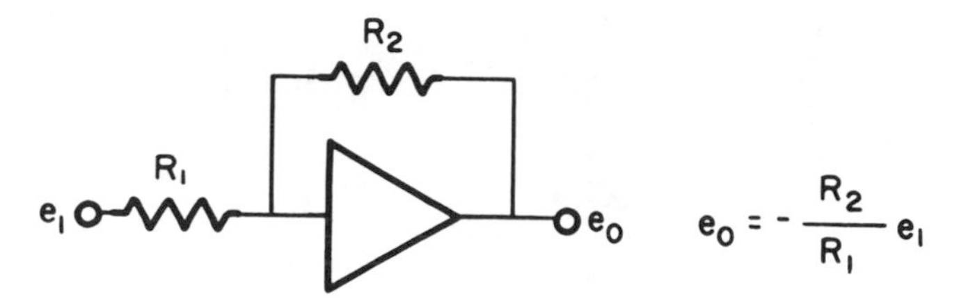


Fig. 6. Multiplication by a constant.

Note in Figs. 5-17 that only the circuit component of the input network and the feedback network are shown and the amplifier is represented simply by a triangle symbol. All other circuitry not involved in the operational transforms such as balancing circuits, power leads, ground connections, etc., are omitted but, of course, understood to be present in the actual circuit.

## 1. Multiplication by a Constant

When the input and feedback impedances of an operational amplifier both are pure resistances, as shown in Fig. 6, the system behaves as a simple inverting amplifier of gain equal to the ratio $R_2/R_1$. The accuracy of multiplication is limited by how well the ratio $R_2/R_1$ (the accuracy of the precision resistors employed) is known and by the gain of the amplifier relative to this value. For example, in using the exact expression [Eq. (39)] for this network and with the operational amplifier having a gain of $10^5$, the use of $R_2/R_1 = 100$ will result in an actual multiplication of 99.9, an error of only 0.1%. As can be seen from Eq. (39), however, the use of higher ratios of feedback-to-input impedances leads to higher errors, and it is also obvious that small changes in the gain of the amplifier (resulting from power supply variations, etc.) do not cause appreciable changes in the multiplication factor for the operational circuit, provided that the value of A is much larger than the quantity, $1 + R_2/R_1$.

## 2. Summation (with Variable Coefficients)

The circuits in Fig. 7(A) differ from that of Fig. 6 in that several input impedances and potential sources are employed in the input network shown in Fig. 7(A). Note that the current flowing from the input circuit into the S. P. is the sum of the individual currents through resistors $R_1$-$R_n$ and is equal, of course, to the current flowing through the single feedback resistor $R_f$. Thus, the output voltage is then equal to

$$e_0 = -\left(\frac{R_f}{R_1}e + \frac{R_f}{R_2}e_2 + \frac{R_f}{R_3}e_3 + \cdots + \frac{R_f}{R_n}e_n\right). \tag{42}$$

If the values of the input resistances $R_1$ to $R_n$ are equal to each other and to the value of $R_f$ ($R_1 = \cdots = R_n = R_f$), the output voltage will be equal to the

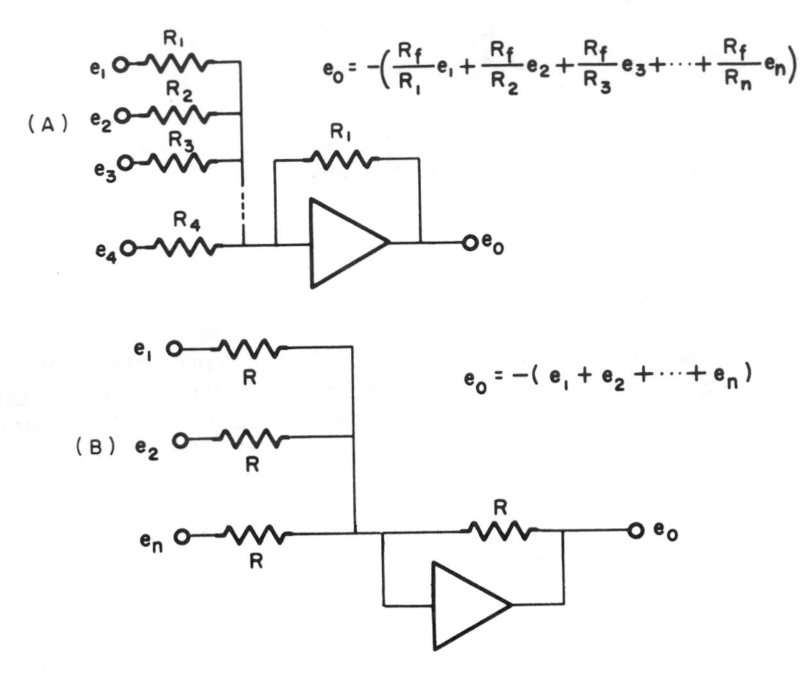


Fig. 7. Weighted adder (and summer) circuit.

sum of the input voltages $e_1$ to $e_n$ (although opposite in sign), as each coefficient is equal to 1 as shown in Fig. 7(B). The circuit may be employed for subtraction if one of the input voltages is made negative. Equation (42) is the general form of a weighted adder or summer.

3. Integration

The operational amplifier of Fig. 8(A), which has a capacitor in the feedback loop, will have a response determined by the ratio of the corresponding transfer impedances of Table 4,

$$\bar{e}_o(s) = -\frac{[1/sC]}{R} \times \bar{e}_i(s) = -\frac{1}{sRC} \times \bar{e}_i(s) \tag{43}$$

where the equation is written in operational form.

Equation (43) may also be expressed in differential equation form as

$$e_o(t) = -\frac{1}{RC}\int e_i \, dt. \tag{44}$$

Thus, the output is the negative integral of $e_i$ multiplied by a coefficient $1/RC$. If the true integral alone is wanted, it is only a matter of making $1/RC = 1$.

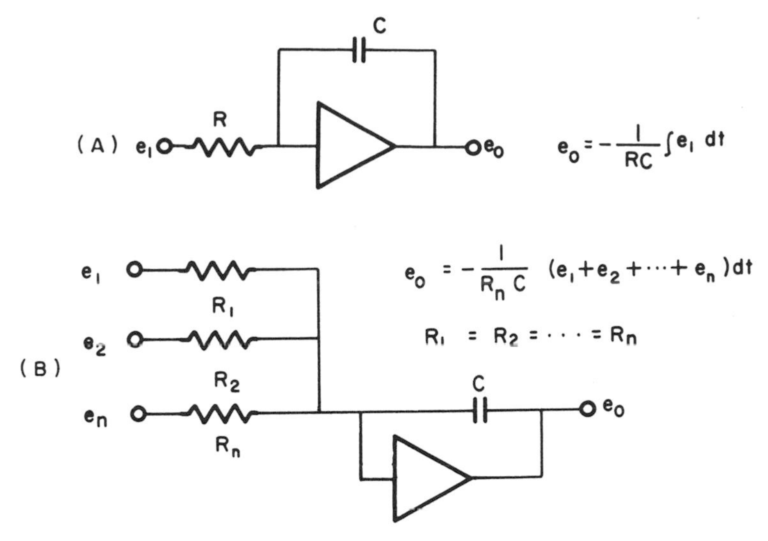


Fig. 8. (A) Integrator circuit. (B) Summing integrator.

Equation (43) may also be derived by substituting into Eq. (36) the appropriate expressions for current flowing in the resistor and capacitor yielding

$$\frac{e_i}{R} = -RC\,\frac{de_o}{dt} \tag{45}$$

and by integrating this relationship under the assumption of zero initial charge on the capacitor.

For sinusoidal inputs of frequency $\omega$ (= $2\pi f$), one substitutes the quantity of $j\omega$ for s in Eq. (43) to obtain the steady-state response

$$\frac{e_o}{e_i} = -\frac{1}{j\omega RC}. \tag{46}$$

Thus, the response is a sine wave with a $90^\circ$ phase angle relative to the output.

Note in Fig. 8(B), that by adding resistors to the input network a summing integrator is obtained.

## 4. Differentiation

By simply reversing the placement of the capacitance and resistance of the input feedback positions, as shown in Fig. 9, one obtains a differentiating

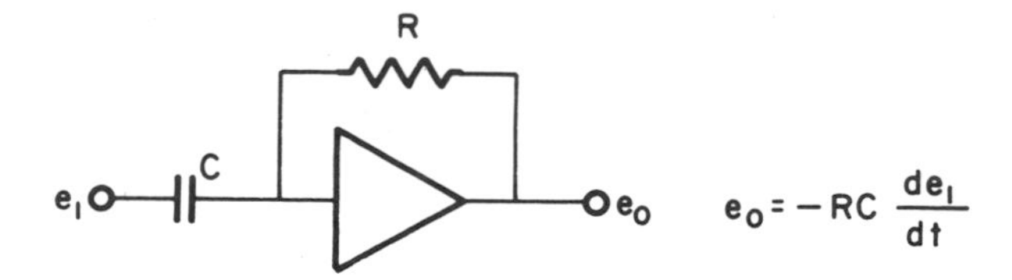


Fig. 9. Simple differentiator circuit.

response as can be shown by making the appropriate substitution from Table 1 into Eq. (40),

$$\bar{e}_o(s) = sRC \times \bar{e}_i(s), \tag{47}$$

which in the time domain is given on taking the inverse transformation by

$$e_o(t) = -RC \frac{de_i}{dt}. \tag{48}$$

This is a circuit of "perfect" differentiating action but is seldom used as such in actual practice. This is because all circuits and/or signals contain a finite level of noise — extraneous signals, some of which are generally of a high-frequency type — which means that the value of $de_i/dt$ for the noise will be extremely large. Thus, this circuit is in practice a potential "super-noise generator" and only special care in circuit design allows it to be used in this "perfect differentiator" form [16]. In order to discriminate against such unwanted response, the circuit shown in Fig. 10 is usually employed in practice. From Table 4 one can obtain the response of this circuit which is given by

$$\bar{e}_o(s) = -\frac{sR_2C_1}{(1 + sR_1C_1)(1 + sR_2C_2)} \times \bar{e}_i(s). \tag{49}$$

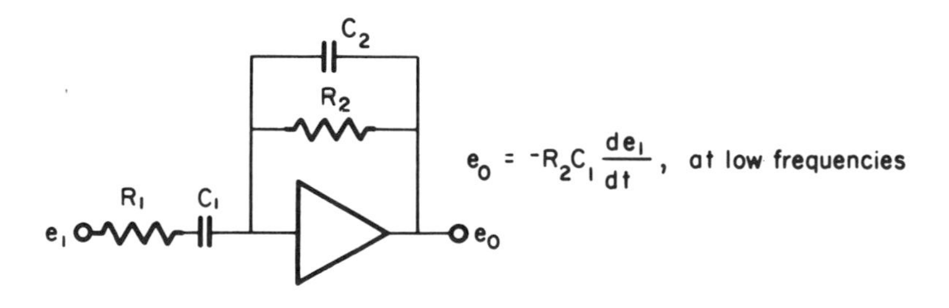


Fig. 10. Practical differentiator circuit.

(Usually the value of product $R_2C_2$ is made equal to that of $R_1C_1$.) Thus, the steady-state response to sinusoidal inputs, found by the substitution of $j\omega$ for s, is

$$e_o(j\omega) = -\frac{j\omega R_1 C_1}{(1 + j\omega R_1 C_1)^2} \times e_i(j\omega). \tag{50}$$

At low frequencies (where $\omega R_1 C_1$ is much less than unity), the response is that given by Eq. (47) or (48), i.e., excellent differentiation action for low-frequency signals. As indicated in Eq. (50), however, the circuit gain decreases rapidly beyond a critical frequency ($\omega = 1/R_1C_1$). At high frequencies ($\omega > 1/R_1C_1$), the differentiating action decreases 100-fold for each 10-fold increase of frequency. One can see that, at high frequency, Eq. (50) reduces to

$$e_o(j\omega) = -\frac{1}{(j\omega)RC} e_i(j\omega) \tag{51}$$

or

$$e_o(t) = -\frac{1}{RC}\int e_i \, dt. \tag{52}$$

Thus, for high frequencies, the circuit is acting as an "integrator" and, of course, the integral of random noise is equal to zero.

## 5. Double Integrator

The operational response for the circuit shown in Fig. 11, obtained from the transfer impedances of Table 1, is

$$\bar{e}_o(s) = -\frac{4}{R^3C^3s^3} \times \bar{e}_1(s). \tag{53}$$

The response in differential form is given by

$$e_o = -\frac{4}{R^2C^2}\iint e_i \, dt \, dt \tag{54}$$

where the circuit behaves as a double integrator (with the coefficient $4/R^2C^2$). Note that a sinusoid input will lead to a sinusoid output of the same phase. If the output is connected to the input, the circuit will usually oscillate at that frequency where the input signal is identical in magnitude to the output signal; thus $\omega = 2/RC$. If only a fraction of the output $\rho$ is fed back to the input, i.e., by a potentiometer, as shown in Fig. 11(B), the frequency of oscillation will be given by $\omega = 2\sqrt{\rho/RC}$.

## 6. Variable Coefficient Multipliers

Often it is useful to have a variable gain (multiplication by a variable factor) amplifier in computation and instrumentation applications. It is a simple matter just to include a precision multiturn potentiometer in the

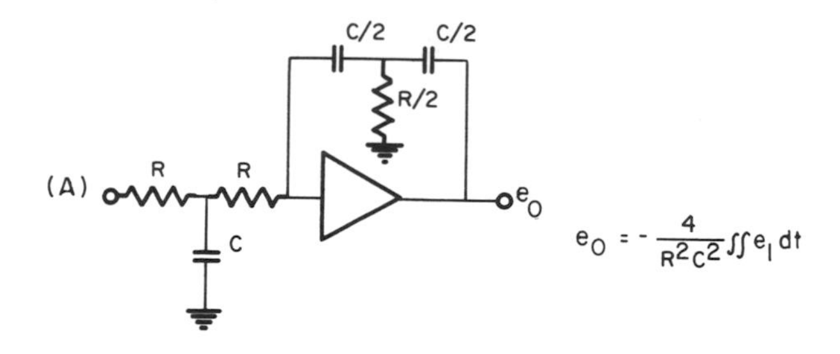


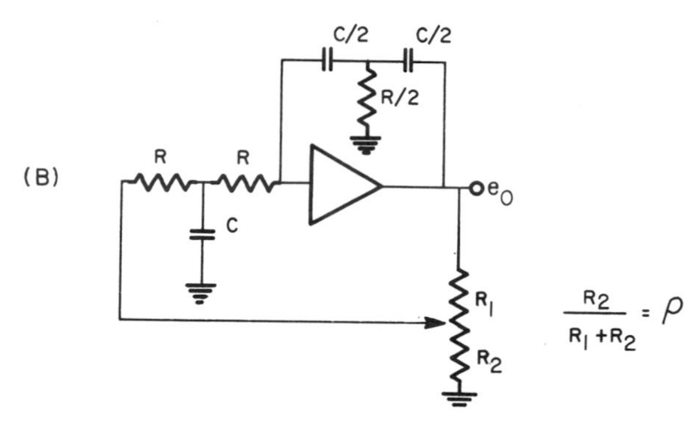


Fig. 11. (A) Double integrator circuit. (B) Oscillator circuit.

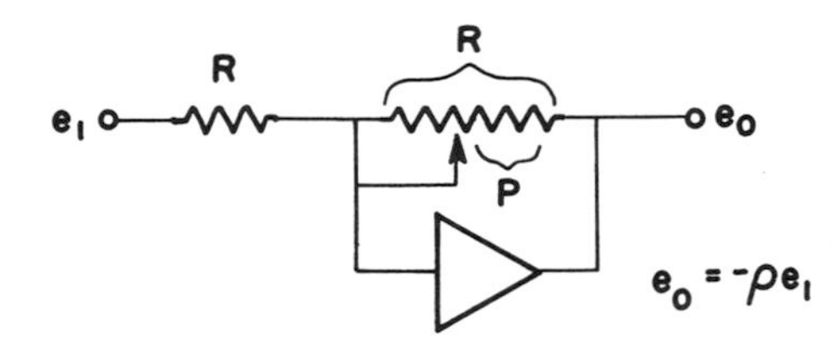


Fig. 12. Variable coefficient multiplier (ρ varies from 0 to 1).

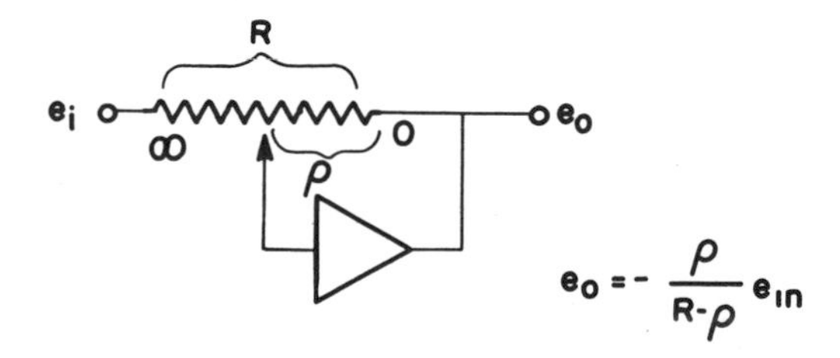


Fig. 13. Variable coefficient multiplier (coefficient varies from 0 to ∞).

circuit of the operational amplifier. Two circuit configurations are generally used. By means of a simple variable resistance in the feedback loop, Fig. 12, one obtains an amplifier whose coefficient can vary from 0 to 1.00. Figure 13 shows a configuration which gives a coefficient which can be varied from 0 to ∞ (in theory only — in practice the current and potential limitations of an operational amplifier will limit this range).

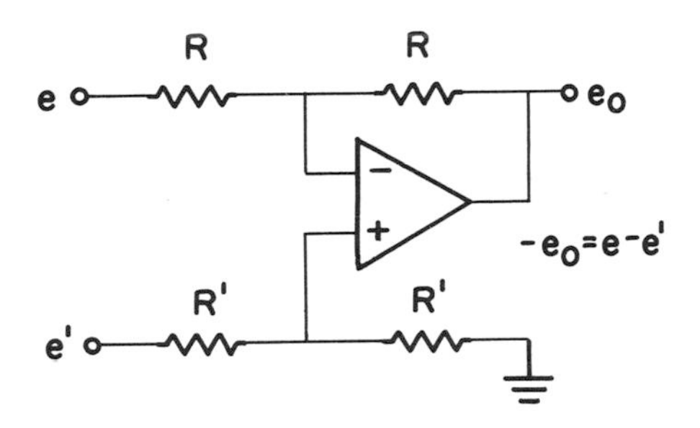


Fig. 14. Subtractor circuit.

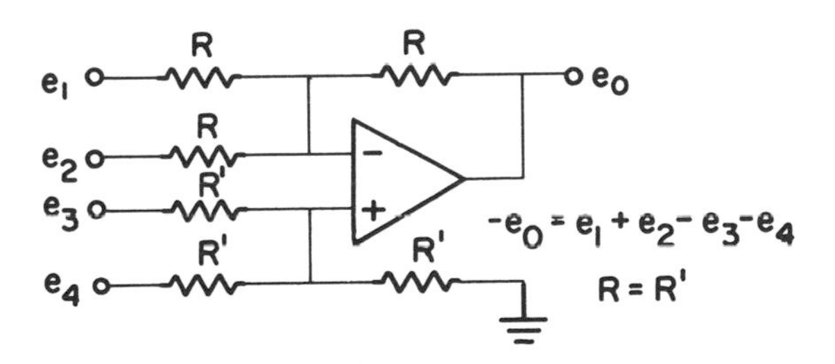


Fig. 15. Adder-subtractor circuit.

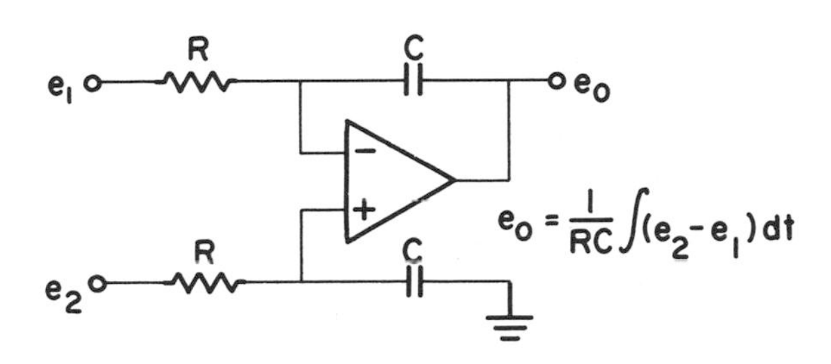


Fig. 16. Subtractor-integrator circuit.

## 7. Subtraction Operations

Operational amplifiers may be obtained with a differential input [5-7, 12-15]. This means that there is the usual inverting (negative) input and a noninverting (positive) input with respect to a common ground as shown in the amplifiers in Figs. 14-16. This type of operational amplifier can be thus employed for subtracting two signals (the signal to the inverting input will be subtracted from that applied to the noninverting input). Three typical subtraction applications are illustrated in Figs. 14-16 along with the equations describing their response functions.

## 8. Logarithmic Operations

It should be pointed out that it is not necessary for the elements in the input and/or feedback networks of an operational amplifier to be linear. For example it is possible to take advantage of the exponential (nonlinear) vs. E response of certain transistors and/or diodes to obtain logarithmic and antilogarithmic operations. Typical logarithmic- and antilogarithmic-response operational amplifier circuits are shown in Figs. 17 and 18. Such

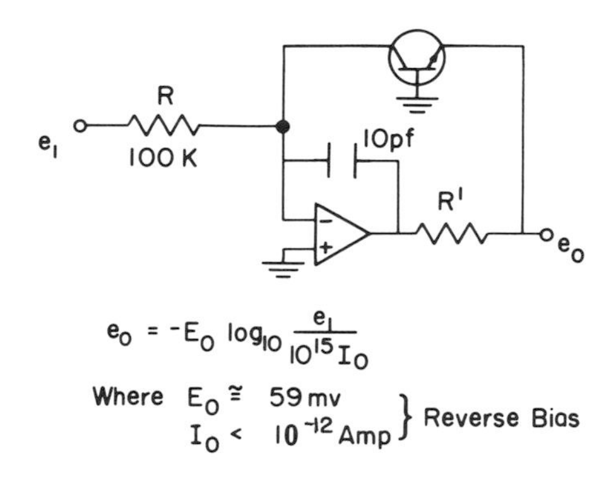


Fig. 17. Logarithmic circuit.

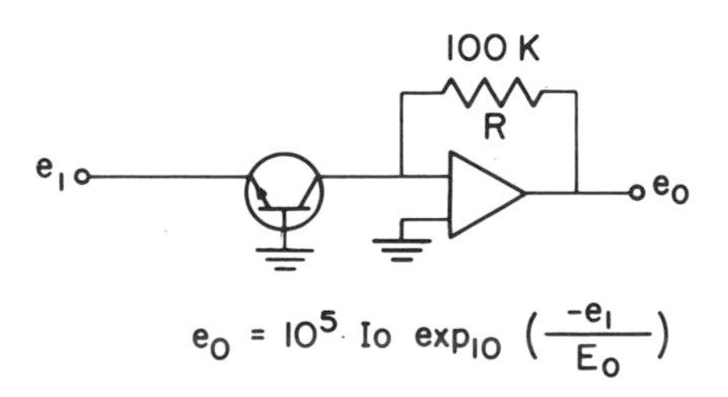


Fig. 18. Antilog circuit.

circuits are very useful in that they are the bases for many time-function multiplier-divider circuits

$$e_o = e^1_{in} \times e^2_{in} \quad \text{or} \quad e^1_o = e^2_{in}/e^2_{in}.$$

(In order to perform time-function multiplication and division it is necessary to convert the signals to log functions, then sum, and then perform an anti-log operation [12-15]. Applications of log circuits and function multiplier-dividers are discussed in general [14, 15] and in specific chapters on general spectrophotometric apparatus [16, 17] for kinetics [18, 19], to be discussed in a chapter on instrumentation for kinetic studies in a subsequent volume, etc.) Other examples of typical nonlinear operation modules are discussed in refs. [12-15].

## IV. THE ANALOG COMPUTER

As any algebraic or differential equation is in reality a collection of individual mathematical operations, it is a relatively simple matter to interconnect individual operational amplifiers which have the appropriate operational transform circuitry representing each specific operation of the equation in a sequence which is the exact analog of the equation. Actually the logic employed in the interconnection of the individual modules is quite straightforward and simple. A few algebraic and differential equations representing typical chemical and physical systems will be described in detail below. The reader can find numerous other examples in the books of Korn and Korn [6] and Huskey and Korn [8]. Many more specific and complex applications are presented in the series edited by Mattson et al. [20]. The material presented here is intended to be an introduction to these more complex applications in instrumentation, simulation, and computation by analog devices.

### A. Algebraic Equations

The mathematical task of solving two simultaneous algebraic equations in two unknowns is perhaps the simplest possible example to illustrate the principle of setting up individual operational modules to be the electronic analog of an equation(s) or problem. Consider the problem of finding the values of x and y in the following equations:

$$x - 2y = 3, \tag{55}$$

$$\frac{x}{3} + y = 3. \tag{56}$$

The first step in approaching the electronic solution of Eqs. (55) and (56) is to solve one equation for x and one equation for y as illustrated below:

$$x = 3 + 2y, \tag{57}$$

$$y = 3 - \frac{x}{3}. \tag{58}$$

At this point it is convenient to recognize in advance that the operational amplifier is a voltage-inverting device. Thus, it is a simple matter of rewriting the above equations in terms of a sign-inverted expression for x and y:

$$x = -(-2y - 3), \tag{59}$$

$$y = -\left(\frac{x}{3} - 3\right). \tag{60}$$

Equation (59) shows that the unknown x is mathematically equal to the negative sum of the two negative quantities -2y and -3. Thus, a summing operational amplifier can be connected to give x at the output, provided that the appropriate inputs can be made available as shown in Fig. 19. The coefficients or gain factors for the two inputs are conveniently obtained by proper selection of the input and feedback resistors. The same approach can be employed to obtain y as the output of another summing amplifier as shown in Fig. 20.

Now it is only necessary to connect the x output from the first amplifier to the x input of the second amplifier (Fig. 20) and then take the y output of the second amplifier, pass it through an inverter to obtain -y and connect this to the -y input of the first amplifier. The overall circuit thus obtained is shown in Fig. 21 and represents the exact electrical analog of Eqs. (59) and (60). The solutions to these equations (the values of x and y) are obtained as voltages (1 V = unity) at the outputs of amplifiers 1 and 2 of Fig. 21.

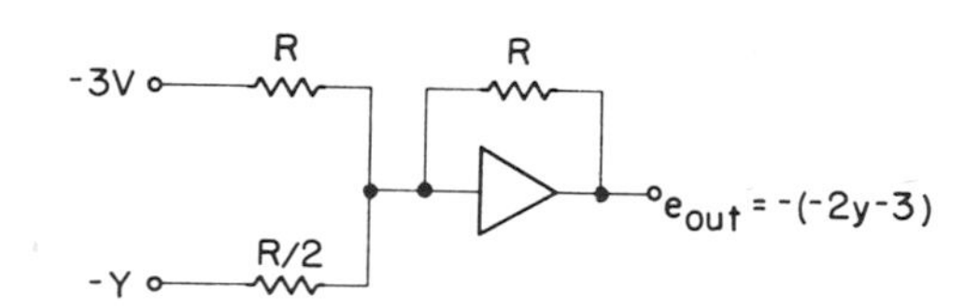


Fig. 19. Analog generation of x [Eq. (59)].

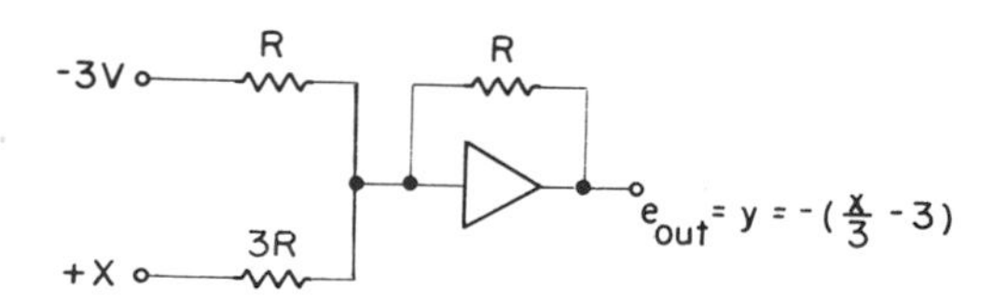


Fig. 20. Analog generation of y [Eq. (60)].

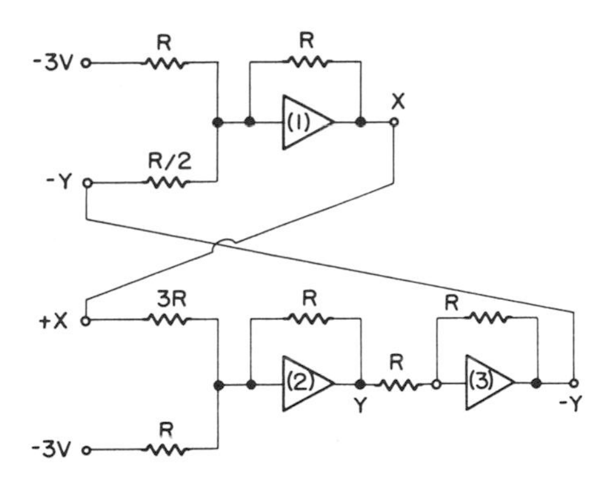


Fig. 21. Computer simulation of Eqs. (59) and (60).

1. Oscillation Considerations (Stability)

a. Closed-Loop Gain. It is important to note here that the loop gain of the three amplifiers of Fig. 21 is less than 1. The net gain of several amplifiers connected together is given by the product of the gain of each individual amplifier in the closed loop. In this particular case, the over-all gain would be $2 \times 1/3 \times 1 = 2/3$ for the loop starting with the -y input of amplifier 1. As the over-all gain is less than 1, the over-all configuration is stable and cannot oscillate. But if the following two equations are solved simultaneously:

$$x - 2y = 3 \tag{61}$$

$$3x + y = 3 \tag{62}$$

the computer network needed for their solution is given in Fig. 22. Note that the absolute closed-loop gain in this case is greater than 1 $(2 \times 3 \times 1 = 6)$. If the amplifiers were connected to solve this equation, as

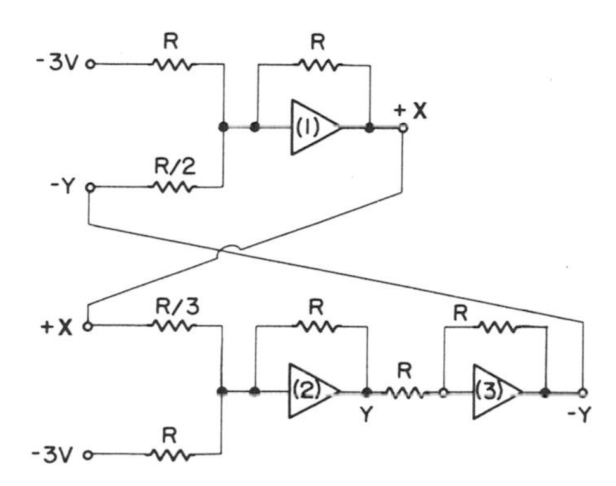


Fig. 22. Computer simulation of Eqs. (61) and (62).

shown in Fig. 22, the system would undoubtedly oscillate at some frequency and not give the correct answers.

This oscillation arises from positive feedback which occurs at high frequencies. For example, if a positive voltage is applied to amplifier 1 at the y input and, after passing through amplifiers 1, 2, and 3, it is amplified (made larger in magnitude) and is fed back to the starting point with the same sign as it had originally, it will enhance the total signal at this input. This means that the signal will continue to grow in magnitude until the amplifiers are driven to the limit (saturation). The question arises: with an odd number of inverting amplifiers, how could the signal be fed back with the same sign? This is easily shown by considering the gain vs. frequency and the phase shift vs. frequency of the computing loop.

b. Phase Shift and Gain Characteristics. Figure 23 is a schematic plot of the logarithm of the gain A of amplifiers 1, 2, and 3 of Fig. 21 vs. the logarithm of the frequency f and a corresponding plot of phase shift vs. logarithm of frequency. The total gain for the three amplifiers is the product of the individual gains of the amplifers; but only at low frequencies, it is given by $2 \times 3 \times 1 = 6$. At high frequencies, the gain of each individual amplifier begins to roll off [14, 21, 22]. This is typical behavior for any amplifier system. Also, the phase angle begins to shift [14, 21, 22]. At low frequencies the net phase shift for the three amplifier loops is $180^\circ$, i. e., sign inversion of the input. But at $f_1$, a certain frequency at the

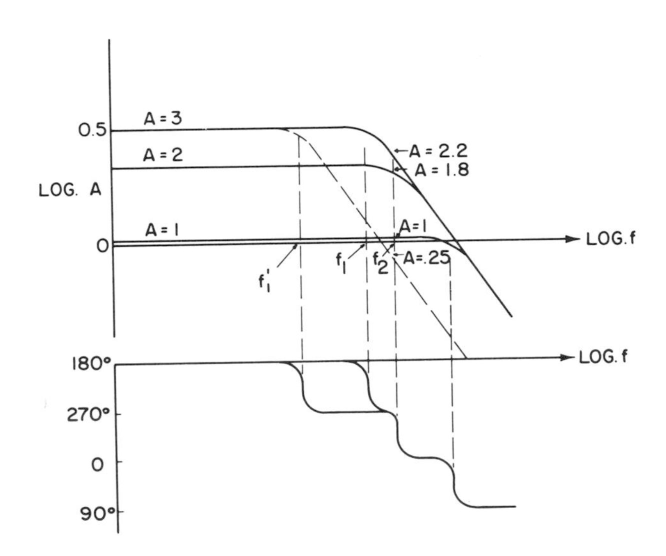


Fig. 23. Gain and phase shift of Fig. 22 computer loop as a function of frequency.

roll-off point of amplifier 2, there is a $90^\circ$ phase shift in amplifier 2 giving a net $270^\circ$ phase angle for the system, and at $f_2$ there is another $90^\circ$ shift in amplifier 1 giving a net $360^\circ$ phase angle for the loop. Thus, there is no net change in the phase or sign of the feedback signal if, from some stray voltage source, a transient noise of $f_2$ is accidentally fed into the input of this closed loop of the network; and if the over-all gain is greater than 1 at that frequency (as shown for this system in Fig. 22), the signal will be amplified and fed back to the input with the same phase. Since $f_2$ is the frequency at which the phase shift becomes $360^\circ$, or no net change in sign of the input, the network will oscillate at that frequency. It will then be amplified still further, until finally the amplifiers will reach their limit. This results in a serious error on the dc outputs, representing the solution to the simultaneous equations.

c. Stabilization. If the gain of the system at the frequency of oscillation is made to be less than 1, the oscillation will cease. However, it should be pointed out that the system may still be used with an over-all gain greater than 1 for low frequencies. This type of stabilization is accomplished by placing a capacitor in parallel with the feedback resistor of one of the amplifiers. As shown in Fig. 23, if a capacitor is placed in parallel with the feedback resistor of amplifier 2 a phase shift of $90^\circ$ for this amplifier occurs at a lower frequency $f_1'$, but now at $f_2$ where the total phase shift is still $360^\circ$, the gain of amplifier 2 is only about 0.25. Although the gain of amplifier 1 is still about 1.8 and the gain of amplifier 3 is still 1.0, the over-all loop gain is now $0.25 \times 1.8 \times 1.0 = 0.45$ at $f_2$, which is less than 1. Thus the system will no longer oscillate, for at frequencies where the gain is greater than 1 the feedback signal is not of the same phase or sign as the input.

The capacitor used to prevent oscillation should be as small as possible, as the larger the capacitor, the more the frequency response is limited. The steady-state transfer function for the three amplifiers of Fig. 22, with a capacitor in parallel with the feedback resistor of amplifier 2, is given by

$$A = \frac{E_o}{E_i} = -\left(\frac{R_{o1} \times R_{o2} \times R_{o3}}{R_{i1} \times R_{i2} \times R_{i3}}\right)\left(\frac{1}{R_{o2} j\omega C + 1}\right). \tag{63}$$

This relationship allows one to approximate the size of the capacitor to be used. If one wishes A, the gain, to be less than 1 at $f_2$, and if the value of $(R_{o1} \times R_{o2} \times R_{o3})/R_{i1} \times R_{i2} \times R_{i3})$ is known, and $\omega$, the frequency of oscillation, can be measured, then the value of C may be chosen such that the right-hand side of Eq. (63) is less than 1.

## B. Differential Equations

A linear differential equation, such as the example given by Eq. (64), can easily be solved by an electronic analog computer network, using a combination of integration and summation operations, with operational amplifier elements. As will be shown later, this fundamental type differential equation,

$$a_2 \frac{d^2x}{dt^2} + a_1 \frac{dx}{dt} + a_0 x = f(t) \tag{64}$$

with appropriate values for the coefficients $a_2$, $a_1$, $a_0$, and f(t), can be used to describe many chemical and physical systems.

As a first step in setting up such a differential equation in a form most suitable for analog computer programming, it is first solved for the highest derivative. Thus, the right-hand side of the resulting equation will then be a summation of terms

$$\frac{d^2x}{dt^2} = \frac{f(t)}{a_2} - \frac{a_1}{a_2}\frac{dx}{dt} - \frac{a_0}{a_2}x. \tag{65}$$

### 1. Chemical Applications

a. Simple Irreversible First-Order Kinetics. The rate of a chemical reaction [Eq. (66)] which follows irreversible first-order or pseudofirst-order kinetics can be expressed by a simple, linear differential equation,

$$A \xrightarrow{k_f} B. \tag{66}$$

The rate of change of concentration of A with respect to time can be expressed mathematically by the differential equation

$$\frac{dx}{dt} = k_f(A^o - x) \tag{67}$$

where $A^o$ is the initial concentration of A, x is the concentration of A that has reacted at any time t, and $k_f$ is the rate constant of the reaction. Note that Eq. (67) is simply (64) obtained when $a_2 = 0$, $a_0/a_1 = k_f$, and $f(t)/a_1 = k_f A^o$. Equation (67) is thus already arranged with only the highest derivative on the left-hand side of the equation. One of several possible computer networks which may be used to solve this equation is shown in Fig. 24. Amplifier 1 is a summing amplifier, and since it is inverting, the inputs +x and $-A^o$ (from a precision dc voltage source) give the sum $(A^o - x)$ at the output of amplifier 1. This is then multiplied by a factor $-k_f$ by amplifier 2, as $R''/R' = k_f$. Now $x = \int_0^t k_f(A^o - x)dt$, which means that if one integrates the quantity $-k_f(A^o - x)$ with respect to time, using the integrating amplifier 3, the output will be +x. The output of amplifier 3, +x, is then fed back to the appropriate input of amplifier 1 and a solution is obtained which is recorded as a function of time.

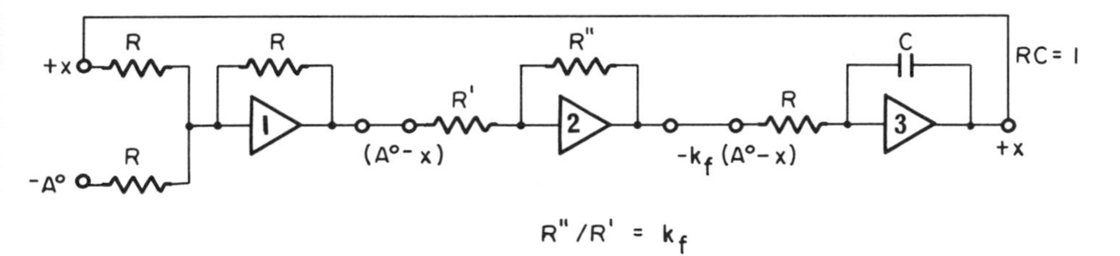


Fig. 24. Computer simulation of simple irreversible first-order reaction kinetics [solution of Eq. (67)].

b. Reversible First-Order Kinetics. The reversible first-order case may be solved, using four operational amplifiers:

$$A \underset{k_b}{\overset{k_f}{\rightleftarrows}} B. \tag{68}$$

The differential equation describing this reaction is

$$\frac{dx}{dt} = +k_f(A^o - x) - k_b(B^o + x) \tag{69}$$

where $A^o$ and $B^o$ are the initial concentrations of the species A and B, respectively, and x corresponds to the amount of B formed from A as the

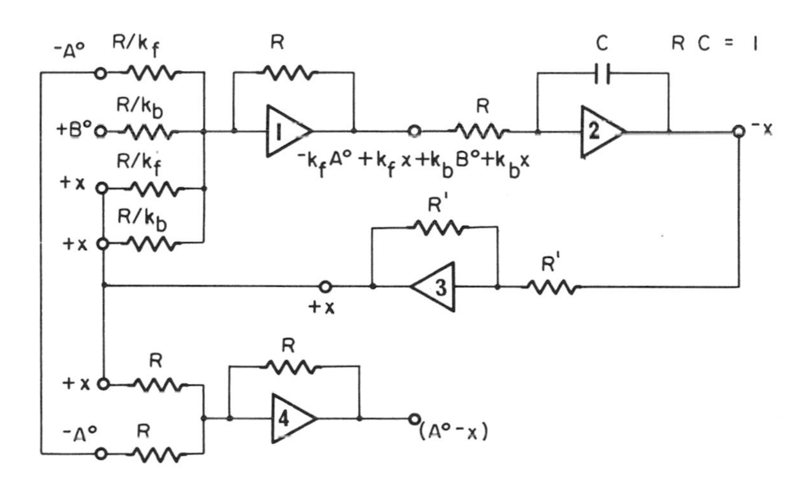


Fig. 25. Computer simulation of reversible first-order reaction kinetics [solution of Eq. (70)].

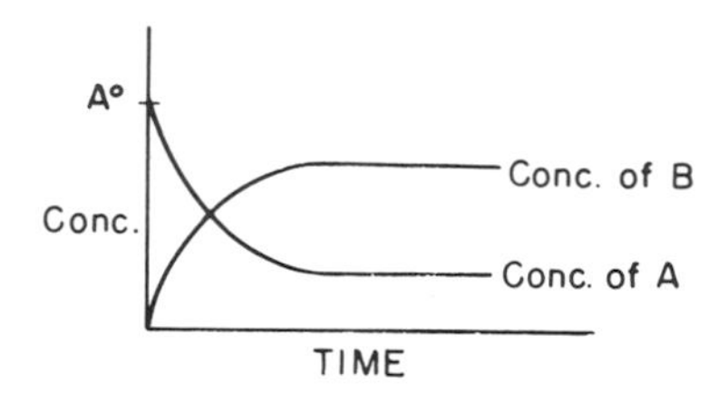


Fig. 26. Example output of simulation of Eq. (70).

reaction proceeds. Expanding the right-hand side of the above equation yields a weighted summation of terms:

$$\frac{dx}{dt} = -(-k_f A^o + k_f x + k_b B^o + k_b x). \tag{70}$$

A network may now be set up on the basis of this inverted sum (or difference) as shown in Fig. 25. Amplifier 1 gives the weighted (multiplied by $k_f$ or $k_b$) inverted sum of the quantities $-k_f A^o$, $+k_b B^o$, $+k_f x$, and $+k_b x$. Integration of this sum with respect to time by an integrating amplifier gives the quantity $-x$, which is then fed through an inverter to get $+x$ and then to the appropriate inputs of amplifier 1.

Usually in such a problem or simulation it is desirable to follow the change in concentration of A and B as a function of time. This requires a summing amplifier for each of the two quantities. Such an arrangement to follow the change in concentration of A is illustrated by amplifier 4 in Fig. 25 if the initial concentration of B is zero and the outputs of amplifiers 3 and 4 are hooked to a dual-beam oscilloscope or Y, Y′, X Recorder. A typical output would be that illustrated by Fig. 26 and which represents the simultaneous change in the concentrations of A and B.

c. Consecutive First-Order Reactions. Another interesting chemical kinetic case which may readily be solved or simulated by a computer network, if the initial concentrations of the reactants and the rate constants are known, is that of the consecutive irreversible first-order reactions,

$$A \xrightarrow{k_1} B \xrightarrow{k_2} C. \tag{71}$$

The solution will be programmed for a case where there is no initial concentration of B and C. Therefore, the differential rate expressions for the reactions are described by

$$\frac{dx}{dt} = k_1(A^o - x) = -k_1(x - A^o) \tag{72}$$

$$\frac{dy}{dt} = k_2(x - y) = -k_2(y - x) \tag{73}$$

where $A^o$ is the initial concentration of A, x is the amount of A which has reacted to form B, and y is the amount of B which has reacted to form C. As there is no initial concentration of B or C, then at any time t we have $B = x - y = -(y - x)$.

In setting up previous computer networks, there was no concern for the number of amplifiers employed. Separate amplifiers were used to sum and integrate. Often, however, the number of amplifiers of a computer is limited. This is of course not necessary as one amplifier can perform both operations, as shown in Fig. 8(B). The network shown in Fig. 27 can be used to record the change in the concentration of each component simultaneously.

Amplifier 1 integrates and inverts the weighted (multiplied by $k_1$) sum $(x - A^o)$. The output is, therefore, +x. Amplifier 2 inverts +x to give -x. This is then fed to amplifier 3, which integrates and inverts the weighted (multiplied by $k_2$) sum $(y - x)$. The output of this amplifier is then C. Amplifiers 4 and 5 give the change in the concentration of A and B, respectively, as shown in Fig. 27. The simultaneous recording of the outputs of amplifiers 3, 4, and 5 would give the curves of the shape shown in Fig. 28.

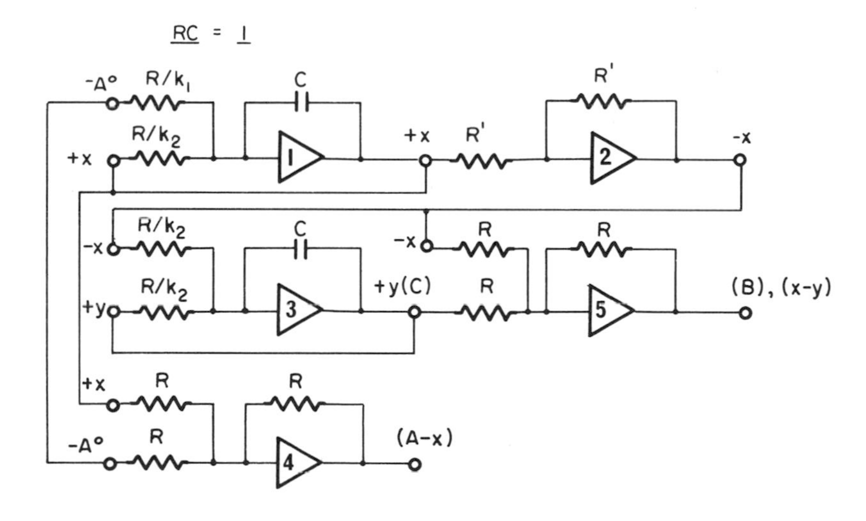


Fig. 27. Computer simulation of consecutive first-order reactions [solution of Eqs. (72) and (73)].

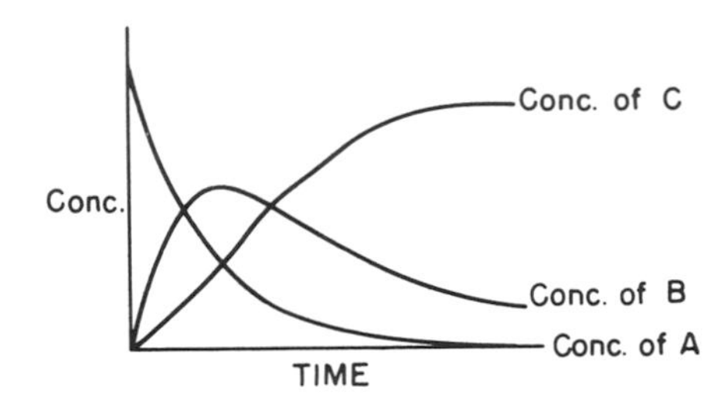


Fig. 28. Example output of the simulation of Eqs. (72) and (73).

2. Scaling Factors

a. Amplitude Scaling. Thus far, in this discussion of analog computer networks, the correct mathematical relations were established by allowing 1 V in the computer simulation to equal one unit in the original equations. This is not always desirable and in some cases it is not possible. It is important to keep in mind that there is a limit to operating voltage of the amplifiers themselves. Thus, one must consider that there is a limit to the magnitude of the machine variables. In order to utilize the optimum voltage range of the amplifiers and avoid the saturation limit it may be desirable, for example, to have 1 V correspond to 0.001 M concentration, 10 m, etc. Thus, in setting up the computer network, it is often necessary to scale the variables so that they are brought within the operating voltage of the amplifiers by multiplying the real variable by some appropriate factor, called the scale factor. Therefore, for a real variable x, the corresponding machine variable X is expressed by

$$X = a_x x \tag{74}$$

where $a_x$ is called the scale factor. Each voltage in the simulation may be given, numerically and dimensionally, by the product of the scale factor and variable. Conversely, the correct (actual) value of any variable can be determined from the corresponding voltage by dividing the machine variable by the scale factor.

The choice of scale factors is governed by two conflicting considerations: (i) scale factors should be as large as possible in order to minimize percent errors owing to stray voltages, drift, etc., and (ii) machine variable(s) or amplifier outputs must not exceed the voltage limit of the amplifier.

As a simple example, suppose that one wished to solve the following simultaneous algebraic equations:

$$x - 2y = 300, \tag{75}$$

$$\frac{x}{3} + y = 300. \tag{76}$$

The network necessary for the solution of these two simultaneous equations is the same as that shown in Fig. 21, except that the dc voltage inputs of amplifiers 1 and 2 would be -300 V instead of -3V. Solving Eqs. (74) and (75) simultaneously will yield the answers x = +540 and y = +120. As these answers must be given in terms of voltages by the computer network, the amplifiers would be driven to saturation and the solution would be incorrect. Therefore, the use of amplitude scaling is necessary.

By letting $a_x = 1/100$, then $X = (1/100)x$ or $x = 100\,X$, which simply means that the machine voltage representing x is multiplied by 100 to get the actual value of x. By letting $a_y = 1/10$, then $Y = (1/10)y$ or $y = 10Y$, and the voltage representing Y is multiplied by 10 to get the true value of y. Before actually programming the computer, these new values for x and y are substituted into the original equations to obtain the "machine" equations:

$$100X - 2(10Y) = 300 \quad \text{or} \quad X = 3 + 0.20Y, \tag{77}$$

$$\frac{100X}{3} + 10Y = 300 \quad \text{or} \quad Y = 30 - 3.33X. \tag{78}$$

The analogous computer network is given by Fig. 29 and it will give an output of 12 V representing y and an output of 5.4 V representing x.

b. Time Scaling. In the same manner, it may not be possible to let one unit of magnitude of the variable time equal to 1 V in those problems involving time. The computer time, of course, can be made any definite fraction or multiple of real time so that computer solutions can be obtained in a convenient time span. This process is called time scaling.

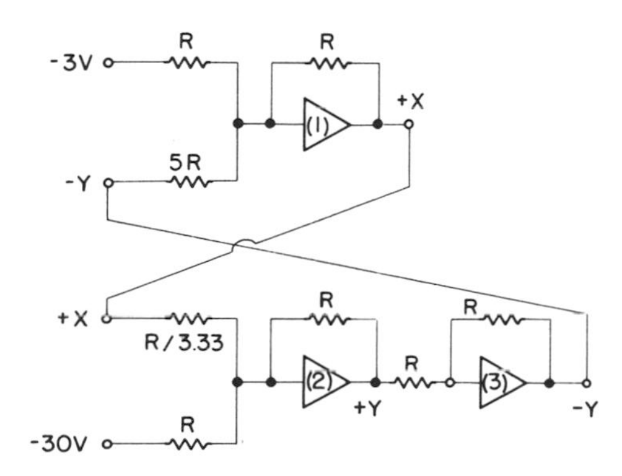


Fig. 29. Computer simulation of Eq. (78).

The time scale is established by writing a transformation equation relating the computer time T to the actual time t, much the same as was done in amplitude scaling. Thus,

$$T = a_t t \tag{79}$$

where $a_t$ is the time-scale factor. Similarly,

$$p = \frac{d}{dt} = a_t \frac{d}{dT} = a_t P \tag{80}$$

where for convenience p is an operator that denotes differentiation with respect to real time and P is an operator that denotes differentiation with respect to computer time. Again, the value of real time at any instant in the computation is obtained by dividing the elapsed computing time by the time-scaling factor.

For example, consider the simple differential

$$\frac{dy}{dt} = 50 \qquad \text{or} \qquad py = 50. \tag{81}$$

If it is desired that computer or machine time be 100 times real time, then $a_t = 100$, or $T = 100t$, and then $p = 100P$. Substituting 100P for p in Eq. (81) gives

$$Py = 0.50. \tag{82}$$

If Eq. (82) is integrated, one obtains $Y = 0.50T$ or $T = y/0.50$. Thus, if a value for y of 1.0 V is obtained from the computer in 1 sec, the real time required to give a value of 1 for y would have been 0.01 sec.

To further illustrate how time and amplitude scale factors are substituted into the actual equation to obtain the machine equation (on which the computer network is based), consider the general differential Eq. (64). For example, using $a_x = 0.10$ and $a_t = 100$, then $t = T/100$, $p = 100P$, and $x = X/0.1$. In terms of the operator p, Eq. (63) is expressed as

$$a_2 p^2 x + a_1 px + a_0 x = f(t). \tag{83}$$

Substituting the chosen scale factors for p and x gives

$$a_2(100P)^2(10X) + a_1(100P)(10X) + a_0(10X) = f(T). \tag{84}$$

This can be simplified to the form

$$a_2 P^2 X + 0.01a_1 PX + 10^{-4} a_0 X = 10^{-5} f(T). \tag{85}$$

The computer simulation network is then set up on the basis of Eq. (85).

It is obvious from Eq. (83) that both time and amplitude scaling affect the computer output together, and a poorly chosen combination of time and amplitude scale factors may still cause the amplifiers to limit. Practice and a familiarity with the type of problem to be solved makes the proper selection of scale factors much easier.

c. Frequency Scaling. Frequently, the output of a computer solution, which is usually to be recorded, is an oscillating signal. Naturally, there is a limit to the frequency response of any recording device. This means that it is often necessary to have some idea of the frequency of the computer output for a particular problem in order to choose the appropriate scale factors.

For an equation such as $a_2(d^2x/dt) + a_1(dx/dt) + a_0 x = f(t)$, the undamped natural frequency (really the angular velocity) is given by $\omega_n = \sqrt{a_0/a_2}$.

Consider, for example, the equation

$$\frac{d^2x}{dt^2} + \frac{dx}{dt} + 2500x = 0. \tag{86}$$

The undamped natural frequency of this equation is given by $\omega_n = \sqrt{2500/1} = 50$ or $50/2\pi = 7.96$ Hz. For a typical slow recorder, this frequency may be too high to follow. If we let $a_t = 10$, then $T = 10t$ and $p = 10P$. Equation (85) expressed in terms of p is

$$p^2x + px + 2500x = 0. \tag{87}$$

Substituting $p = 10P$ gives

$$100P^2x + 10Px + 2500x = 0 \tag{88}$$

or

$$P^2x + 0.1Px + 25x = 0. \tag{89}$$

Now $\omega_n = \sqrt{25/1} = 5$ or $5/2\pi = 0.796$ Hz which would be well within the range of response of most recorders.

## 3. Initial Conditions

In all previous discussion, the initial values (values at $t = 0$) of the functions of x were taken to zero. However, in many practical problems, it is more common for these parameters to have some finite initial values. To apply such initial conditions to the computer simulation is quite simple. Voltages corresponding in magnitude to initial conditions are applied in parallel to feedback capacitors of the amplifiers generating the specific parameter which has an initial value in the equation. In order to put an initial charge, which results in a voltage corresponding to the magnitude of the initial condition, on a feedback capacitor, initially the input terminal of the particular amplifier in question must be disconnected from the input signal lead and grounded by a switch or relay. At the start of the experiment ($t = 0$), the initial input-condition voltages must be disconnected from across the capacitors [the moment f(t) is applied to the input of another amplifier in the loop] by the same switch. This same switch that disconnects the initial

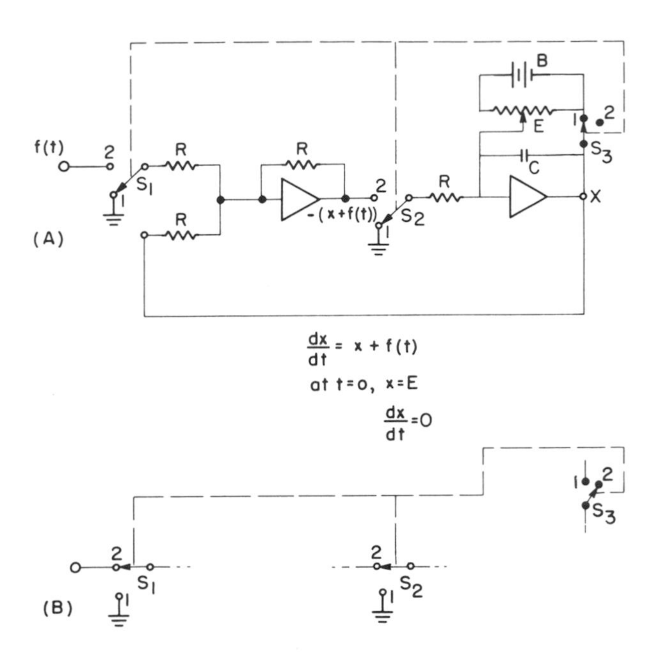


Fig. 30. Initial conditions circuits: (A) switch positions at $t = 0$; (B) switch positions at $t > 0$.

condition voltages at $t = 0$ also connects the input terminal of the f(t) input amplifier, allowing f(t) to be applied and the amplifiers generating the parameters which have an initial condition to be functional at the same instant. A typical initial-condition circuit is illustrated in the computer solution given in Fig. 30. The application of initial conditions in this problem is discussed in the next section.

It must be pointed out that if the particular problem requires scaling, the initial conditions also must be scaled accordingly. Thus, if $X = a_X(x)$, then $X_0 = a_X(x_0)$; and if $T = a_t(t)$ or $dx/dt = a_t(dx/dT)$, then $dx/dT$ (at $t = 0$) is equal to $1/(a_t)[dx/dt \text{ (at } t = 0)]$.

## 4. General Solution to Fundamental Differential Equations

As was mentioned earlier, the fundamental differential equation $a_2(d^2x/dt^2) + a_1(dx/dt) + a_0x = f(t)$ [Eq. (64)] may be used to solve many chemical and physical systems. It is thus worthwhile to consider a general analog computer program in solving this equation and to compare the computer solution to another commonly used method of solving differential equations by Laplace transforms.

Expressing Eq. (64) in the Laplace plane yields

$$a_2\left[s^2\bar{x} - s(x_{t=+0}) - \frac{\partial x}{\partial t}\ (\text{at } t = +0)\right] + a_1(s\bar{x} - x_{t=+0}) + a_0\bar{x} = \bar{f}(t) \qquad (90)$$

and expanding and simplifying gives

$$\bar{x} = \frac{\bar{f}(t) + a_2 s(x_{t=+0}) + a_1(x_{t=+0}) + a_2\dfrac{\partial x}{\partial t}\ (\text{at } t = +0)}{a_2 s^2 + a_1 s + a_0}. \qquad (91)$$

The form of Eq. (64) to be solved depends of course upon which of the coefficients $a_2$, $a_1$, and $a_0$ have real, finite values and on the nature of f(t).

For this case where f(t) = H sin $\omega$t (a coupled system), Eq. (64) becomes

$$a_2\frac{d^2x}{dt^2} + a_1\frac{dx}{dt} + a_0 x = H\ \sin\omega t. \qquad (92)$$

There are many coupled systems that are frequently encountered in physical sciences.

One example illustration is the response of the simple system illustrated in Fig. 31 coupled to an impulse source described by f(t) = H sin $\omega$t. This is the physical problem represented by Eq. (92) and shows how the dis-displacement of the mass m is affected by the frequency and amplitude of the vibrations from some other system causing the function f(t) = H sin $\omega$t. The term C is the friction damping coefficient, k is the spring constant, and x is the displacement of m.

a. <u>Steady-State Solution Using Laplace Transform and j$\omega$ Methods.</u> Substituting the condition t = 0 (x = 0, dx/dt = 0) into Eq. (91) gives

$$\bar{x} = \frac{f(t)}{a_2 s^2 + a_1 s + a_0}. \qquad (93)$$

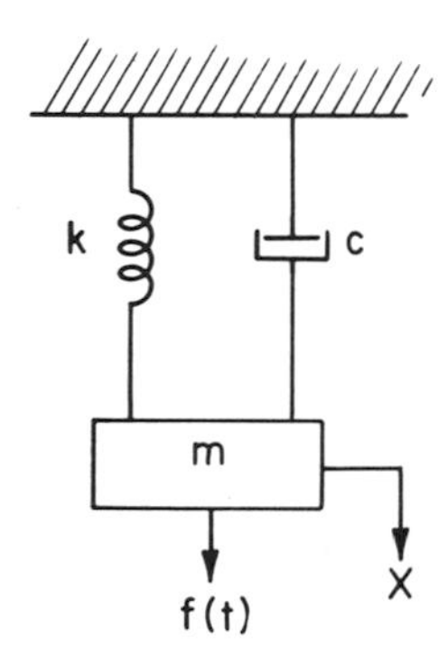


Fig. 31. Mass on a damped spring problem.

From this expression, one can obtain an expression for the ratio of output to the input in the "imaginary" plane ($j\omega$ plane) by substituting $j\omega$ for s:

$$\frac{\bar{x}}{\bar{f}}(j\omega) = \frac{1}{a_2(j\omega)^2 + aj\omega + a_0}$$

$$\frac{\bar{x}}{\bar{f}}(j\omega) = \frac{1}{a_0 - a_2\omega^2 + aj\omega}. \tag{94}$$

The numerator and denominator are now multiplied by the complex conjugate $(a_0 - a_2\omega^2) - a_1 j\omega$:

$$\frac{\bar{x}}{\bar{f}}(j\omega) = \frac{(a_0 - a_2\omega^2) - a_1 j\omega}{(a_0 - a_2\omega^2)^2 + (a_1\omega)^2}. \tag{95}$$

Rearranging Eq. (95) into the "real" and "imaginary" parts gives

$$\frac{\bar{x}}{\bar{f}(t)}(j\omega) = \frac{a_0 - a_2\omega^2}{(a_0 - a_2\omega^2)^2 + (a_1\omega)^2} + j\frac{-a_1\omega}{(a_0 - a_2\omega^2)^2 + a_1\omega^2}. \tag{96}$$

(a) "real" (b) "imaginary"

Then $x = HC' \sin(\omega t + \varphi)$, where H = amplitude of the input, $C' = \sqrt{a^2 + b^2}$, and $\varphi = \tan^{-1}(b/a)$:

$$C' = \left(\frac{(a_0 - a_2\omega^2)^2 + (-a_1\omega)^2}{[(a_0 - a_2\omega^2)^2 + (a_1\omega)^2]^2}\right)^{1/2}, \tag{97}$$

$$C' = \frac{1}{[(a_0 - a_2\omega^2)^2 + (a_1\omega)^2]^{1/2}}, \tag{98}$$

$$\varphi = \tan^{-1}\left(\frac{a_1\omega}{a_0 - a_2\omega^2}\right). \tag{99}$$

Therefore,

$$x = \frac{H}{[(a_0 - a_2\omega^2)^2 + (a_1\omega)^2]^{1/2}} \sin\left[\omega t + \tan^{-1}\left(\frac{-a_1\omega}{a_0 - a_2\omega^2}\right)\right]. \tag{100}$$

It can be seen from Eq. (97) that as the frequency of the signal, $\sin \omega t$, increases the amplitude or the displacement decreases and the phase angle is shifted more and more towards $90^\circ$. Also increasing the damping $a_1$ decreases the amplitude and increases the phase shift. If the damping is zero, there is no phase shift.

b. Analog Computer Solution. To obtain the computer solution of Eq. (92), solve for the highest derivative which gives

$$\frac{d^2x}{dt} = \frac{H}{a_2}\sin \omega t - \left(\frac{a_1\,dx}{a_2\,dt}\right) - \frac{a_0\,x}{a_2}. \tag{101}$$

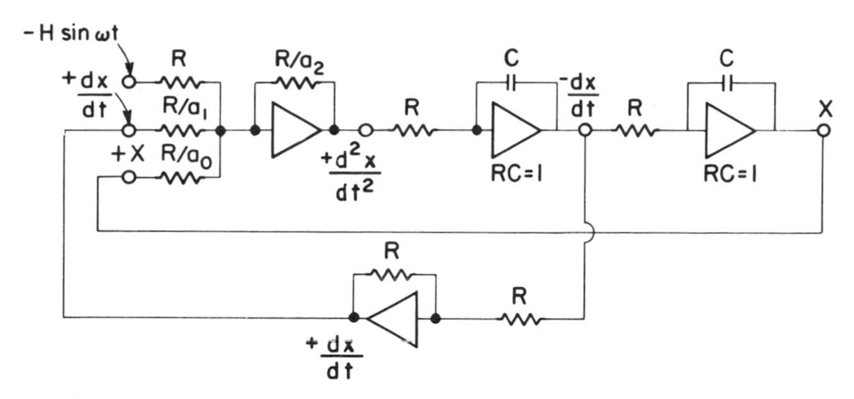


Fig. 32. Computer simulation of the mass on a damped spring problem.

The computer network for the solution of this equation is given by Fig. 32. Note how simply the general solution to this problem is generated by the analog computer system as compared to the steady-state only solution obtained by Laplace transform method.

## REFERENCES

[1]. F. W. Sears and M. W. Zemansky, College Physics, 2nd ed., Addison-Wesley, Reading, Mass., 1956, pp. 513-538.

[2]. J. J. Brophy, Basic Electronics for Scientists, McGraw-Hill, New York, 1966, pp. 20-28.

[3]. R. V. Churchill, Operational Mathematics, McGraw-Hill, New York, 1958.

[4]. B. J. Ley, A. G. Lutz, and C. F. Rehberg, Linear Circuit Analysis, McGraw-Hill, New York, 1959.

[5]. R. W. Tandee, D. C. Davis, and A. P. Albrecht, Electronic Designer's Handbook, McGraw-Hill, New York, 1957.

[6]. G. A. Korn and T. M. Korn, Electronic Analog Computers, 2nd ed., McGraw-Hill, New York, 1956.

[7]. J. A. Aseltine, Transform Method in Linear Systems Analysis, McGraw-Hill, New York, 1958.

[8]. H. D. Huskey and G. A. Korn, Computer Handbook, Analog Computers, McGraw-Hill, New York, 1962.

[9]. A. Erdelyi, W. Magnus, F. Oberhettinger, and F. Tricomi, Tables of Integral Transforms, Vols. 1 and 2, McGraw-Hill, New York, 1954.

[10]. Handbook of Chemistry and Physics, Chemical Rubber Co., Cleveland, Ohio (any recent editions).

[11]. R. W. Murray and C. N. Reilley, J. Electroanal. Chem., 3:182 (1962).

[12]. H. V. Malmstadt, C. G. Enke, and E. C. Toren, Jr., Electronics for Scientists, W. A. Benjamin, New York, 1963.

[13]. R. W. Benedict, Electronics for Scientists and Engineers, Prentice-Hall, Engelwood Cliffs, New Jersey, 1967.

[14]. Application Manual for Operational Amplifiers, G. A. Philbrick Researchers, Dehdam, Mass.; B. Seddon, 12th Ann. Symp. Computers Process Ind., N. J. Section, Inst. Soc. of Amer. Inc., April 1960; Handbook of Operational Amplifier Applications, Burr-Brown Res., Tuscon, Arizona, 19 .

[15]. C. N. Reilley, J. Chem. Ed., 39:A855, A933 (1962).

[16]. T. E. Weichselbaum, W. H. Plumpe, Jr., and H. B. Mark, Jr., Anal. Chem., 41:103A (1969); T. E. Weichselbaum, W. H. Plumpe, Jr., R. E. Adams, and H. B. Mark, Jr., Anal. Chem., 41:725 (1969); H. V. Malmstadt and S. R. Crouch, J. Chem. Ed., 43:340 (1966).

[17]. G. E. James and H. L. Pardue, Anal. Chem., 40:796 (1968); J. W. Strojek, G. A. Gruver, and T. Kuwana, Anal. Chem., 41:481 (1969).

[18]. J. Janata and H. B. Mark, Jr., in Electroanalytical Chemistry, Vol. 3 (A. J. Bard, ed.), Marcel Dekker, New York, 1968.

[19]. H. B. Mark, Jr., L. J. Papa, and C. N. Reilley, in Advances in Analytical Chemistry and Instrumentation, Vol. 2 (C. N. Reilley, ed.), Wiley-Interscience, New York, 1963.

[20]. J. S. Mattson, H. B. Mark, Jr., and H. MacDonald, eds., Computer Applications in Analytical Chemistry, Marcel Dekker, New York, to be published.

[21]. S. Sealy, Electron Tube Circuits, 2nd ed., McGraw-Hill, New York, 1958.

[22]. E. J. Angelo, Jr., Electronic Circuits, McGraw-Hill, New York, 1958.

Chapter 3

# BASIC PRINCIPLES OF DIGITAL CIRCUITRY

Richard D. Sacks

Department of Chemistry
University of Michigan
Ann Arbor, Michigan



## I. INTRODUCTION

All physical quantities are either analog or digital in nature. A familiar example of an analog measurement is that of temperature. An analogy is formed between the height of a mercury column in a tube and the average velocity of atoms or molecules in the system being measured. Although these velocities are not measured directly, the use of the analogy with the column height allows the straightforward measurement of temperature.

The height of the mercury column is not restricted to any finite number of values. A change in temperature, no matter how small, will produce a corresponding change in the height of the column. Thus we can speak of the column height as a continuous function of temperature.

While many physical quantities are analog in nature, they are not easily communicated in that form. It is first necessary to convert the analog quantity into a digital form which can be expressed as a simple number. For temperature measurements, the mercury column is divided into a number of segments. The segments are numbered consecutively, and the temperature is expressed as the segment number at the top of the column. The accuracy of the temperature measurement is limited only by the length of each segment. This is an example of digitizing an analog quantity. It is important to remember that the analog quantity is exact but of limited utility. The digital representation of the analog quantity is only an approximation but is easily communicated and manipulated.

An analog computer makes use of physical analogies to solve problems. While many types of problems are conveniently solved in this way, analog computers which are useful in solving a wide variety of problems are difficult to construct. In addition, accuracy is limited to the production tolerances of the computer components. For example, if an analog computer performs additions and subtractions by summing the currents through a resistive network, the accuracy of the computation is limited by the tolerances of the resistors, their temperature coefficients, and the temperature control in the computer.

In the digital computer, all quantities, analog and digital, are converted into a numerical representation. All problems, no matter how complex, are solved by the manipulation of numbers. In addition, these manipulations are few in number and easily implemented electronically. The accuracy of digital computations is limited only by the maximum number of digits or places in the numbers processed by the computer.

The nature of these mathematical operations and their electronic implementation will be discussed in this chapter.

## II. NUMBER SYSTEMS, CODES, AND BINARY ARITHMETIC

### A. Decimal Numbers

The familiar decimal number system is an example of a weighted base or radix numerical notation. To avoid needing a large number of symbols, the position of a symbol in a number conveys information as well as the type of symbol. In the decimal system, any number, no matter how large, is expressed using only the ten symbols 0 through 9. If the number to be expressed is larger than 9, this is indicated by shifting an appropriate symbol one or more decimal places to the left. For a number less than 1, an appropriate symbol is shifted to the right. The direction of shifting is indicated by the use of a decimal point. Each shift to the left results in multiplying the symbol value by 10. A shift to the right divides the symbol value by 10. Here 10 is called the base or radix of the number system. In general, the base of a number system is one unit more than the magnitude of the largest symbol in the system.

Consider how a decimal number such as 2947 might be generated. Each time a symbol is shifted to the left, a zero is inserted to the right of the symbol to indicate that a shift has been performed. We begin by generating the number 2000 by shifting the symbol 2 three places to the left. This is equivalent to multiplying the symbol value by $10 \times 10 \times 10$ or $10^3$. Note that this is just the base of the number system raised to a power equal to the number of shifts. Next we generate 900 by shifting 9 twice to the left or multiplying it by $10^2$. The number 40 is generated by multiplying by $10^1$, and 7 is generated by multiplying by $10^0$ or 1. The final number is formed by adding these four results:

$$2947 = 2000 + 900 + 40 + 7 = 2 \times 10^3 + 9 \times 10^2 + 4 \times 10^1 + 7 \times 10^0.$$

Any decimal number can be expressed in this way.

For numbers less than 1, the symbol values are multiplied by negative powers of 10. For a number such as 37.946, we have

$$37.946 = 3 \times 10^1 + 7 \times 10^0 + 9 \times 10^{-1} + 4 \times 10^{-2} + 6 \times 10^{-3}.$$

Decimal numbers in general are not processed directly in digital computing systems. The base 2 or binary system and the base 8 or octal system are the most frequently used.

## B. Binary Numbers

In the binary system only two symbols are used. These are 0 and 1. This system is conveniently used in electronic computing systems since the two symbols can be represented by the open and closed states of an electronic switch. Thus all operations on binary numbers can be performed using proper combinations of switches. This in fact is the basic operating feature of most digital devices.

In the decimal system, each decimal digit or place is weighted by a factor of 10 greater than the next decimal place to the right. In the binary system, each binary digit is weighted by a factor of 2 greater than the next place to the right. Thus the binary number 10 is a factor of 2 greater than the binary number 1, and the binary number 100 is a factor of 4 or $2^2$ greater than binary 1. Any binary number, $a_n a_{n-1} \ldots a_1 a_0$, is evaluated in the following general form:

$$a_n 2^n + a_{n-1} 2^{n-1} + \cdots + a_1 2^1 + a_0 2^0.$$

Here $a_i$ is the binary coefficient of the i-th binary digit or place.

### 1. Binary-to-Decimal Conversion

The decimal equivalent of any binary number is evaluated directly from the general form of a binary number. For example, the binary number

1011001 is converted to decimal in the following way:

$$1011001 = 1 \times 2^6 + 0 \times 2^5 + 1 \times 2^4 + 1 \times 2^3 + 0 \times 2^2 + 0 \times 2^1 + 1 \times 2^0$$

$$= 64 + 0 + 16 + 8 + 0 + 0 + 1$$

$$= 89.$$

Notice that the decimal number required only two decimal places or digits for its representation while the binary number required seven binary places. Thus the ease of processing binary numbers through electronic switches is achieved only at the expense of a more unwieldy representation.

Fractional numbers are represented using negative powers of 2. For example, the binary number 101.011 is converted to decimal as follows:

$$101.011 = 1 \times 2^2 + 0 \times 2^1 + 1 \times 2^0 + 0 \times 2^{-1} + 1 \times 2^{-2} + 1 \times 2^{-3}$$

$$= 4 + 0 + 1 + 0 + 0.25 + 0.125$$

$$= 5.375.$$

## 2. Decimal-to-Binary Conversion

Decimal numbers are readily converted to binary using a division process. Consider the decimal number 89. The largest power of 2 that can be divided into 89 is $2^6$ or 64. This leaves a remainder of 25. The largest power of 2 that can be divided into 25 is $2^4$ or 16. This leaves a remainder of 9 which can be divided by $2^3$ or 8. The remainder of 1 can only be divided by $2^0$ or 1 leaving a remainder of 0. Thus 89 can be represented as $1 \times 2^6 + 1 \times 2^4 + 1 \times 2^3 + 1 \times 2^0$. To write the complete binary number, the 0 coefficients of $2^5$, $2^2$, and $2^1$ must be introduced as place holders. So we have

$$89 = 1 \times 2^6 + 0 \times 2^5 + 1 \times 2^4 + 1 \times 2^3 + 0 \times 2^2 + 0 \times 2^1 + 1 \times 2^0.$$

The binary number is just the coefficients of this expression or 1011001.

A more systematic approach is to repeatedly divide the decimal number by 2. The remainders are then the coefficients of the binary number. This procedure is outlined below.

| | | | | | | | |
|---|---|---|---|---|---|---|---|
| 0 | 1 | 2 | 5 | 11 | 22 | 44 | |
| 2⟌1 ← | 2⟌2 ← | 2⟌5 ← | 2⟌11 ← | 2⟌22 ← | 2⟌44 ← | 2⟌89 | Start |
| 0 | 2 | 4 | 10 | 22 | 44 | 88 | |
| 1 | 0 | 1 | 1 | 0 | 0 | 1 | Binary number |

Fractional decimal numbers are converted to binary by dividing by successive negative powers of 2. Consider for example the decimal number 5.375. The integer and fractional parts are treated separately. Using the division procedure outlined above, decimal 5 is converted to binary 101. The fraction 0.375 is not divisible by $2^{-1}$, but it can be divided by $2^{-2}$ with a remainder of 0.125. This remainder can be divided by $2^{-3}$ with a new remainder of 0. Thus we have

$$0.375 = 0 \times 2^{-1} + 1 \times 2^{-2} + 1 \times 2^{-3}.$$

The binary fraction is just the sequence of coefficients of this expansion or 0.011. The complete binary representation of 5.375 is then 101.011.

## C. Octal Numbers

The more compact decimal representation and the more easily processed binary representation are compromised in octal representation. Octal numbers, while less compact than decimal, are much more readily converted to binary since the base 8 is exactly $2^3$. The eight symbols 0 through 7 used in the octal system can be represented using the eight possible combinations of three binary digits. These are shown in Table 1.

TABLE 1

Binary Representation of Octal Numbers

| Octal | Binary |
|---|---|
| 0 | 000 |
| 1 | 001 |
| 2 | 010 |
| 3 | 011 |
| 4 | 100 |
| 5 | 101 |
| 6 | 110 |
| 7 | 111 |

### 1. Octal-Binary Conversion

To convert from binary to octal, the binary digits are collected in groups of three working in both directions from the binary point. The octal equivalent of each group of three binary digits is then evaluated by inspection. For example, the binary number 1011001 is grouped to give 1 011 001. These groups correspond to the octal number 131. The binary number 1101.011 is grouped to give 1 101 • 011 or octal 15.3.

The conversion from octal to binary is also straightforward. Here each octal digit is assigned a corresponding three-digit binary number. Thus the octal number 47.106 is expressed as 100 111 • 001 000 110 in binary representation.

2. Octal-Decimal Conversion

Octal numbers are converted to decimal using the same procedure as for converting from binary to decimal. The general form for evaluating an octal number, $a_n a_{n-1} \cdots a_1 a_0 a_{-1} \cdots a_{-m}$, is

$$a_n 8^n + a_{n-1} 8^{n-1} + \cdots + a_1 8^1 + a_0 8^0 + a_{-1} 8^{-1} + \cdots + a_{-m} 8^{-m}.$$

Here $a_i$ is the octal coefficient of the i-th octal digit or place. The octal number 247.36 is converted to decimal as follows:

$$\begin{aligned} 247.36 &= 2 \times 8^2 + 4 \times 8^1 + 7 \times 8^0 + 3 \times 8^{-1} + 6 \times 8^{-2} \\ &= 128 + 32 + 7 + 0.375 + 0.009375 \\ &= 167.384375. \end{aligned}$$

To convert a decimal number to octal, it is usually easier to first convert the decimal number to binary. The octal number is then obtained directly from inspection of the binary representation. For example, the decimal number 85.375 is converted to binary giving 1010101.011. This is grouped to give 1 010 101 • 011 which by inspection corresponds to the octal number 125.3.

## D. Binary-Number Arithmetic

1. Binary Addition

Since the binary-number system uses only two symbols, the rules for binary arithmetic are much simpler than those in the decimal system. Each digit or place in a binary number is called a bit. Consider the addition of two 1-bit binary numbers A and B. Since each number has only two possible values, 0 and 1, there are only four possible combinations of A and B. These combinations and the resulting sum are shown in Table 2.

TABLE 2

Rules for Binary Addition

| A | B | A + B |
|---|---|---|
| 0 | 0 | 0 |
| 0 | 1 | 1 |
| 1 | 0 | 1 |
| 1 | 1 | (1) 0 |

These results are readily verified by considering the decimal equivalent of each binary number. Thus 1 + 1 = 2 in the decimal system, but decimal 2 is equivalent to binary 10. The carry bit generated when both A and B are binary 1 is added to the next column of bits to the left. Again this is analogous to decimal arithmetic.

Once proficiency is gained in manipulating binary numbers, binary addition can be performed as easily as decimal addition; however, the systematic approach described below is useful in understanding the steps involved. In this approach, carry bits are not added directly but are later added to the partial sum generated from the addition of the two numbers. Only the four addition rules from Table 2 are required. Consider the addition of two binary numbers A and B:

$$\begin{array}{rrl} A = & 10110011 & \\ +B = & 1111010 & \\ \hline S_1 = & 11001001 & = \text{first partial sum} \\ C_1 = & 1100100 & = \text{first carry.} \end{array}$$

Next, the first partial sum is added to the first carry using the same four addition rules:

$$\begin{array}{rrl} S_1 = & 11001001 & \\ +C_1 = & 1100100 & \\ \hline S_2 = & 10101101 & = \text{second partial sum} \\ C_2 = & 10000000 & = \text{second carry.} \end{array}$$

The process is repeated until the carry is 0:

$$\begin{array}{rr} S_2 = & 10101101 \\ +C_2 = & 10000000 \\ \hline S_3 = & 00101101 \\ +C_3 = & 100000000 \\ \hline S_4 = & 100101101 \\ C_4 = & 0. \end{array}$$

This gives $S_4$ as the final result.

## 2. Binary Subtraction

Binary subtraction is similar to binary addition except that borrow bits must be considered rather than carry bits. The four rules for binary subtraction are given in Table 3. The (-1) in Table 3 indicates a borrow from the next column to the left. The subtraction procedure is outlined below for two binary numbers M and N:

$$\begin{array}{rrl} M = & 100101101 & \\ -N = & 10110011 & \\ \hline D_1 = & 110011110 & = \text{first partial difference} \\ B_1 = & 100100100 & = \text{first borrow.} \end{array}$$

The first borrow $B_1$ is subtracted from the first partial difference $D_1$ to give a second partial difference and a second borrow:

TABLE 3

Rules for Binary Subtraction

| A | B | A - B |
|---|---|---|
| 0 | 0 | 0 |
| 0 | 1 | (-1) 1 |
| 1 | 0 | 1 |
| 1 | 1 | 0 |

$$\begin{array}{rl} D_1 = & 110011110 \\ -\ B_1 = & 100100100 \\ \hline D_2 = & 010111010 = \text{second partial difference} \\ B_2 = & 1000000 = \text{second borrow.} \end{array}$$

The process is repeated until the borrow is 0:

$$\begin{array}{rl} D_2 = & 010111010 \\ -\ B_2 = & 1000000 \\ \hline D_3 = & 011111010 \\ -\ B_3 = & 010000000 \\ \hline D_4 = & 001111010 \\ B_4 = & 0. \end{array}$$

This gives $D_4$ as the final answer.

a. Complement Decimal Subtraction. While electronic circuits have been developed for performing binary addition and subtraction, it is often more practical to use the same circuit for both operations. This is easily accomplished by using complement subtraction.

The 9's complement of a decimal number is formed by subtracting every decimal digit from 9. It should be noted that no borrowing is needed to form a 9's complement. To show how the complement is used in subtraction, consider the following example:

$$\begin{array}{rl} A = & 436 \\ -\ B = & 187 \\ \hline D = & 249 = \text{difference.} \end{array}$$

Note that borrowing was required in this subtraction. We now compute the 9's complement of B:

$$\begin{array}{rl} & 999 \\ - & 187 = B \\ \hline & 812 = 9\text{'s complement of B.} \end{array}$$

The complement of B is then added to A.

```
A = 436              436
  + 812      or   | + 999 |
   1248           | - 187 |  = 9's complement of B.
                    1248.
```

From this we see the answer is 999 (or 1000 - 1) larger than it should be. To obtain the final answer we subtract 1000 and add 1:

```
  1248
- 1000
   248
+    1
   249.
```

The same result is obtained if the leftmost carry is brought around and added to the least significant digit. This is called an end-around carry:

```
  1248 ─┐
+    1 <┘
   249.
```

The end-around carry is eliminated by using 10's complement subtraction. The 10's complement of a decimal number is formed by adding 1 to the 9's complement. The 10's complement is formed from the 9's complement rather than directly to avoid borrowing. Consider the previous example using 10's complement subtraction:

B = 187
9's complement of B = 812
10's complement of B = 813

```
   436 = A
+  813 = 10's complement of B
  1249
- 1000
   249.
```

If the subtrahend is larger than the minuend, the difference must be a negative number. In complement subtraction, this is indicated by the lack of a leftmost carry digit. However, the magnitude of the difference is obtained only after complementing. In addition, a 1 must be added after complementing if 10's complement subtraction is used. The following example is worked using both 9's and 10's complement subtraction:

```
  A =  187
- B =  436
  D = -249
```

9's complement of B = 563
10's complement of B = 564

```
  187                     187
+ 563     999    or     + 564      999
  750 → - 750             751  → - 751
          249                      248
                                 +   1
                                   249.
```

These procedures give only the magnitude of the difference. The negative sign must be supplied.

b. Complement Binary Subtraction. In binary subtraction, 1's and 2's complements are used rather than 9's and 10's complements. The 1's complement is formed by subtracting each binary bit from 1. This corresponds to changing all 0 bits to 1 and all 1 bits to 0. The 2's complement is formed by adding 1 to the 1's complement.

The algebraic sign of a binary number is usually indicated by the value of an extra sign bit placed to the left of the number. A 0 is used for positive numbers and a 1 for negative numbers. For example,

```
0   1001 = +9
1   1001 = -9.
```

In addition to identifying the sign of a binary number, this sign bit is very useful in subtraction and in the addition of negative numbers. When the subtrahend is complemented for subtraction, the sign bit is also complemented. Consider the following examples using 1's complement subtraction:

A = 21 = 0 10101
B = 12 = 0 01100
1's complement of B = 1 10011

```
  0 10101            21
+ 1 10011          - 12
1 0 01000          +  9
└────→ +1
  0  1001 = +9
```

A = 12 = 0 01100
B = 21 = 0 10101
1's complement of B = 1 01010

```
  0 01100       12
+ 1 01010     - 21
  1 10110 = -9  -  9.
```

In the first example, the 0 sign bit indicates a positive answer. Also, an end-around carry is generated. In the second example, the 1 sign bit indicates a negative result. Here, no end-around carry is generated, and the magnitude of the difference is obtained only after complementing.

The same procedure is used in 2's complement subtraction with two important differences: (1) the end-around carry is ignored for positive answers, and (2) a 1 must be added after complementing to find the magnitude of a negative answer. These points are illustrated in the following examples.

```
                A = 21 = 0 10101
                B = 12 = 0 01100
                2's complement of B = 1 10100

                  0 10101              21
                + 1 10100            - 12
                -----------          ----
                 10 01001 = +9        + 9

                A - 12 - 0 01100
                B = 21 = 0 10101
                2's complement of B = 1 01011

                  0 01100              12
                + 1 01011            - 21
                ---------            ----
                  1 10111             - 9
    complement    0 01000
                      + 1
                  -------
                  0 01001 = 9.
```

In the first example, the positive answer is obtained directly, and the end-around carry is ignored. In the second example, the magnitude of the negative answer is obtained only after it is complemented and increased by 1.

Some electronic devices to be discussed later produce two complementary outputs. One output can be used for addition and the other for complement subtraction. This is convenient since the same binary adding circuit can be used for both operations.

## 3. Binary Multiplication

TABLE 4

Rules for Binary Multiplication

| A | B | A × B |
|---|---|---|
| 0 | 0 | 0 |
| 0 | 1 | 0 |
| 1 | 0 | 0 |
| 1 | 1 | 1 |

The rules for binary multiplication are given in Table 4. Multiplication of binary numbers is performed in one of three ways. The simplest method is by repeated addition. Here the multiplicand is added to itself the number of times equal to the magnitude of the multiplier.

```
  5       101 =  5
 x 4    + 101 =  5
 ---
 20     + 101 =  5
        + 101 =  5
       ----------
        10100 = 20.
```

While this procedure can be carried out using a simple binary adder, it is far too slow for most data processing and control systems.

The second approach is identical to the common method used for decimal multiplication.

```
  13 =    1101
 x 5 =     101
 -------------
          1101
         0000
        1101
 -------------
  65 = 1000001.
```

Here a series of partial products is formed, and each one is shifted relative to the previous one to indicate the weight of the corresponding bit in the multiplier. These partial products then are added to obtain the final product.

In the previous example, it is seen that each partial product is either all zeros or identical to the multiplicand. This is used in the third method, multiplication by shifting. Here the multiplier is inspected one bit at a time starting with the lowest-order bit. For each 1-bit in the multiplier, the multiplicand is shifted one place to the left and added to the previous partial product. For each 0-bit in the multiplier, the multiplicand is shifted but not added. This procedure is outlined in the following example.

```
  13 =    1101
 x 5 =     101
 -------------
  65 = 1000001

Initial partial product =      0000
First multiplier bit = 1     + 1101    Multiplicand added
                             ------
                               1101
Second multiplier bit = 0     1101     Multiplicand shifted
Third multiplier bit = 1       1101
                             1101      Multiplicand shifted and added
                            -------
                            1000001
```

Multiplication by shifting and adding is the method most frequently used in high-speed computing and control systems.

## 4. Binary Division

Binary division is quite similar to multiplication. Here either repeated subtractions or subtraction and shifting is used. In a computer, these subtractions are usually performed by 1's or 2's complement addition. Consider the following example.

```
Dividend = 27 = 11011
Divisor  =  5 =   101.
```

The divisor is repeatedly subtracted from the dividend until the new dividend is smaller than the divisor. The quotient is equal to the number of subtractions performed.

```
  11011 = 27
-   101 =  5
  -----
  10110 = 22
-   101
  -----
  10001 = 17
-   101
  -----
   1100 = 12
-   101
  -----
    111 =  7
-   101
  -----
     10 =  2 = remainder.
```

Since 5 subtractions were required, the quotient is equal to 5 with a remainder of 2.

While division by repeated subtraction can be implemented with simple binary adding circuits, it is quite slow, and division by shifting is usually preferred. Here the divisor is subtracted from the highest-order bits of the dividend. If the divisor is smaller than the corresponding bits of the dividend, the difference is computed, and a 1 is entered for the partial quotient. The difference is then shifted one place to the left and the process repeated. If after any shift the divisor is larger than the corresponding bits in the new difference, the difference is shifted but no subtraction is performed. In this case, a 0 is entered into the partial quotient. This procedure is outlined in the following example.

```
  11011 = 27 = dividend
- 101   =  5 = divisor
  -----
  00111 = difference — enter 1 in quotient
  <----
  0111    difference shifted left
- 101     divisor larger than difference — enter 0 in quotient
  ----
  <---
  111     difference shifted left
- 101
  ---
  010     new difference — enter 1 in quotient.
```

The quotient is then 101 or decimal 5, and the final difference gives a remainder of 10 or decimal 2. In a computer, the subtractions would be performed using 1's or 2's complementation.

Only the salient features of binary arithmetic have been presented. The operations performed in a large computing machine may be considerably more complex, particularly when signed numbers are processed using a floating decimal or binary point format. These topics are discussed in more detail in the references given at the end of this chapter.

## E. Binary-Coded Decimal Numbers

In Section II. C. 1, it was shown how binary-octal conversion can be implemented by inspection. Since decimal numbers are more suitable for human processing, it would be advantageous to develop a simplified binary-decimal conversion procedure. This can be accomplished if decimal numbers are represented as a binary code rather than as true binary numbers. Three binary digits were required to represent the eight octal symbols. The ten decimal symbols can be represented using four binary digits. It should be noted that this conversion is quite wasteful since 16 unique symbols can be represented with four binary digits. Six of these are not used in binary-coded decimal or BCD representation. In addition to being wasteful, the normal counting sequence is not preserved. That is, if a BCD number is converted to decimal or octal using the previously discussed procedures, an erroneous answer will result. This is because six binary numbers are skipped in the coded representation of any one decimal digit.

### 1. The 8421 Code

There are many possible schemes for coding decimal numbers in binary form. Several of these have found wide application. The most common code uses the normal binary counting sequence. This is called the 8421 or natural code where the numbers refer to the numerical weights of the four binary places in the BCD representation. Examples of nonweighted codes will be considered later. The 8421 coding for the decimal numbers 0 through 9 is given in Table 5. From this table, the BCD representation for any decimal number can be determined directly. For example, the decimal number 472 is expressed as 0100 0111 0010 in 8421 BCD.

It must be remembered that this is not a binary number, but rather a binary-coded decimal number. The binary number for decimal 472 is 111011000. The binary number contains 9 bits while the binary-coded decimal number required 12 bits.

It is often necessary to convert a BCD number to its binary equivalent. Often the arithmetic processing unit of a computer uses binary numbers while the input and output units require BCD. Thus BCD-binary conversion

TABLE 5

The 8421 Binary-Coded Decimal

| Decimal | 8421 BCD |
|---|---|
| 0 | 0000 |
| 1 | 0001 |
| 2 | 0010 |
| 3 | 0011 |
| 4 | 0100 |
| 5 | 0101 |
| 6 | 0110 |
| 7 | 0111 |
| 8 | 1000 |
| 9 | 1001 |

is an important feature of many digital computing systems. This conversion is a straightforward process of alternate addition and multiplication. The highest-order coded decimal number is first multiplied by the binary equivalent of decimal 10. The next-highest-order coded decimal digit is then added to this product, and the sum is again multiplied by decimal 10. This process is continued until the lowest-order decimal digit has been added. The multiplication is omitted after this final addition. The procedure is outlined below for the decimal number 472.

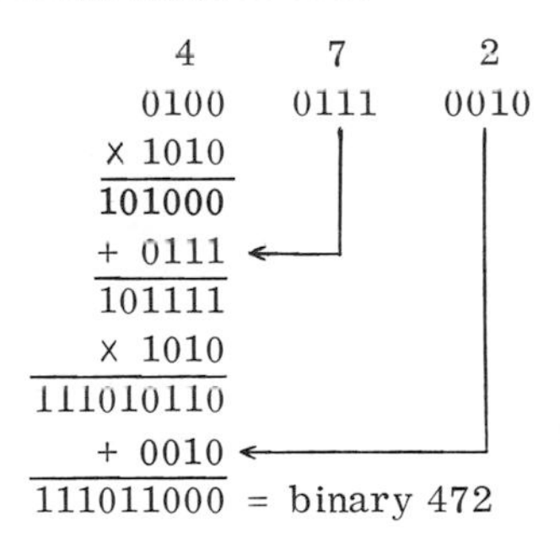

4 7 2
0100 0111 0010
× 1010
101000
+ 0111
101111
× 1010
111010110
+ 0010
111011000 = binary 472


While the 8421 code is most frequently used, other codes have been developed which have attractive properties for certain computing operations.

## 2. The Excess-3 Code

The excess-3 code is formed by adding binary 3 to every coded decimal digit in the 8421 code. Excess-3 representation for the decimal numbers 0 through 9 is given in Table 6. Again, 16 symbols could have been represented using four binary bits, but only 10 were required. In the 8421 code, the last

TABLE 6

Excess-3 Binary-Coded Decimal

| Decimal | Excess-3 BCD |
|---|---|
| 0 | 0011 |
| 1 | 0100 |
| 2 | 0101 |
| 3 | 0110 |
| 4 | 0111 |
| 5 | 1000 |
| 6 | 1001 |
| 7 | 1010 |
| 8 | 1011 |
| 9 | 1100 |

six combinations corresponding to decimal 10 through 15 were skipped. In the excess-3 code, the first three and the last three are skipped. This is a direct result of adding decimal 3 (binary 0011) to each 8421 code. It should be noted that this is not a weighted code. The bit positions do not have a fixed weight or numerical value.

The useful feature of the excess-3 code is its self-complementing property. That is, if two decimal numbers are 9's complements, their excess-3 codes are 1's complements. Referring to Table 6, the excess-3 code for decimal 6 is 1001. The 9's complement of 6 is 3 which has an excess-3 code of 0110 which is just the 1's complement of 1001. This self-complementing property is used in binary subtraction and in BCD addition.

### 3. The 2421 Code

The 2421 code is both weighted and self complementing. The numerical weights of the four binary places are indicated in the code name. The 2421 coding of the decimal numbers 0 through 9 is given in Table 7. Since two bits of the 2421 code have numerical weights of decimal 2, there is more than one representation for certain decimal numbers. In fact, the six decimal numbers 2 through 7 each have two possible representations in this code. Either representation can be used without affecting the self-complementing properties of the code if the representations are consistent. For example, if 0011 is used for decimal 3, then 1100 must be used for decimal 6. On the other hand, if 1001 is used for decimal 3, then 0110 must be used for decimal 6. The 2421 code is frequently used in subtraction procedures.

TABLE 7

The 2421 Binary-Coded Decimal

| Decimal | 2421 BCD |
|---|---|
| 0 | 0000 |
| 1 | 0001 |
| 2 | 0010 |
| 3 | 0011 |
| 4 | 0100 |
| 5 | 1011 |
| 6 | 1100 |
| 7 | 1101 |
| 8 | 1110 |
| 9 | 1111 |

## 4. The Gray Code

A code seldom used in computer hardware but frequently used with input transducers is the reflected or Gray code. This is often referred to as a minimum-change code since only one binary bit changes between any two consecutive decimal numbers. The Gray code for decimal numbers 0 through 15 is given in Table 8.

TABLE 8

Gray-Code Binary-Coded Decimal

| Decimal | Gray Code BCD | |
|---|---|---|
| 0 | 0000 | |
| 1 | 0001 | |
| 2 | 0011 | |
| 3 | 0010 | |
| 4 | 0110 | |
| 5 | 0111 | |
| 6 | 0101 | |
| 7 | 0100 | |
| - - - - - | - - - - - | Reflection plane |
| 8 | 1100 | |
| 9 | 1101 | |
| 10 | 1111 | |
| 11 | 1110 | |
| 12 | 1010 | |
| 13 | 1011 | |
| 14 | 1001 | |
| 15 | 1000 | |

The Gray code is an example of a reflected code. That is, reflections through a mirror plane are used to generate the code. For example, the lower-order three bits in every number above the reflection plane are the mirror images of the same three bits in every number below the reflection plane. Only the highest-order bits are different above and below the plane. To generate the code for decimal numbers up to 31, a new reflection plane is drawn under the code for decimal 15. The mirror image of all entries above the plane is then constructed below the plane. Finally, a new 1-bit is placed to the left of all entries below the plane, and a new 0-bit is placed to the left of all entries above the plane. The process is repeated as many times as necessary to generate the codes for larger decimal numbers.

### 5. The 2-out-of-5 Code

A code frequently used in error detection is the 2-out-of-5 code. Here, five binary bits are required rather than four to code the decimal numbers 0 through 9. The code is shown in Table 9. The code for every decimal number contains two 1-bits and three 0-bits. This property is very useful in detecting errors when binary information is transmitted from one part of a computing or control system to another. After transmission, the number of 1-bits is counted in the code for each decimal digit. If the code for any decimal digit contains anything other than two 1-bits, a transmission error must have occurred. It should be noted here that two errors in the same decimal digit may not be detected.

Only a few of the more frequently used codes have been discussed. Numerous others have been developed, and the interested reader is referred to the references at the end of this chapter.

TABLE 9

The 2-out-of-5 Binary-Coded Decimal

| Decimal | 1-out-of-5 BCD |
|---|---|
| 0 | 00011 |
| 1 | 00101 |
| 2 | 00110 |
| 3 | 01001 |
| 4 | 01010 |
| 5 | 01100 |
| 6 | 10001 |
| 7 | 10010 |
| 8 | 10100 |
| 9 | 11000 |

## III. COMPUTER LOGIC

The strength of a digital computing or control system is in its decision making capability. For example, the arithmetic control section of a computer may have to decide whether or not to complement a certain number in performing a computation. Outside the computer, a decision in a control loop may be required to change the temperature of a reaction vessel based on the continuous analysis of the reaction products.

A series of electronic switches is used to produce binary numbers in a digital computing system. The voltages corresponding to the open and closed states of the switch are used to represent the 0 or 1 values of the bits in a binary number. In the same way, the states of a switch can be used to represent the true or false conditions of a logic statement. The choice of switch states corresponding to true and false is arbitrary. For this discussion, we will assume a closed switch corresponds to a true statement, and an open switch corresponds to a false statement.

### A. The Logical OR Operation

Consider the following example. Suppose a chemical reaction is proceeding in a closed vessel. Assume the reaction is to be terminated when the temperature exceeds a certain critical value OR when the pressure exceeds a certain critical value. Although both temperature and pressure can have an infinite number of possible values, they can all be grouped into only two meaningful classes, those below the critical value and those above the critical value.

The state of the temperature will be designated by the variable A, and the state of the pressure by the variable B. Since we are interested only in whether a variable is above or below its critical value, only two values are needed for each of the variables. Such two-state variables will be called logic variables. Let us assign a logic value of 1 to these variables if the corresponding temperature or pressure is above the critical value and a logic value of 0 if the temperature or pressure is below the critical value. Thus, the statement $A = 1$ implies the temperature is above the critical value, and $B = 0$ implies the pressure is below the critical value. This is equivalent to the statement, "The temperature is above its critical value," being true and the statement, "The pressure is above its critical value," being false. It is important to note that the 0's and 1's used here are not binary numbers but rather states or values of a logic variable. They have no direct numerical significance.

We will now assign a third logic variable C to correspond to the state of the chemical reaction. If it is proceeding then $C = 0$; if it has been terminated then $C = 1$. We can now write a logic function to describe the state of the reaction:

A OR B = C.

From the description of the problem, it is apparent that C = 1 if A = 1 OR if B = 1 OR if they are both equal to 1. Only if both A and B are 0 will C be 0. These results are summarized in Table 10. This table lists all possible combinations of A and B together with the resulting values of C. This is referred to as a table of combinations. Such tables are used frequently in computer logic design.

TABLE 10

Logical OR Operation

| A | B | C = A + B |
|---|---|---|
| 0 | 0 | 0 |
| 1 | 0 | 1 |
| 0 | 1 | 1 |
| 1 | 1 | 1 |

The OR operation introduced in the previous example is one of the three basic logic operations performed in digital computing and control systems. These logic operations are not restricted to two input variables. For example, A OR B OR C = D is a logic function which states that the dependent or output variable D is at logic 1 if any one or more of the independent or input variables A, B, and C are in the logic 1 state. The sumbol + is used to represent the OR operation. Thus, A + B is equivalent to A OR B.

The electronic circuit used to implement the OR operation is called an OR gate. This term comes from the gate-like action of the electronic switch being either open or closed. The logic symbol for a 3-input OR gate is shown in Fig. 1. Its table of combinations is Table 11.

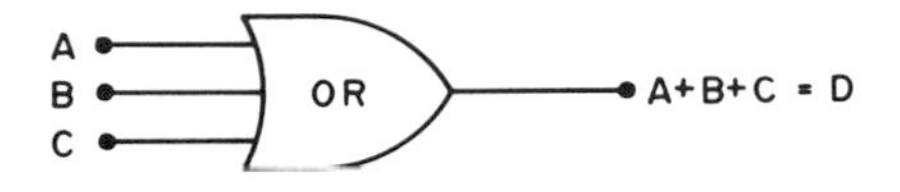


Fig. 1. The OR gate logic symbol.

It should be noted that there are eight possible combinations for the three logic variables A, B, and C. Thus eight rows or entries are needed in the table of combinations. For the 2-input OR operation, only four entries were required. In general, a logic function of n independent variables will require $2^n$ entries or rows in its table of combinations.

TABLE 11

Combinations for a 3-Input OR Gate

| A | B | C | D = A + B + C |
|---|---|---|---|
| 0 | 0 | 0 | 0 |
| 1 | 0 | 0 | 1 |
| 0 | 1 | 0 | 1 |
| 0 | 0 | 1 | 1 |
| 1 | 1 | 0 | 1 |
| 1 | 0 | 1 | 1 |
| 0 | 1 | 1 | 1 |
| 1 | 1 | 1 | 1 |

## B. The Logical AND Operation

In the previous example, the chemical reaction was to be terminated if either the temperature OR the pressure exceeded certain critical values. If both temperature AND pressure had to exceed their critical values before the reaction would be terminated, the OR operation would be logically invalid. Here a logic function is needed which is true only if A AND B are true. This AND operation is the second of the three basic computer logic operations. It describes the situation where two or more things must be true simultaneously. In other words, all independent logic variables must be in the logic 1 state to produce a logic 1 state in the dependent variable. The symbol • is used to represent the AND operation. Thus A • B is equivalent to A AND B.

The electronic circuit used to implement the AND operation is called an AND gate. The logic symbol for a 3-input AND gate is shown in Fig. 2. Its table of combinations is Table 12.

TABLE 12

Combinations for a 3-Input AND Gate

| A | B | C | D = A • B • C |
|---|---|---|---|
| 0 | 0 | 0 | 0 |
| 1 | 0 | 0 | 0 |
| 0 | 1 | 0 | 0 |
| 0 | 0 | 1 | 0 |
| 1 | 1 | 0 | 0 |
| 1 | 0 | 1 | 0 |
| 0 | 1 | 1 | 0 |
| 1 | 1 | 1 | 1 |

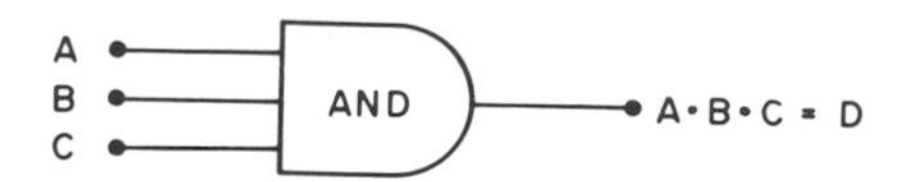


Fig. 2. The AND gate logic symbol.

C. The NOT Operation

If the logic variable A is used to test the truth of the statement, "The temperature is above the critical value," then A = 1 implies the statement is true, and A = 0 implies the statement is false. We can define a new variable S to test the truth of the statement, "The temperature is below the critical value." If S = 1 the statement is true, and the temperature is below the critical value; if S = 0 the statement is false, and the temperature is above the critical value. It should be obvious that the variables A and S are describing the same thing but in inverse or complementary ways. Since the temperature has only two logically significant values, above and below the critical value, if A is true (A = 1) then S must be false (S = 0). Conversely, if A is false (A = 0) then S must be true (S = 1). Thus we see that A and S are not independent variables but are related through the logic operation of complementation or inversion.

Since the variables A and S are not independent, it is not necessary to consider them both in a logic statement. In fact, this could lead to confusion and ambiguity. A more satisfactory approach is to define a new logic operation which we will call NOT. The NOT operation simply inverts or complements the states of a logic variable or function. Thus if A = 0, NOT A = 1. The term NOT A is usually represented by $\bar{A}$. The electronic circuit used to implement the NOT operation is called an inverter gate. The symbol for an inverter gate is shown in Fig. 3, and its table of combinations is Table 13. More accurately, the circle on the right side of Fig. 3 is used to

Fig. 3. Inverter gate logic symbol.

TABLE 13

Combinations for an Inverter Gate

| A | $\bar{A}$ |
|---|---|
| 0 | 1 |
| 1 | 0 |

represent logical inversion. The triangle indicates that the inversion operation is performed by an inverting amplifier. Since most inverter gates are amplifiers, we will include the triangle as part of the inverter-gate notation.

## D. The NAND and NOR Operations

There are two other logic operations which, while not independent of the three already discussed, are so widely used in digital systems that they are usually considered as separate self-consistent operations. The NAND operation (NOT AND) is just the AND operation followed by logical inversion. The NOR operation (NOT OR) is the OR operation followed by logical inversion. Since inversion follows AND in the NAND operation, the results of NAND and AND operations are always complementary. That is, if the AND operation produces a logic 1 for a given set of logic variables, the result for a NAND operation must be a logic 0. Conversely, if the AND operation produces a logic 0, the NAND operation will produce a logic 1. A completely analogous relationship exists between the OR and NOR operations.

The electronic circuit used to implement the NAND operation is called a NAND gate, and the circuit used to implement the NOR operation is called a NOR gate. The symbols for 3-input NAND and NOR gates are shown in Fig. 4. Their tables of combinations are given in Table 14.

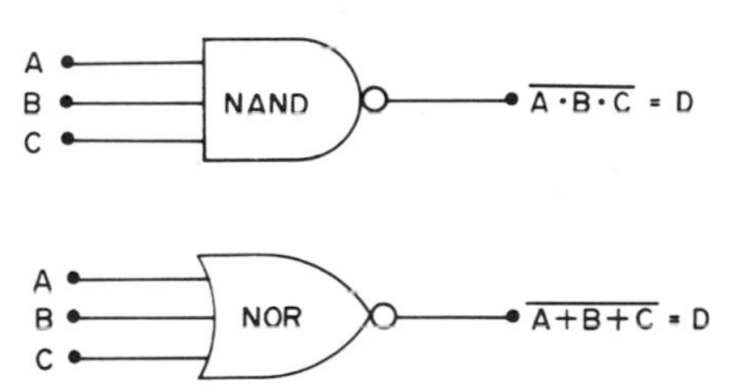


Fig. 4. The NAND and NOR gate logic symbols.

TABLE 14

Combinations for 3-Input NAND and NOR Gates

| A | B | C | $D = \overline{A \cdot B \cdot C}$ | $D = \overline{A + B + C}$ |
|---|---|---|---|---|
| 0 | 0 | 0 | 1 | 1 |
| 1 | 0 | 0 | 1 | 0 |
| 0 | 1 | 0 | 1 | 0 |
| 0 | 0 | 1 | 1 | 0 |
| 1 | 1 | 0 | 1 | 0 |
| 1 | 0 | 1 | 1 | 0 |
| 0 | 1 | 1 | 1 | 0 |
| 1 | 1 | 1 | 0 | 0 |

Complex logic expressions can be implemented using combinations of logic gates. For example, consider the implementation of the following logic expression,

$$\overline{(\bar{A} \cdot \bar{B}) + (\bar{C} \cdot \bar{D})} = E.$$

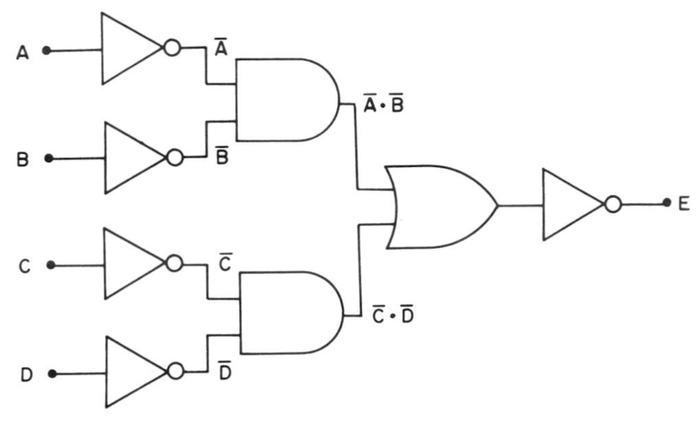


Fig. 5. Implementation of $\overline{(\bar{A} \cdot \bar{B}) + (\bar{C} \cdot \bar{D})} = E$.

TABLE 15

Combinations for $\overline{(\bar{A} \cdot \bar{B}) + (\bar{C} \cdot \bar{D})} = E$

| A B C D | $\bar{A}\ \bar{B}\ \bar{C}\ \bar{D}$ | $\bar{A} \cdot \bar{B}$ | $\bar{C} \cdot \bar{D}$ | $(\bar{A} \cdot \bar{B}) + (\bar{C} \cdot \bar{D})$ | $E = \overline{(\bar{A} \cdot \bar{B}) + (\bar{C} \cdot \bar{D})}$ |
|---|---|---|---|---|---|
| 0 0 0 0 | 1 1 1 1 | 1 | 1 | 1 | 0 |
| 1 0 0 0 | 0 1 1 1 | 0 | 1 | 1 | 0 |
| 0 1 0 0 | 1 0 1 1 | 0 | 1 | 1 | 0 |
| 0 0 1 0 | 1 1 0 1 | 1 | 0 | 1 | 0 |
| 0 0 0 1 | 1 1 1 0 | 1 | 0 | 1 | 0 |
| 1 1 0 0 | 0 0 1 1 | 0 | 1 | 1 | 0 |
| 1 0 1 0 | 0 1 0 1 | 0 | 0 | 0 | 1 |
| 1 0 0 1 | 0 1 1 0 | 0 | 0 | 0 | 1 |
| 0 1 1 0 | 1 0 0 1 | 0 | 0 | 0 | 1 |
| 0 1 0 1 | 1 0 1 0 | 0 | 0 | 0 | 1 |
| 0 0 1 1 | 1 1 0 0 | 1 | 0 | 1 | 0 |
| 1 1 1 0 | 0 0 0 1 | 0 | 0 | 0 | 1 |
| 1 0 1 1 | 0 1 0 0 | 0 | 0 | 0 | 1 |
| 1 1 0 1 | 0 0 1 0 | 0 | 0 | 0 | 1 |
| 0 1 1 1 | 1 0 0 0 | 0 | 0 | 0 | 1 |
| 1 1 1 1 | 0 0 0 0 | 0 | 0 | 0 | 1 |

We will begin by inverting A, B, C, and D using inverter gates. Next, $\bar{A}$ and $\bar{B}$ will be processed through an AND gate as will $\bar{C}$ and $\bar{D}$. The outputs of these AND gates will then be processed by an OR gate, the output of which will be complemented by an inverter gate. The resulting gating network is shown in Fig. 5, and the table of combinations is Table 15.

Careful inspection of the table of combinations reveals that the final output E is at logic 1 when either A OR B or both are at logic 1 and simultaneously either C OR D or both are at logic 1. Thus an equivalent expression can be formulated,

$$(A + B) \cdot (C + D) = E.$$

The table of combinations for this expression is Table 16. The logic gating network is shown in Fig. 6.

TABLE 16

Combinations for $(A + B) \cdot (C + D)$

| A B C D | A + B | C + D | E = (A + B) • (C + D) |
|---|---|---|---|
| 0 0 0 0 | 0 | 0 | 0 |
| 1 0 0 0 | 1 | 0 | 0 |
| 0 1 0 0 | 1 | 0 | 0 |
| 0 0 1 0 | 0 | 1 | 0 |
| 0 0 0 1 | 0 | 1 | 0 |
| 1 1 0 0 | 1 | 0 | 0 |
| 1 0 1 0 | 1 | 1 | 1 |
| 1 0 0 1 | 1 | 1 | 1 |
| 0 1 1 0 | 1 | 1 | 1 |
| 0 1 0 1 | 1 | 1 | 1 |
| 0 0 1 1 | 0 | 1 | 0 |
| 1 1 1 0 | 1 | 1 | 1 |
| 1 0 1 1 | 1 | 1 | 1 |
| 1 1 0 1 | 1 | 1 | 1 |
| 0 1 1 1 | 1 | 1 | 1 |
| 1 1 1 1 | 1 | 1 | 1 |

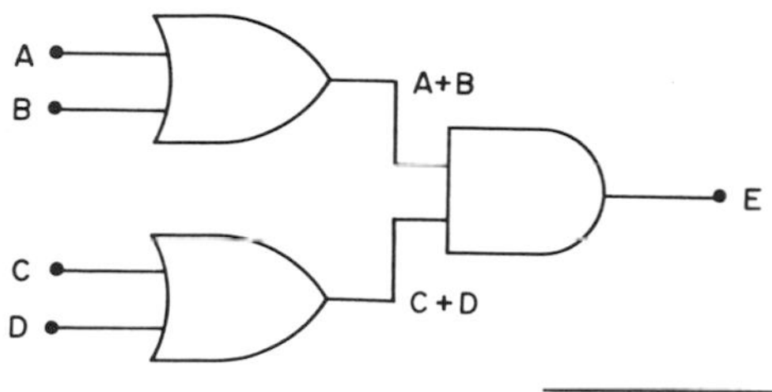


Fig. 6. Simplified implementation of $\overline{(\bar{A} \cdot \bar{B}) + (\bar{C} \cdot \bar{D})} = E.$

Since the rightmost columns of Tables 15 and 16 are identical, the logic expressions for the two logic networks must be logically equivalent:

$$\overline{(\bar{A} \cdot \bar{B}) + (\bar{C} \cdot \bar{D})} = (A + B) \cdot (C + D).$$

However, one required eight logic gates for its implementation while the other required only three gates. The simplification of logic expressions to minimize the number of gates required for implementation is an important aspect of logic circuit design.

## E. Boolean Algebra and Minimization Techniques

The most systematic approach for simplifying logic expressions is through Boolean algebra. Here a series of definitions and postulates are used to define the various properties of two-state logic variables. These simple mathematical relations allow us to express the logic properties of complex switching networks in a concise and unambiguous way. The two values of a Boolean variable are usually represented by the symbols 1 and 0. Alternative representations are true and false or the on and off states of an electronic switch.

### 1. Basic Boolean Relations

Since the Boolean variable A has only two mutually exclusive states, by definition we have

$$\text{if } A \neq 1, \text{ then } A = 0$$
$$\text{if } A \neq 0, \text{ then } A = 1.$$

Three basic operations are defined in Boolean algebra. These are (1) addition (+) which is logically equivalent to OR, (2) multiplication (•) which is logically equivalent to AND, and (3) complementation or inversion. Postulates for these three operations are formulated by considering all possible combinations of the appropriate Boolean variables:

| Addition | Multiplication | Complementation |
|---|---|---|
| $0 + 0 = 0$ | $0 \cdot 0 = 0$ | $\bar{0} = 1$ |
| $0 + 1 = 1$ | $0 \cdot 1 = 0$ | $\bar{1} = 0$ |
| $1 + 0 = 1$ | $1 \cdot 0 = 0$ | |
| $1 + 1 = 1$ | $1 \cdot 1 = 1$ | |

A number of other important relations can be derived from the basic definitions and postulates:

| | | |
|---|---|---|
| Unit rule: | $1 + A = 1$ | $1 \cdot A = A$ |
| Zero rule: | $0 + A = A$ | $0 \cdot A = 0$ |
| Complementerity: | $A + \bar{A} = 1$ | $A \cdot \bar{A} = 0$ |
| Idempotency: | $A + A = A$ | $A \cdot A = A$ |

Commutativity: $A + B = B + A$ $\quad A \cdot B = B \cdot A$

Associativity: $A + (B + C) = (A + B) + C$ $\quad A \cdot (B \cdot C) = (A \cdot B) \cdot C$

Distributivity: $A + B \cdot C = (A + B) \cdot (A + C)$ $\quad A \cdot (B + C) = A \cdot B + A \cdot C$

Involution: $\bar{\bar{A}} = A$

Absorption: $A \cdot (A + B) = A$

$A + \bar{A} \cdot B = A + B$

$A + A \cdot B = A$

DeMorgan's theorems: $\overline{(A + B)} = \bar{A} \cdot \bar{B}$

$\overline{(A \cdot B)} = \bar{A} + \bar{B}$

To illustrate the use of Boolean relations, consider the following examples.

$(A + B) \cdot \overline{(A \cdot B)}$

$= (A + B) \cdot (\bar{A} + \bar{B})$ DeMorgan's theorem

$= A \cdot \bar{A} + A \cdot \bar{B} + B \cdot \bar{A} + B \cdot \bar{B}$ Distributivity

$= A \cdot \bar{B} + B \cdot \bar{A}$ Complementarity

The table of combinations for this expression is given in Table 17. A result of logic 1 is obtained if A OR B is at logic 1 but not if A AND B are at logic 1. This is known as the exclusive OR operation since it excludes the case where A AND B are simultaneously at logic 1. This case is not excluded in the normal OR operation. The logic symbol for exclusive OR is $\oplus$. Thus $A \oplus B = A \cdot \bar{B} + B \cdot \bar{A}$. Implementation of exclusive OR using NAND and NOR gates is shown in Fig. 7.

TABLE 17

Combinations for $A \cdot \bar{B} + B \cdot \bar{A}$

| A | B | $A \cdot \bar{B}$ | $B \cdot \bar{A}$ | $A \cdot \bar{B} + B \cdot \bar{A}$ |
|---|---|---|---|---|
| 0 | 0 | 0 | 0 | 0 |
| 1 | 0 | 1 | 0 | 1 |
| 0 | 1 | 0 | 1 | 1 |
| 1 | 1 | 0 | 0 | 0 |

The complement of $A \oplus B$ is obtained as follows:

$\overline{A \oplus B} = \overline{(A \cdot \bar{B}) + (B \cdot \bar{A})}$

$= \overline{(A \cdot \bar{B})} \cdot \overline{(B \cdot \bar{A})}$ DeMorgan's theorem

$= (\bar{A} + \bar{\bar{B}}) \cdot (\bar{B} + \bar{\bar{A}})$ DeMorgan's theorem

$= (\bar{A} + B) \cdot (\bar{B} + A)$ Involution

$= \bar{A} \cdot \bar{B} + A \cdot \bar{A} + \bar{B} \cdot B + A \cdot B$ Distributivity

$= \bar{A} \cdot \bar{B} + A \cdot B$ Complementarity

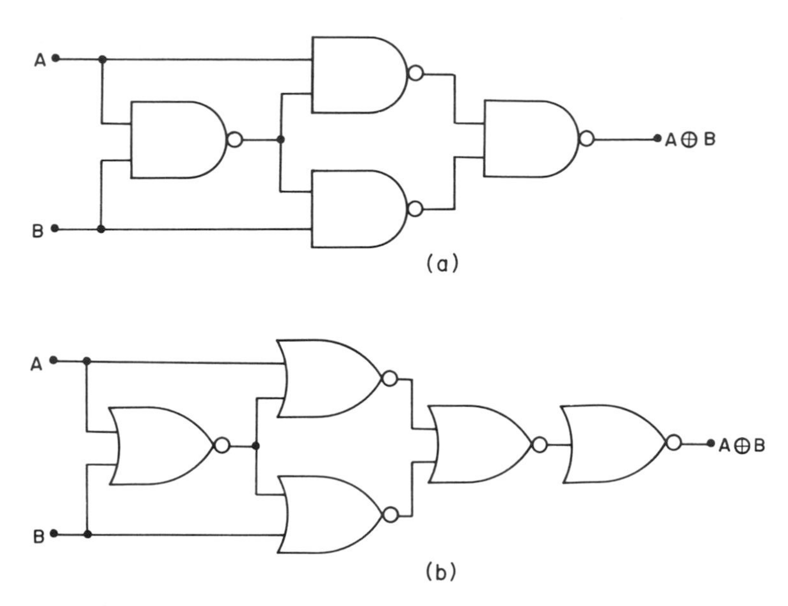


Fig. 7. Implementation of exclusive OR, (a) using NAND gates and (b) using NOR gates.

The table of combinations for $\overline{A \oplus B}$ is given in Table 18. This logic function is useful in equality testing since the result is a logic 1 only when A = B.

The equivalence of logic expressions can be determined using appropriate tables of combinations. If the rightmost or result columns in the corresponding tables are identical, the logic statements must be equivalent. This is a convenient method for verifying the results of a series of Boolean operations.

An alternative logic gate notation for NAND and NOR gates based on DeMorgan's theorems is frequently used in logic circuit design and analysis.

TABLE 18

Combinations for $\overline{A} \cdot \overline{B} + A \cdot B$

| A | B | $\overline{A} \cdot \overline{B}$ | $A \cdot B$ | $\overline{A} \cdot \overline{B} + A \cdot B$ |
|---|---|---|---|---|
| 0 | 0 | 1 | 0 | 1 |
| 1 | 0 | 0 | 0 | 0 |
| 0 | 1 | 0 | 0 | 0 |
| 1 | 1 | 0 | 1 | 1 |

Consider the two logic variables A and B. From DeMorgan's theorems we have

$$\overline{A + B} = \overline{A} \cdot \overline{B} \quad \text{and} \quad \overline{A \cdot B} = \overline{A} + \overline{B}.$$

Since $\overline{A + B}$ is the OR operation followed by inversion, this complete operation can be performed with a NOR gate. However, $\overline{A} \cdot \overline{B}$ is implemented by complementing A and B and then processing the complements through an AND gate. Thus, complementation followed by AND is logically equivalent to NOR. In the same way, complementation followed by OR is logically equivalent to NAND.

The equivalence of these operations using a simplified notation is shown in Fig. 8. Here the circles at the OR and AND gate inputs indicate that the input variables are complemented before being processed by the gate. This alternative notation together with the logical equivalence of these operations often can be used to simplify logic networks. In Fig. 9(a), the logic network of Fig. 5 has been redrawn using this alternative notation. In Fig. 9(b), the network has been redrawn again using the equivalent notation for NAND and NOR gates. Here two consecutive inversions are observed. From the involution law, $\overline{\overline{X}} = X$, so the network can be simplified to give Fig. 9(c) which is the same network as shown in Fig. 6. In this way, logic network simplification involving DeMorgan's theorems often can be accomplished by inspection.

## 2. Synthesis of Boolean Functions from Tables of Combinations

Another important use of tables of combinations is the synthesis of Boolean functions. Often the results of a logic or gating operation are known but the simplest means of implementation is not clear. If the table of combinations for the gating operation is constructed, the corresponding Boolean function can be readily synthesized. The synthesis is performed by adding a series of terms in the independent logic variables. One term is required for each table entry producing an output of logic 1. Each term is formed from the Boolean product of all independent variables, used directly if in the logic-1 state, and complemented if in the logic-0 state.

Consider, for example, Table 17 for the exclusive OR operation. The logic operation produces a logic-1 output for the second and third entries in the table. Thus two terms involving the product of the independent variables A and B must be summed. For the second entry in the table, A is logic 1 so it is used directly, but B is logic 0 so it must be complemented. Thus the first term in the sum is of the form, $A \cdot \overline{B}$. For the third entry in the table, A is logic 0 and B is logic 1. Thus the second term in the sum is $\overline{A} \cdot B$. When the two terms are logically added, the familiar result for $A \oplus B$ is obtained:

$$A \cdot \overline{B} + \overline{A} \cdot B = A \oplus B.$$

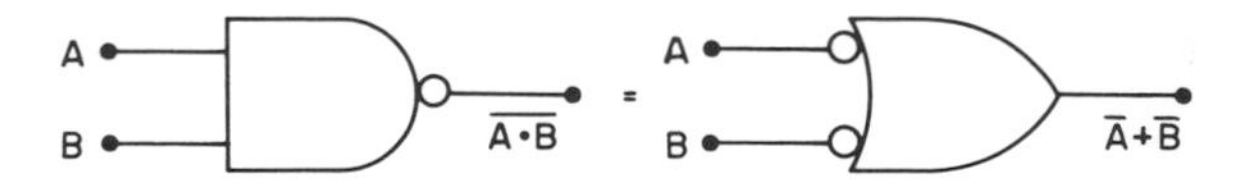


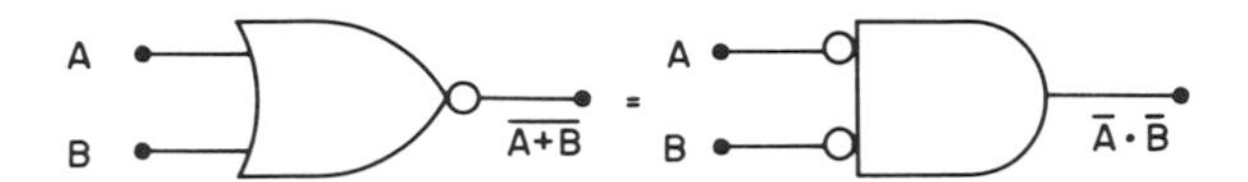


Fig. 8. Equivalent logic symbols for NAND and NOR gates.

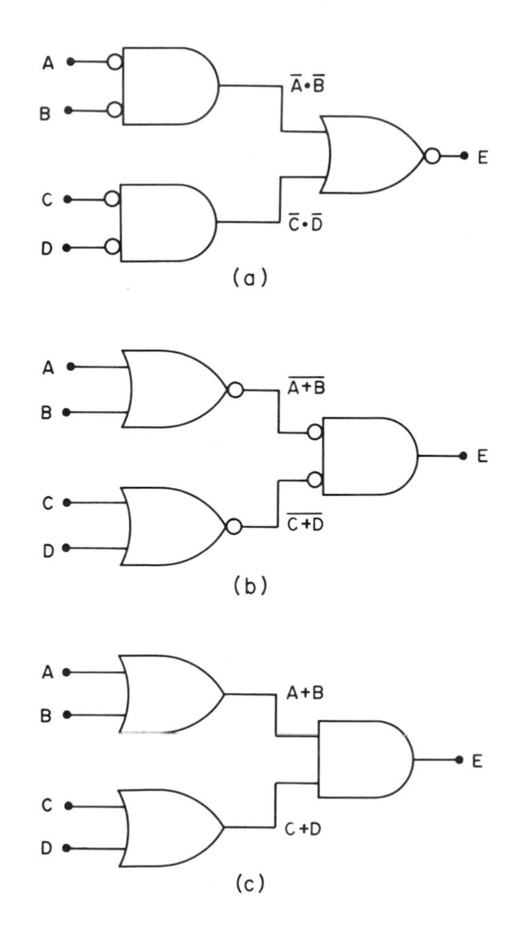


Fig. 9. Simplification of logic networks, using equivalent logic symbols: (a) original network; (b) equivalent logic network; (c) final simplified network.

An alternative approach is to sum all terms where the function has a value of 0. However, here the sum must be complemented. For $A \oplus B$ we have

$$\overline{(\bar{A} \cdot \bar{B}) + (A \cdot B)}$$

| | |
|---|---|
| $= \overline{(\bar{A} \cdot \bar{B})} \cdot \overline{(A \cdot B)}$ | DeMorgan's theorem |
| $= (\bar{\bar{A}} + \bar{\bar{B}}) \cdot (\bar{A} + \bar{B})$ | DeMorgan's theorem |
| $= (A + B) \cdot (\bar{A} + \bar{B})$ | Involution |
| $= A \cdot \bar{A} + A \cdot \bar{B} + B \cdot \bar{B} + \bar{A} \cdot B$ | Distributivity |
| $= A \cdot \bar{B} + \bar{A} \cdot B = A \oplus B$ | Complementarity and commutativity |

The method of choice usually depends on the number of 1's and 0's in the table of combinations for the dependent variable. Consider the table of combinations given in Table 19. Using the entries where the function value is logic 1, three terms are required:

$$\bar{A} \cdot \bar{B} + A \cdot \bar{B} + A \cdot B = C$$

| | |
|---|---|
| $= \bar{B} \cdot (\bar{A} + A) + A \cdot B$ | Distributivity |
| $= \bar{B} + A \cdot B$ | Complementarity |
| $= \bar{B} + A$ | Absorption |

The same result is obtained more directly using the one term with a 0 function value.

$$\overline{(\bar{A} \cdot B)} = C$$

| | |
|---|---|
| $= \bar{\bar{A}} + \bar{B}$ | DeMorgan's theorem |
| $= A + \bar{B}$ | Involution |

TABLE 19

Combinations for $\overline{A + \bar{B}}$

| A | B | C |
|---|---|---|
| 0 | 0 | 1 |
| 1 | 0 | 1 |
| 0 | 1 | 0 |
| 1 | 1 | 1 |

## 3. Synthesis of Boolean Functions from Sets of Binary Numbers

It is often useful to synthesize Boolean functions from sets of binary numbers. Here, the two-state binary numbers are treated as Boolean variables. Consider, for example, the implementation of binary addition. Let $A_i$ and $B_i$ be the binary bits corresponding to the i-th place or digit in the binary numbers A and B. In addition to $A_i$ and $B_i$, we must consider the

TABLE 20

Combinations for Binary Addition

| $A_i$ | $B_i$ | $C_i$ | $S_i$ | $C_{i+1}$ |
|---|---|---|---|---|
| 0 | 0 | 0 | 0 | 0 |
| 1 | 0 | 0 | 1 | 0 |
| 0 | 1 | 0 | 1 | 0 |
| 0 | 0 | 1 | 1 | 0 |
| 1 | 1 | 0 | 0 | 1 |
| 1 | 0 | 1 | 0 | 1 |
| 0 | 1 | 1 | 0 | 1 |
| 1 | 1 | 1 | 1 | 1 |

carry from the addition of $A_{i-1}$ and $B_{i-1}$, the next-lower-order binary digits. This will be called $C_i$. Two answers must be generated, a sum bit $S_i$ and a carry bit $C_{i+1}$, to be added to the next-higher-order binary place. Using the binary addition rules from Section II. D. 1, the table of combinations in Table 20 is readily derived. From this table, Boolean functions for $S_i$ and $C_{i+1}$ are synthesized:

$$
\begin{aligned}
S_i &= A_i \cdot \overline{B}_i \cdot \overline{C}_i + \overline{A}_i \cdot B_i \cdot \overline{C}_i + \overline{A}_i \cdot \overline{B}_i \cdot C_i + A_i \cdot B_i \cdot C_i \\
&= C_i \cdot (A_i \cdot B_i + \overline{A}_i \cdot \overline{B}_i) + \overline{C}_i \cdot (A_i \cdot \overline{B}_i + \overline{A}_i \cdot B_i) \\
&= C_i \cdot \overline{(A_i \oplus B_i)} + \overline{C}_i \cdot (A_i \oplus B_i) \\
C_{i+1} &= A_i \cdot B_i \cdot \overline{C}_i + A_i \cdot \overline{B}_i \cdot C_i + \overline{A}_i \cdot B_i \cdot C_i + A_i \cdot B_i \cdot C_i \\
&= A_i \cdot B_i \cdot \overline{C}_i + A_i \cdot \overline{B}_i \cdot C_i + \overline{A}_i \cdot B_i \cdot C_i + A_i \cdot B_i \cdot C_i \\
&\quad + A_i \cdot B_i \cdot C_i + A_i \cdot B_i \cdot C_i \\
&= (A_i \cdot B_i \cdot \overline{C}_i + A_i \cdot B_i \cdot C_i) + (A_i \cdot \overline{B}_i \cdot C_i + A_i \cdot B_i \cdot C_i) \\
&\quad + (\overline{A}_i \cdot B_i \cdot C_i + A_i \cdot B_i \cdot C_i) \\
&= A_i \cdot B_i \cdot (\overline{C}_i + C_i) + A_i \cdot C_i \cdot (\overline{B}_i + B_i) + B_i \cdot C_i \cdot (\overline{A}_i + A_i) \\
&= A_i \cdot B_i + A_i \cdot C_i + B_i \cdot C_i.
\end{aligned}
$$

The gating circuits used in implementing these functions will be discussed.

## F. Logic-Gate Applications

The variety of logic-gate applications is almost endless. We will consider only some of the more widely used and definitive gating networks. For reasons of cost and interfacing simplicity, some digital systems, particularly

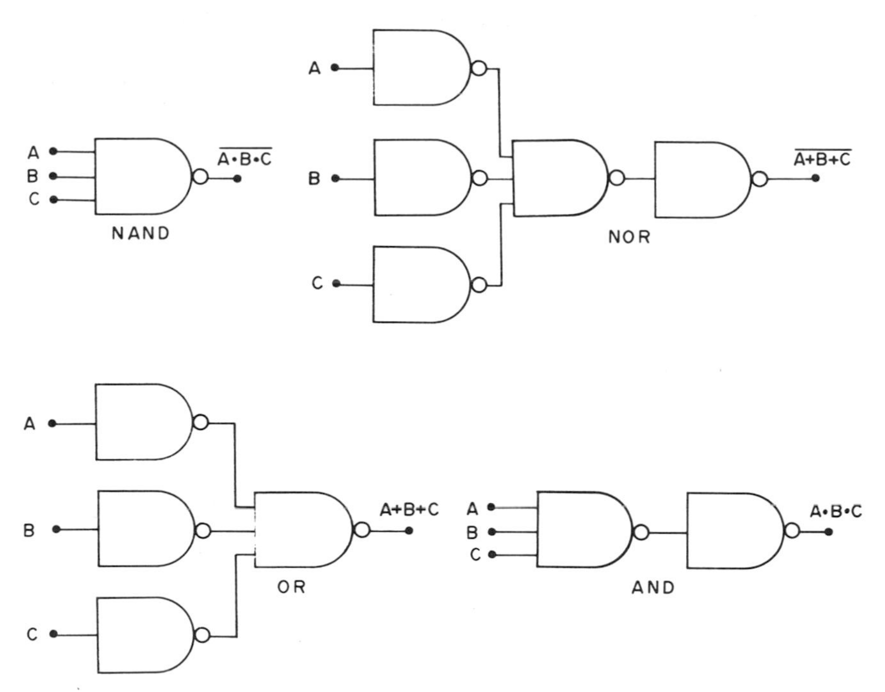


Fig. 10. Implementation of logic operations, using NAND gates.

older ones, are based on a particular logic type. For example, a system might be based on NAND gate operation exclusively. The NOR operation is also common, as is a combination of AND, OR, and NOT. In general, a Boolean function contains more than one type of logic operation. Thus, it would be useful to be able to express all common logic operations in terms of NAND and NOR gate operations. Figure 10 shows the NAND gate equivalents of the various logic operations, and Fig. 11 shows the NOR gate equivalents of these operations. The NAND and NOR implementation of exclusive OR has been discussed. It should be noted here that a 1-input NAND or NOR gate functions as a simple inverter gate.

## 1. Encoding and Decoding Networks

An important application of gating circuits is the coding of decimal numbers into binary or binary-coded decimal. A BCD encoding network will have ten inputs corresponding to the ten decimal symbols 0 through 9, and four outputs for the four binary places required in most BCD codes. The simple OR gate network shown in Fig. 12 serves as a BCD encoder.

The operation of the encoder is straightforward. When any input is raised to logic 1, the outputs of the OR gates to which it is connected also

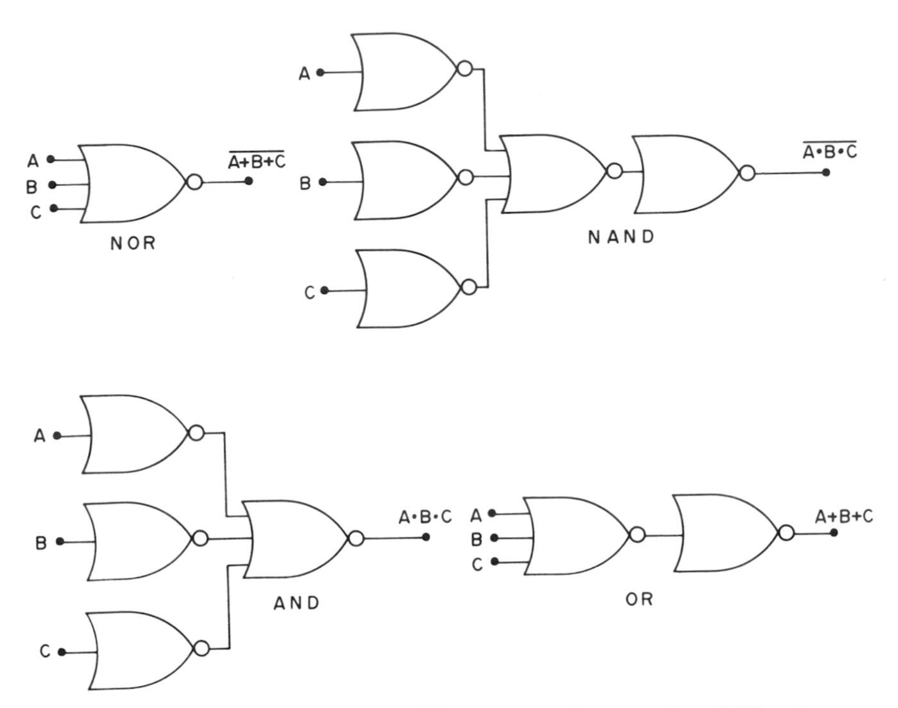


Fig. 11. Implementation of logic operations, using NOR gates.

are raised to logic 1. The outputs of the other gates stay at logic 0. This will generate the proper code. For example, if input 5 is raised to logic 1, the outputs from the 0th and 2nd-order gates will be at logic 1 while the 1st- and 3rd-order gates will be at logic 0. This will produce an output of 0101 which is the 8421 code for decimal 5. Simple modification of the input connections will allow the generation of any of the 4-bit BCD codes.

Binary or binary-coded decimal numbers are converted to decimal using logic-gate decoders. The AND-gate or AND-gate-equivalent logic networks usually are used here. Consider the decoding of a 2-bit binary number AB where A and B correspond to the 1st- and 0th-order binary places, respectively. Since the decimal numbers 0, 1, 2, and 3 can be represented with two binary digits, the decoder will require four outputs. The four combinations of A and B together with the corresponding binary and decimal numbers are given in Table 21. The complements of A and B are generated with inverter gates. The proper combinations of A, B, $\overline{A}$, and $\overline{B}$ are then processed through AND gates. The complete 2-bit decoder is shown in Fig. 13. From the figure, it can be seen that any combination of inputs will result in a logic 1 at one and only one output.

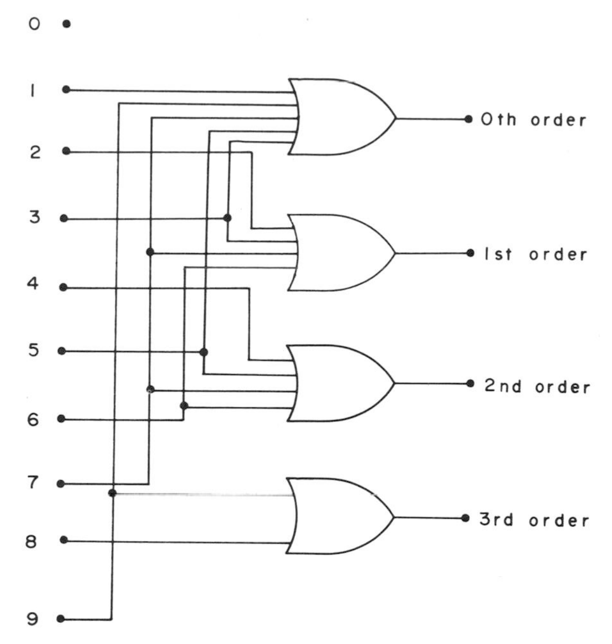


Fig. 12. Decimal-to-binary encoder.

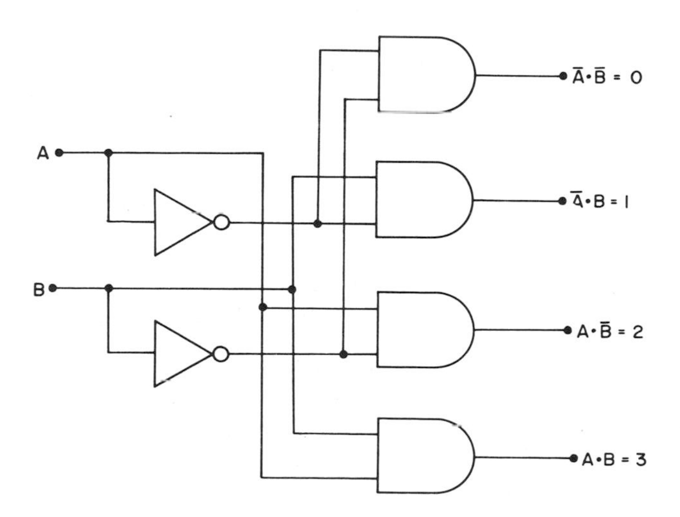


Fig. 13. A 2-bit binary-to-decimal decoder.

TABLE 21

Boolean Functions for Binary Decoding

| Boolean function | Binary | Decimal |
|---|---|---|
| $\bar{A} \cdot \bar{B}$ | 00 | 0 |
| $\bar{A} \cdot B$ | 01 | 1 |
| $A \cdot \bar{B}$ | 10 | 2 |
| $A \cdot B$ | 11 | 3 |

With larger binary numbers, the more systematic approach of pyramid decoding is often preferred. Consider the 3-bit binary number ABC. The two highest-order binary bits are decoded first, producing the four outputs $A \cdot B$, $\bar{A} \cdot B$, $A \cdot \bar{B}$, and $\bar{A} \cdot \bar{B}$. The 2-bit decoder of Fig. 13 can be used here. When these four outputs are combined with C, the odd decimal numbers 1, 3, 5, and 7 are decoded. When these outputs are combined with $\bar{C}$, the even decimal numbers 0, 2, 4, and 6 are decoded. A 3-bit pyramid decoder is shown in Fig. 14.

Another pyramid stage can be connected to the eight outputs of the 3-bit decoder to decode the 4-bit number ABCD. Here D is combined with the eight outputs to decode the odd decimal numbers to 15, and $\bar{D}$ is combined with them to decode the even decimal numbers. While there is no theoretical limit to the number of stages in a pyramid decoder, the number of gates and connections soon becomes prohibitively large. If more than four binary bits are involved, it is often more convenient to use BCD rather than binary. Here a 4-bit decoder is used to decode each decimal digit. If the BCD for each decimal digit is decoded sequentially, one decoder network will suffice. This is known as serial decoding. If a separate decoder is used for each decimal digit, parallel decoding results. This requires considerably more circuitry but is appreciable faster.

In pyramid decoding, two or more steps are required. In the first decoding stage, two binary bits are decoded. These are combined with a third bit and decoded in the second stage. Another decoder state is required for each additional binary bit. In 4-bit representation such as BCD, matrix decoding is often used. Here the two higher-order bits and the two lower-order bits are decoded separately. Thus the binary number ABCD would be decoded to give four outputs in A and B and four outputs in C and D. The 2-bit decoder of Fig. 13 can be used here. These eight outputs are then combined to give the decimal numbers 0 through 15. A 4-bit matrix decoder is shown in Fig. 15. If BCD is being decoded, the output gates for decimal 10 through 15 are eliminated.

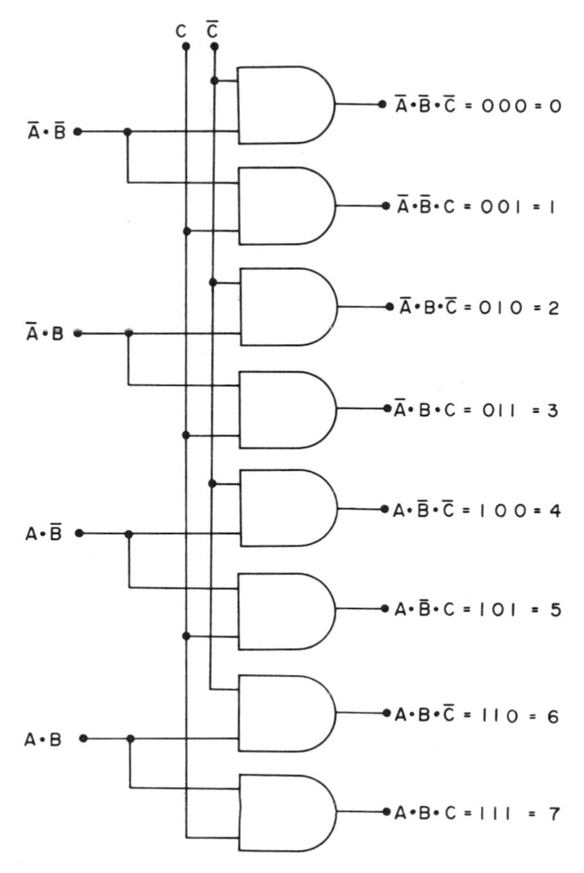


Fig. 14. A 3-bit binary-to-decimal pyramid decoder.

## 2. Signal Multiplexers

Frequently the central processing unit in a computing system will receive signals from and send signals to a variety of other units. This situation is analogous to a railroad switching yard where both incoming and outgoing trains must be switched or gated onto the right track. Gating circuits used to perform these functions are called signal selectors or multiplexers.

A 3-input signal selector is shown in Fig. 16. Here A, B, and C are three logic signal sources. The logic state of one of these three is to appear at the output. The X, Y, and Z are control inputs which are used to select which of the three logic signals is to be gated through the network. For example, if control input Y is raised to logic 1, logic signal B will be gated through the network. That is, the output logic level will be the same as B.

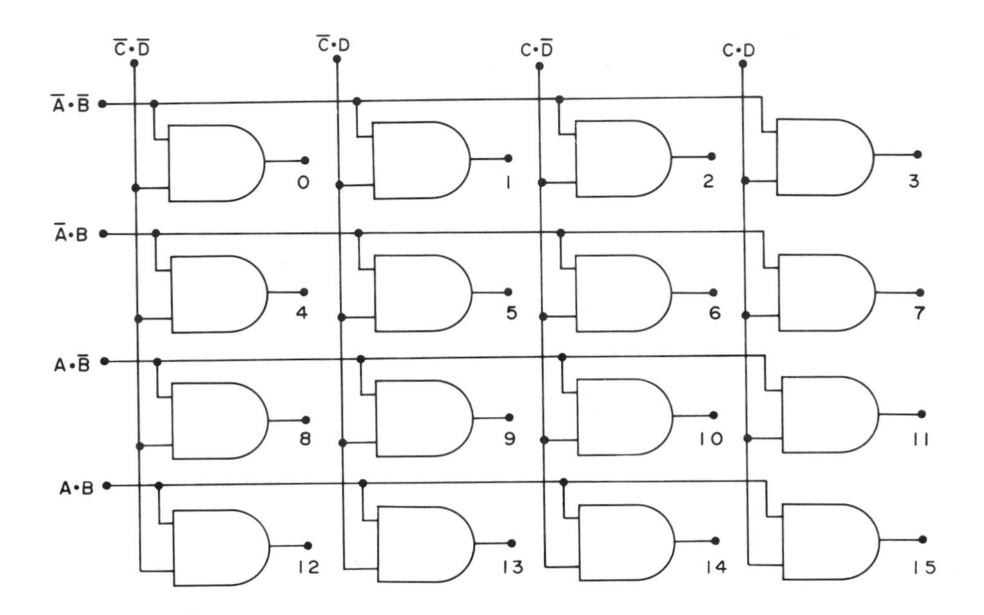


Fig. 15. A 4-bit matrix decoder.

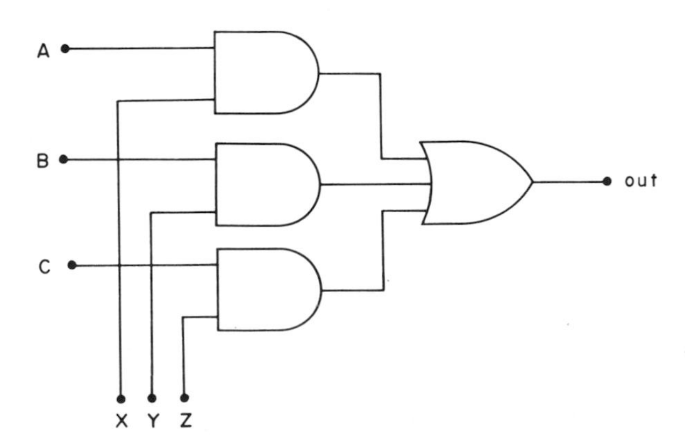


Fig. 16. A 3-input signal selector.

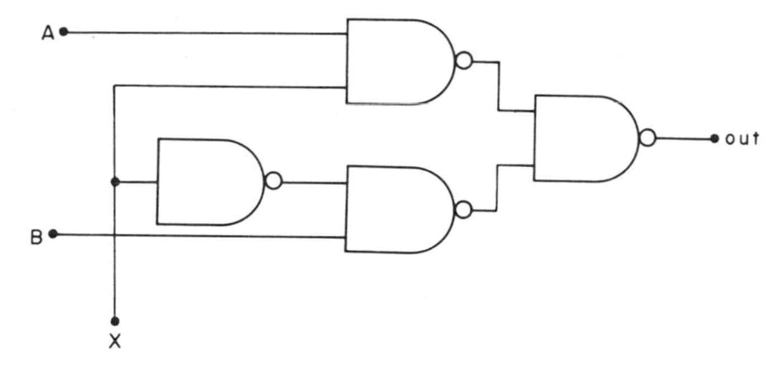


Fig. 17. A 2-input signal selector with one control input.

This simple signal selector can be troublesome since the output is ambiguous if more than one control input is at logic 1. If three of the outputs from a 2-bit decoder are used as control inputs, the problem is eliminated since no two of the decoder outputs can be simultaneously at logic 1. Another approach is to use the signal selector shown in Fig. 17. Here only one control input is used, thus eliminating any chance of ambiguity. If the control input X is at logic 1, the logic state of A is gated through the network. If X is at logic 0, then B appears at the output. Figure 18 shows how three 2-input signal selectors can be connected to make a 4-input selector. Here two control inputs are required. Table 22 gives the control-input states required for gating each channel through the multiplexer.

TABLE 22

4-Channel Multiplexer Coding

| X | Y | Output channel |
|---|---|---|
| 1 | 1 | A |
| 0 | 1 | B |
| 1 | 0 | C |
| 0 | 0 | D |

## 3. Arithmetic Processing Networks

Binary arithmetic is easily implemented using logic gates. The rules for binary arithmetic given in Section II. D can be used to establish tables of combinations from which appropriate Boolean functions can be synthesized and simplified. Since the more rapid method of multiplication and division by shifting is usually used in high-speed digital systems, these arithmetic operations will be discussed later. Addition and subtraction, however, are

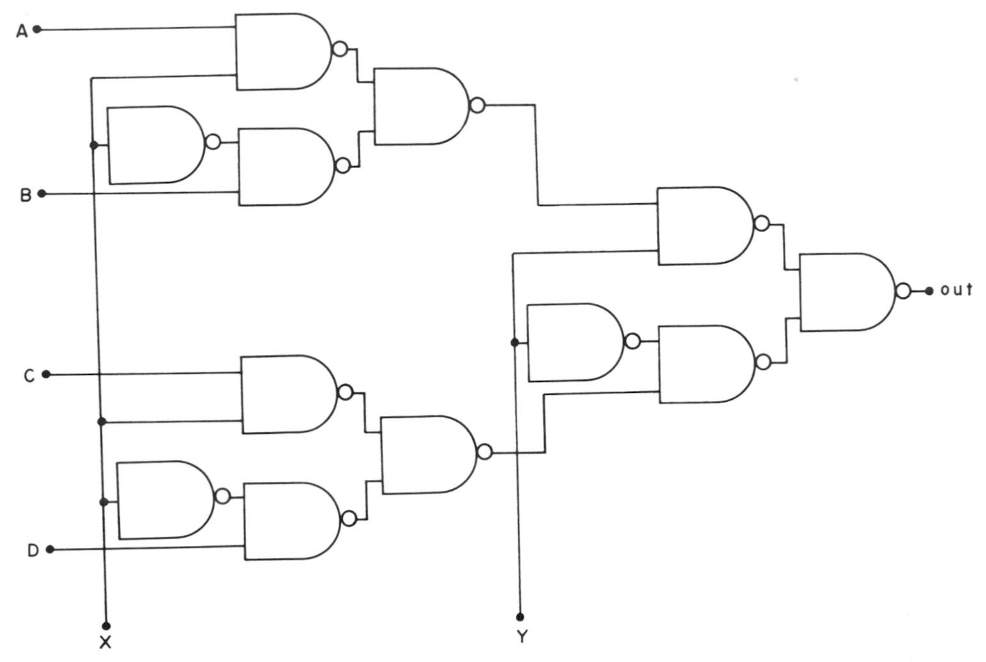


Fig. 18. Three 2-input signal selectors connected for use as a 4-input signal selector.

frequently implemented using logic gates, and these operations will be considered here.

a. Binary Addition Logic Networks. In Section III. E. 3, two functions were synthesized for binary addition. The i-th sum bit had the form

$$S_i = C_i \cdot \overline{(A_i \oplus B_i)} + \overline{C}_i \cdot (A_i \oplus B_i)$$
$$= C_i \oplus (A_i \oplus B_i)$$

where $A_i$ and $B_i$ are the i-th-order bits of the binary numbers A and B, and $C_i$ is the carry from the addition of the next-lower-order bits $A_{i-1}$ and $B_{i-1}$. The carry bit $C_{i+1}$ to be added to the next-higher-order bits $A_{i+1}$ and $B_{i+1}$ has the form

$$C_{i+1} = A_i \cdot B_i + A_i \cdot C_i + B_i \cdot C_i.$$

Logic-gate circuits for implementing these two functions are shown in Fig. 19. The Boolean function for $S_i$ can be rearranged to give a more easily implemented expression,

$$S_i = (A_i \cdot B_i \cdot C_i) + (A_i + B_i + C_i) \cdot \overline{(A_i \cdot B_i + A_i \cdot C_i + B_i \cdot C_i)}.$$

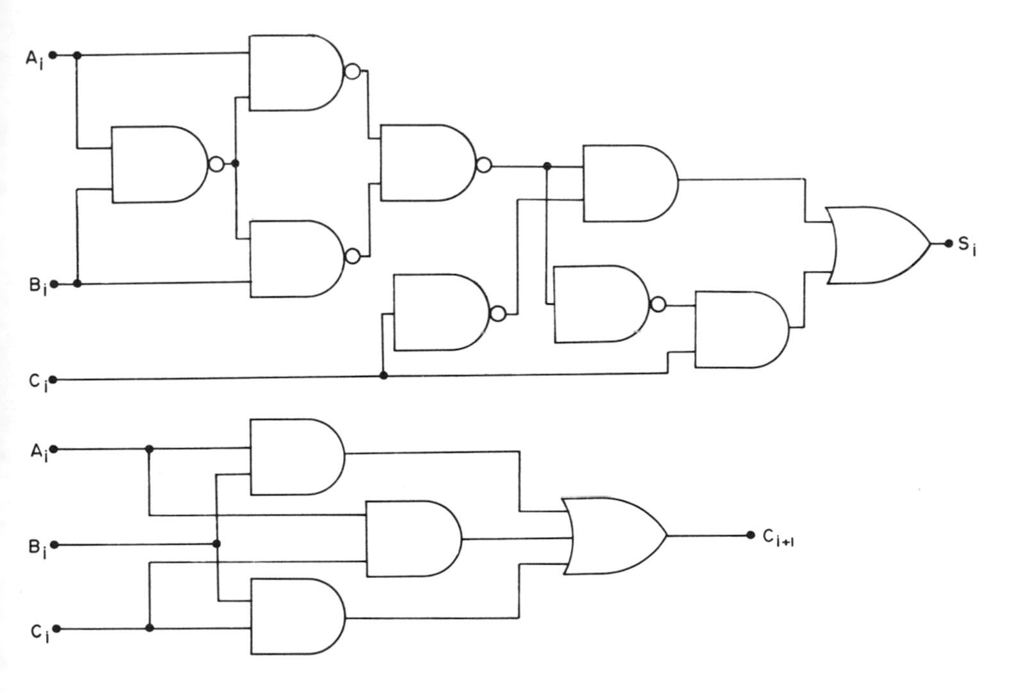


Fig. 19. Logic gate implementation of binary addition.

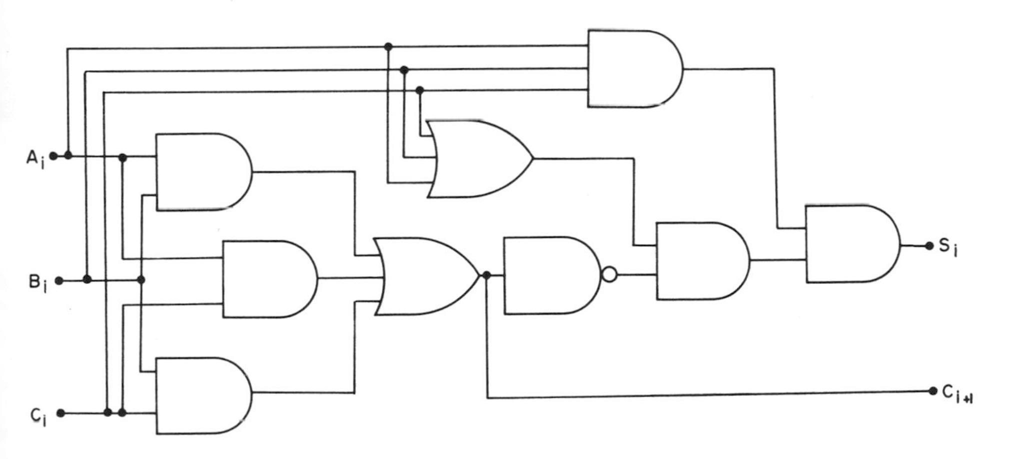


Fig. 20. A binary full-adder.

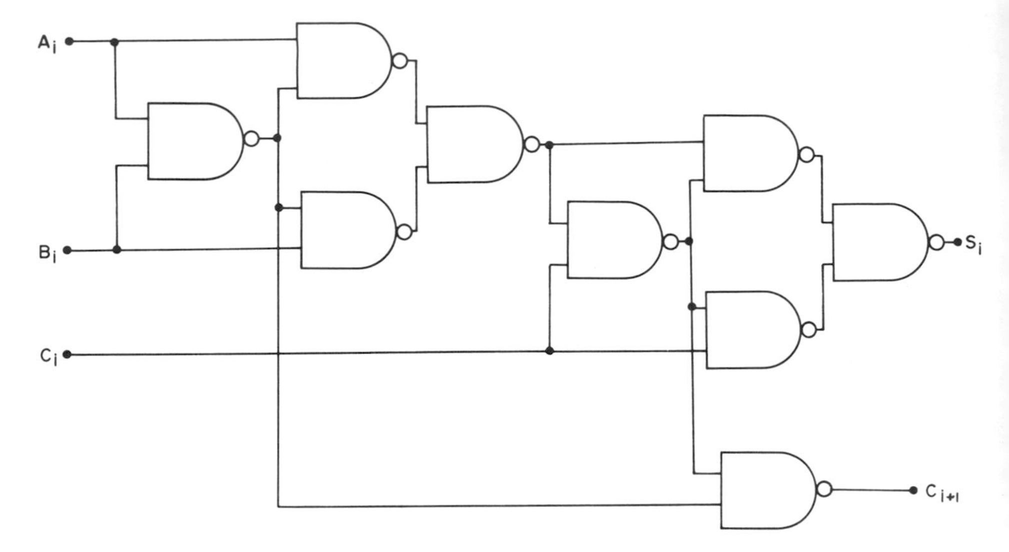


Fig. 21. A binary full-adder using NAND gates.

It should be noted that the last term in parentheses is just the complement of the carry $C_{i+1}$. The complete adding circuit using this function is shown in Fig. 20. An adding circuit using only NAND gates is shown in Fig. 21.

When adding the lowest-order binary bits of the two numbers A and B, the carry $C_i$ from the next-lower-order addition is not present. The Boolean functions for this special case are considerably simpler:

$$S_i = A_i \cdot \overline{B}_i + \overline{A}_i \cdot B_i = A_i \oplus B_i$$

$$C_{i+1} = A_i \cdot B_i.$$

The logic-gate network used to implement these functions is called a half-adder. A NAND-gate half-adder is shown in Fig. 22. It should be noted that the full-adder of Fig. 21 is simply the combination of two half-adders.

b. Binary Subtraction Logic Networks. From the subtraction rules given in Section II.D.2, the table of combinations for logical subtraction shown in Table 23 is readily derived. Here $M_i$ and $S_i$ are, respectively, the i-th bits of the minuend and subtrahend, and $B_i$ is the borrow bit from the subtraction of the next-lower-order bits $M_{i-1}$ and $S_{i-1}$. Two outputs are required, a difference bit $D_i$ and a borrow bit $B_{i+1}$, from the next-higher-order place.

From this table of combinations, Boolean functions are synthesized for $D_i$ and $B_{i+1}$:

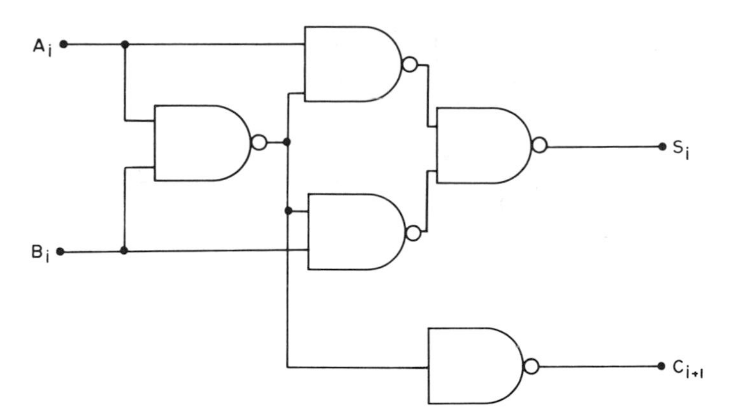


Fig. 22. A binary half-adder using NAND gates.

TABLE 23

Combinations for Binary Subtraction

| $M_i$ | $S_i$ | $B_i$ | $D_i$ | $B_{i+1}$ |
|---|---|---|---|---|
| 0 | 0 | 0 | 0 | 0 |
| 1 | 0 | 0 | 1 | 0 |
| 0 | 1 | 0 | 1 | 1 |
| 0 | 0 | 1 | 1 | 1 |
| 1 | 1 | 0 | 0 | 0 |
| 1 | 0 | 1 | 0 | 0 |
| 0 | 1 | 1 | 0 | 1 |
| 1 | 1 | 1 | 1 | 1 |

$$\begin{aligned} D_i &= \overline{M}_i \cdot S_i \cdot \overline{B}_i + \overline{M}_i \cdot \overline{S}_i \cdot B_i + M_i \cdot \overline{S}_i \cdot \overline{B}_i + M_i \cdot S_i \cdot B_i \\ &= B_i \cdot \overline{(M_i \oplus S_i)} + \overline{B}_i \cdot (M_i \oplus S_i) \\ &= B_i \oplus (M_i \oplus S_i) \end{aligned}$$

$$\begin{aligned} B_{i+1} &= \overline{M}_i \cdot S_i \cdot \overline{B}_i + \overline{M}_i \cdot \overline{S}_i \cdot B_i + \overline{M}_i \cdot S_i \cdot B_i + M_i \cdot S_i \cdot B_i \\ &= \overline{M}_i \cdot (S_i + B_i) + M_i \cdot (S_i \cdot B_i) \end{aligned}$$

A logic-gate full-subtractor is shown in Fig. 23. A full subtractor using only NAND gates is shown in Fig. 24.

When subtracting the lowest-order binary bits of the two numbers M and S, the borrow $B_i$ is not present. As in binary addition, the Boolean functions

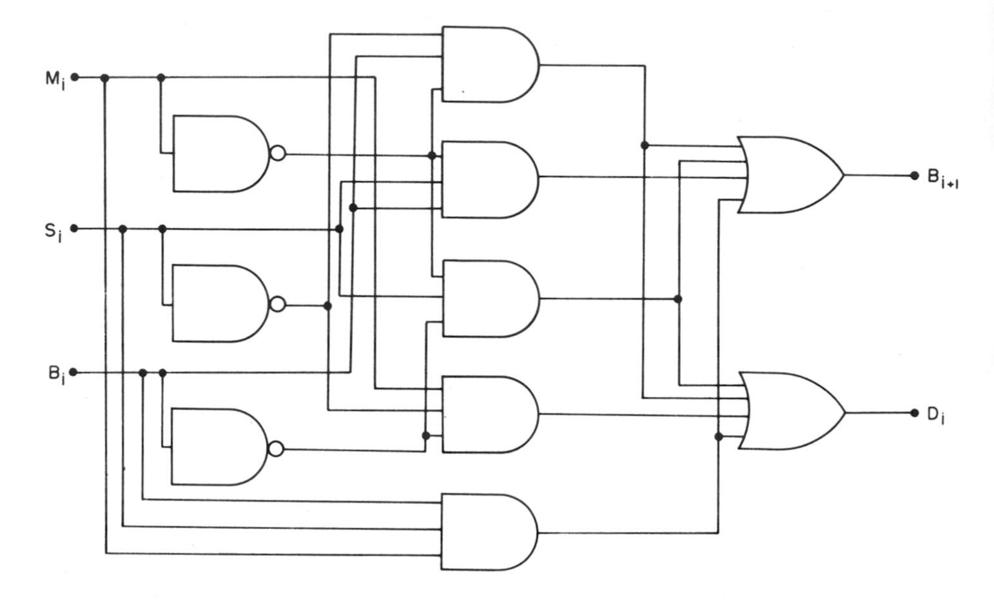


Fig. 23. A binary full-subtractor.

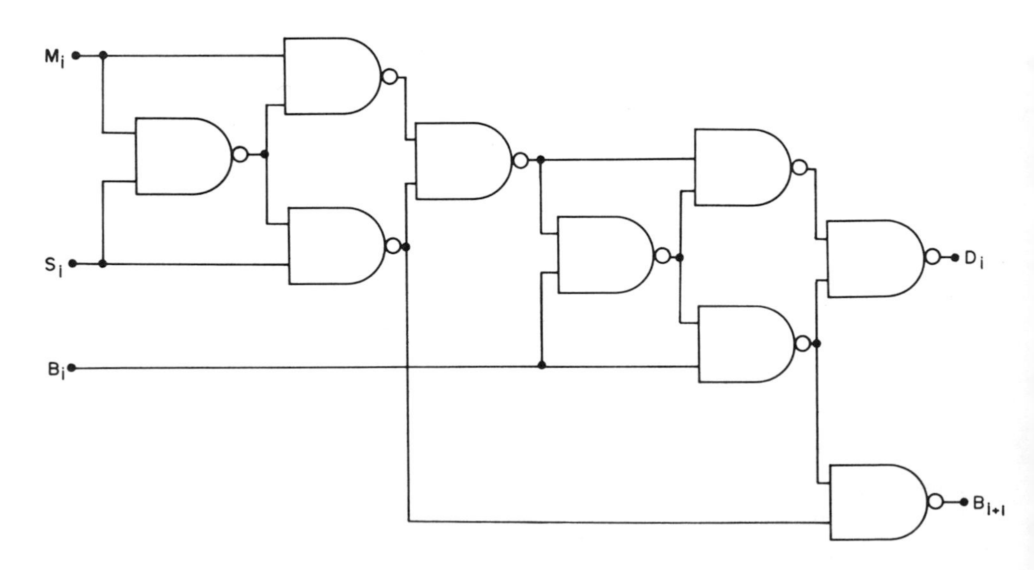


Fig. 24. A binary full-subtractor using NAND gates.

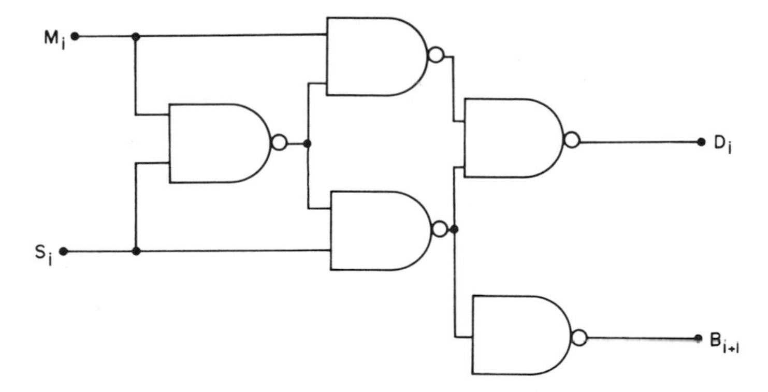


Fig. 25. A binary half-subtractor using NAND gates.

for this special case are much simpler:

$$D_i = \overline{M}_i \cdot S_i + M_i \cdot \overline{S}_i = M_i \oplus S_i$$
$$B_{i+1} = \overline{M}_i \cdot S_i.$$

The logic-gate network used to implement these functions is called a half-subtractor. A NAND-gate half-subtractor is shown in Fig. 25. The full-subtractor of Fig. 24 is just a combination of two half-subtractors.

The circuits presented in this section are used to perform arithmetic operations on individual digits or places in a pair of binary numbers. Combinations of these circuits, along with electronic memory or storage devices known as registers, are used to perform complete addition and subtraction operations on large binary numbers. These topics will be considered in detail in later sections.

Logic gates find numerous other applications in digital computing and control systems. Some of these will be discussed as parts of more complex control and processing networks.

## IV. SOLID-STATE DEVICES AND CIRCUITS

The heart of most modern digital computing systems is the integrated-circuit transistor switch. With modern technology, these devices can be made extremely small, and at relatively low cost. In addition, they require minimal power and usually have very long operating lifetimes. While the integrated-circuit switch is often treated as a black box in logic circuit design, a complex computing or control system may contain numerous interfacing and coupling diodes and transistors. The salient properties and operating parameters of these devices must be understood if an electronic circuit, digital or otherwise, is to be properly designed. In addition, some understanding of the operation of integrated-circuit transistor switches is necessary if the best type of transistor logic is to be chosen for a particular application from the several types commercially available.

## A. Rectification by Diodes

A device which allows current to flow in one direction only is called a rectifier. Most rectifiers are nonlinear resistive elements, the resistance of which depends on the polarity of the potential applied to it. Figure 26 shows a simple model of a rectifier. In this model, the rectifier consists of two resistors and a switch which serves to change the resistance in series with a potential source.

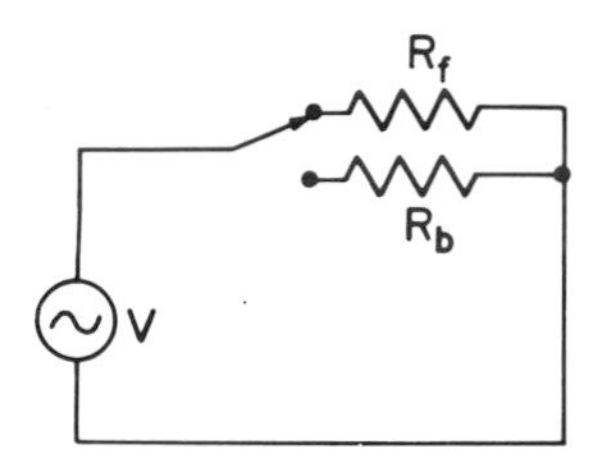


Fig. 26. A resistive model of a rectifier.

The smaller of the two resistors is called the forward resistance $R_f$ and the larger one is called the backward resistance $R_b$. When the input voltage signal V is positive, it is connected to $R_f$. When the input is negative, it is connected to $R_b$. The forward and backward currents are given by

$$I_f = \frac{V}{R_f}, \qquad I_b = \frac{V}{R_b}, \qquad \text{or} \qquad \frac{I_f}{I_b} = \frac{R_b}{R_f}.$$

In a good rectifier, this ratio can be very large.

Most common rectifying devices are diodes, that is, two terminal devices. They are designed so that the flow of electrons from one terminal to the other is easier in one direction than in the other. While several types of diodes are manufactured, we will discuss only the pn-junction diode since this is the type usually associated with modern digital instrumentation. In addition, an understanding of the pn-junction diode is essential to the understanding of the more complex solid-state devices which are responsible for the operation of logic switching circuits.

## B. Semiconductor Materials

Solid-state devices are made from semiconducting materials, silicon and germanium being the most common. A semiconducting material is one that is neither a good conductor nor a good insulator. There are some charge carriers which are free to conduct current if a potential is applied across the material, but the number of charge carriers is small compared to a good conductor. Most of the electrons in a semiconductor are tightly bound to their parent atoms which, in turn, are fixed in the lattice of the material.

Germanium has about $2.5 \times 10^{13}$ charge carriers per cubic centimeter, and silicon has about $1.5 \times 10^{10}$. The resistances for pure germanium and pure silicon are about 60 and 60,000 $\Omega/\text{cm}^3$, respectively.

### 1. Intrinsic Semiconductors

Both silicon and germanium are covalent bonded to four similar atoms with tetrahedral symmetry. The free charge carriers come from the occasional breaking of the covalent bonds. Once an electron is liberated from a bond, it drifts through the crystal until it recombines with another atom which has lost an electron. At equilibrium, the rates of charge carrier production and recombination are equal.

Although the parent atom from which an electron has left is fixed in position in the crystal lattice, the positive charge on this atom is not fixed. An electron from an adjacent atom can fill the vacancy in the original parent atom and thus transfer the positive charge to the adjacent atom. In this way, the positive charges move through the crystal. Since these moving positive charges have no specific particles associated with them, they are called holes. The number and mobility of holes and electrons determine the resistance of the material.

In a pure crystal of silicon or germanium, the number of electrons equals the number of holes. This is called intrinsic semiconductivity. A more useful semiconducting material can be made by adding a small, controlled amount of an impurity to the pure silicon or germanium. This results in a doped semiconductor.

### 2. Doped Semiconductors

Suppose that a small amount of a group V element such as antimony is added to a pure silicon or germanium crystal. The impurity atoms will assume normal lattice positions in the crystal and form four covalent bonds with their nearest neighbors. The fifth valence electron from the impurity atoms cannot form a covalent bond because there are no unbonded valence electrons in the surrounding matrix of silicon or germanium atoms. These excess valence electrons are only weakly bound to their parent impurity atoms, and at room temperature many of them break away and travel freely through the crystal.

After an electron has left an impurity atom, the atom has a positive charge. This charge, however, is fixed in the crystal lattice and does not constitute a hole because the impurity atom and all its neighbors have a full complement of covalent bonds. The number of holes, in fact, is less than in the pure material because the increase in the number of free electrons increases the rate of hole-electron recombination. The doped crystal is called an n-type semiconductor, the n implying a negative sign on the major charge carriers.

If a group III element is added as an impurity to a pure silicon or germanium crystal a p-type semiconductor is produced. Since the group III elements have only three valence electrons, only three covalent bonds can be formed by each impurity atom. This vacancy can be filled by supplying an electron from a neighboring silicon or germanium atom. This results in the propagation of a hole through the crystal lattice. The free-electron concentration in p-type material is very low because the increased hole concentration has increased the rate of hole-electron recombination.

The impurity atoms in n-type semiconducting material are called donors since they donate free electrons. The impurity atoms in p-type material are called acceptors since they accept electrons from the lattice, thus forming holes.

### C. The pn Junction

We now ask what will occur when an n-type and a p-type semiconductor are joined together. We would expect the electrons in the n-type material to recombine with the holes in the p-type material. This is exactly what happens at the interface or junction between the n and p materials. This recombination does not continue through the entire crystal but is confined to a narrow region near the interface. This can be understood by a closer examination of the pn junction. It is important to remember that while the electrons and holes are mobile, the donor and acceptor atoms are locked in the crystal lattice.

After a few recombinations occur at the pn junction, an uncompensated positive charge exists near the junction in the n-type material, and an uncompensated negative charge exists in the p-type material, as shown in Fig. 27. After this initial recombination, a narrow region exists near the junction which has no holes or free electrons. This is known as the depletion region. The uncompensated negative charge on the p side of the junction prevents electrons from crossing the junction and recombining with holes. In the same way, the uncompensated positive charge on the n side of the junction prevents holes from crossing the junction. In essence, the immobile donor and acceptor ions form a potential barrier which prevents further recombination.

The barrier height can be represented as a space-charge equivalent

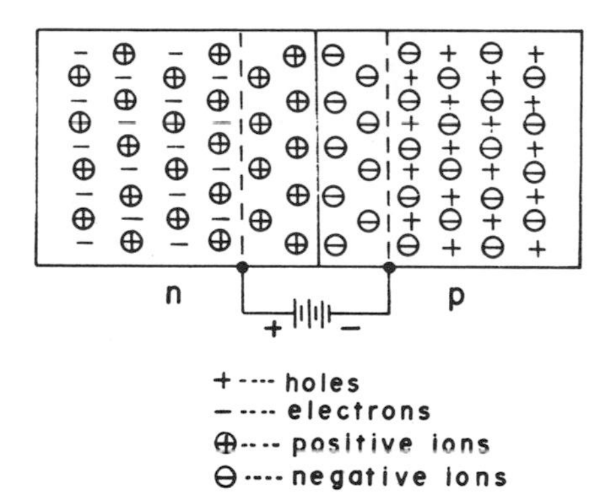


Fig. 27. The pn junction showing depletion region and space-charge equivalent battery.

battery as shown in Fig. 27. It should be noted that free electrons in the n-type material are already at their highest possible potential, that is, near the positive pole of the space-charge equivalent battery. Thus, there would be little tendency for them to move toward the junction.

## 1. Junction-Biasing Configurations

The next question we must answer is what will occur when a potential difference is applied across the pn junction. In Fig. 28(a), a battery $V_{rb}$ has been connected to the pn junction such that the n-type material is made positive with respect to the p-type material. This will cause holes to be attracted to the negative terminal and electrons to the positive terminal. As a result, the width of the depletion region and the height of the potential barrier both will increase. Under these conditions, it would be extremely difficult for holes or electrons to cross the junction, and hence no current should flow across the junction or in the external circuit. This is known as reverse biasing the junction, and $V_{rb}$ is the reverse-bias voltage.

In Fig. 28(b), the p side of the junction has been made positive with respect to the n side. In this case, holes from the p side and electrons from the n side will be accelerated toward the junction. This will result in a lowering of the barrier height and a narrowing of the depletion region. Under these conditions, it will be much easier for electrons and holes to cross the junction and recombine.

Every time an electron and hole are neutralized at the junction, an electron enters the n material from the external circuit, and an electron-pair bond is broken in the p material, and an electron leaves through the external circuit. The resulting hole in the p material and the electron in the n material are accelerated toward the junction under the influence of the uncompensated charge at the junction. Thus we observe a current flowing in

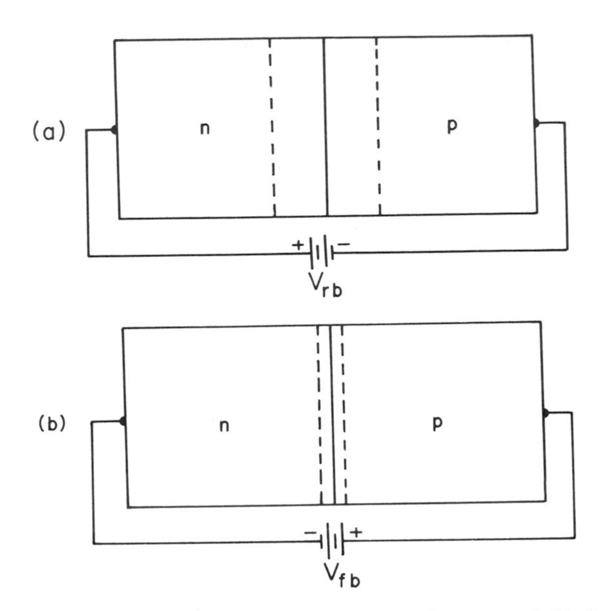


Fig. 28. The pn-junction diode with (a) reverse bias, and (b) forward bias.

the external circuit. This is called forward biasing the junction; $V_{fb}$ is the forward-bias voltage.

2. Diode-Current-Voltage Characteristics

The current-voltage relationship for both forward- and reverse-bias configurations is shown in the diode characteristic curve in Fig. 29. Notice that the reverse current is not zero. This is because the n-type material

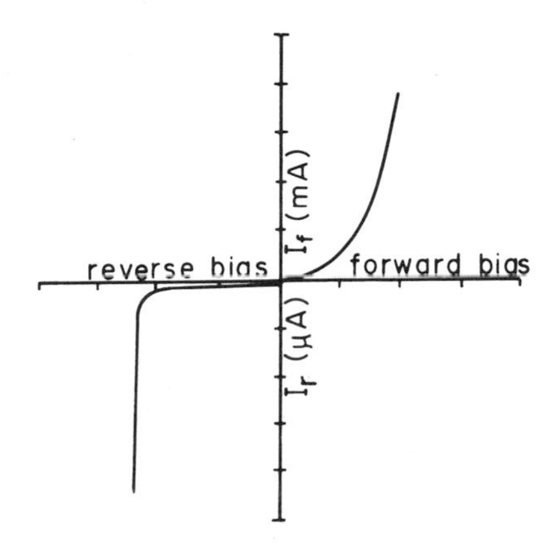


Fig. 29. Current-voltage characteristics of the pn-junction diode.

contains small traces of acceptor atoms, and the p-type material contains small traces of donor atoms. Thus n material contains some holes and p material contains some free electrons. These are called minority carriers. The small reverse current is a result of the fact that a reverse-bias configuration for the majority carriers is a forward-bias configuration for the minority carriers.

Another important feature of the diode characteristic curve is the sudden increase in reverse current at large values of reverse-bias voltage. This is the result of an avalanche or internal breakdown phenomenon in the semiconductor material. With the exception of certain special purpose diodes which are designed to operate with large reverse currents, this region is to be avoided since the excessive power dissipation may destroy the device.

The conventional circuit notation for the pn-junction diode is shown in Fig. 30. Note the direction of forward and reverse currents.

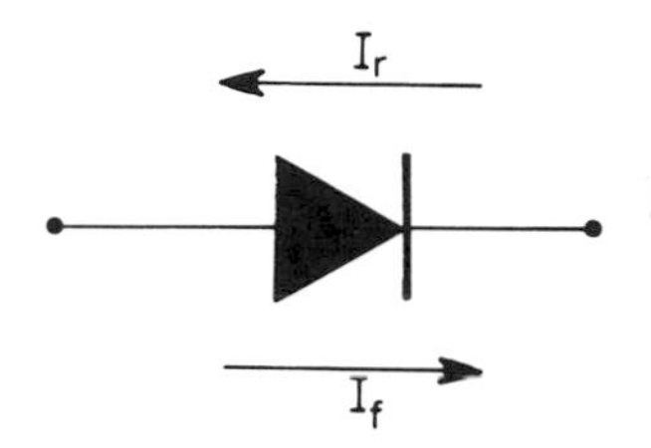


Fig. 30. Circuit symbol for a pn-junction diode.

## D. The Junction Transistor

The junction transistor largely has replaced the vacuum tube in modern electronic circuits. Because of its small size, ruggedness, and lower voltage and power requirements, it has become the basic electronic amplification device. Much of the theory needed to understand basic transistor operation was discussed in the previous section.

A transistor is simply two pn junctions placed back to back, one having a forward bias and the other having a reverse bias. For a given current, the power developed in a larger resistor is greater than that developed in a smaller one. The transistor is designed so that the current carried by the reverse-biased junction is almost the same as that carried by the forward-biased junction. Since the reverse-biased junction has a higher resistance, more power can be developed in an external circuit connected to the reverse-biased junction than in the circuit connected to the forward-biased junction. A small power change in the forward-biased-junction external circuit will produce a larger power change in the reverse-biased-junction external

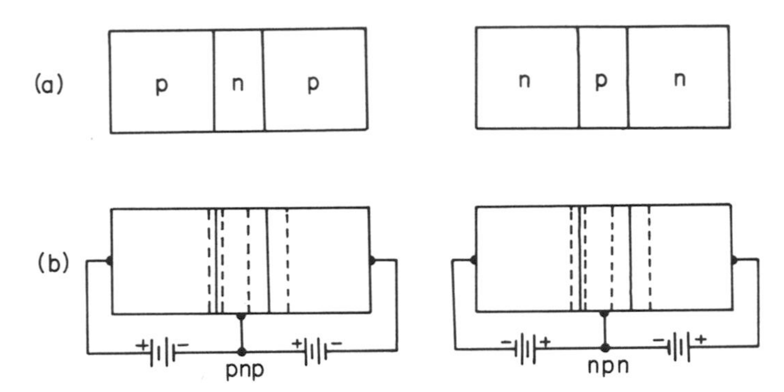


Fig. 31. Junction transistors showing (a) the formation of two pn junctions, and (b) the depletion regions at the forward- and reverse-biased junctions.

circuit. Thus, the transistor can function as a power amplifier.

The transistor is made by sandwiching a thin wafer of n-type or p-type semiconductor material between two pieces of the other type of material. This is shown in Fig. 31(a). If the center section is n-type material, the device is called a pnp transistor. If the center section is p-type material, the device is called an npn transistor. The size of the center section is greatly exaggerated in Fig. 31(a) and, in fact, is usually less than 0.001 in. wide.

The depletion regions under the previously discussed bias configurations are shown in Fig. 31(b). To forward bias the left-hand junction of the pnp transistor, the left section must be made positive with respect to the center section. The opposite polarity is used to reverse bias the right-hand junction. For the npn transistor, the polarities of the bias supplies must be reversed.

Majority carriers, holes in the pnp transistor and electrons in the npn transistor, will accerate toward the forward-biased junction, cross the junction, diffuse across the center section, and finally cross the reverse-biased junction. There they will recombine with carriers of the opposite sign introduced at the right-hand side from the external bias supply. The left-hand side which emits the majority carriers is called the emitter. It is analogous to the cathode in a vacuum tube except that the carriers may be positive or negative. The right-hand side which collects the carriers produced in the emitter is called the collector. It is analogous to the plate in a vacuum tube. The center section is called the base, and in some ways its function is similar to the grid in a vacuum tube. However, the tube grid is a high-impedance, voltage-sensing element while the transistor base is a low-impedance, current-sensing element.

Carriers opposite in sign to those introduced in the emitter from the left will enter the base region from the base-emitter bias supply. These carriers will accelerate toward the forward-biased junction, and will recombine with carriers of opposite sign crossing the junction from the emitter. The extent of this recombination must be kept low so that most of the carriers coming from the emitter will cross into the collector. This is accomplished by making the base region extremely thin so that most of the majority carriers from the emitter will diffuse across the base region without encountering carriers of opposite sign. This means that the current carried by the external base circuit will be only a small fraction of the total current carried across the base-emitter junction.

An important transistor parameter is the fraction of majority carriers which cross the base region without recombining. This parameter, which is called the forward current gain and is symbolized by $\alpha_f$, is between 0.92 and 0.99 for most transistors. If $I_E$ is the total emitter current, then $\alpha_f I_E$ is the fraction of this current going through the collector and $(1 - \alpha_f)I_E$ is the fraction recombining in the base. Thus the following relations exist among the currents in the external emitter, base, and collector circuits:

$$I_E = I_C + I_B, \quad I_C = \alpha_f I_E, \quad I_B = (1 - \alpha_f)I_E.$$

Figure 32 shows the circuit notation used for junction transistors. The emitter arrow points in the direction of positive current flow in the external circuit.

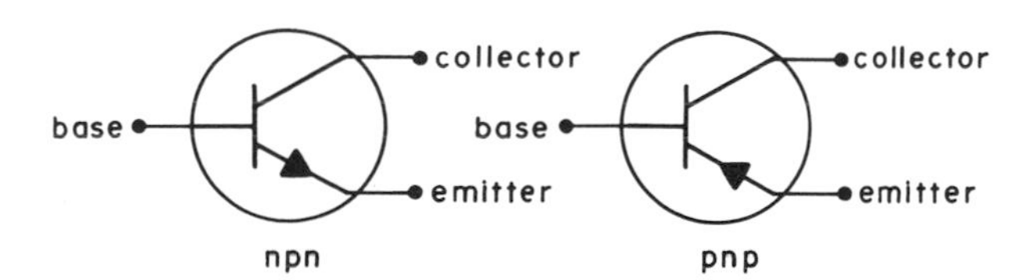


Fig. 32. Circuit notation for junction transistors.

## E. Basic Amplifier Configurations

There are three basic ways of introducing an input signal into a transistor and extracting an output signal from it. In the common base (CB) configuration, the input signal is introduced into the base-emitter circuit, and the output signal is extracted from the base-collector circuit. The base is common to both input and output circuits. A CB configuration for a pnp transistor is shown in Fig. 33(a). The input impedance $R_i$ of a CB amplifier is typically between 30 and 150 Ω, while the output impedance $R_o$ is typically between 300 and 500 kΩ.

In the common emitter (CE) configuration shown in Fig. 33(b), the input

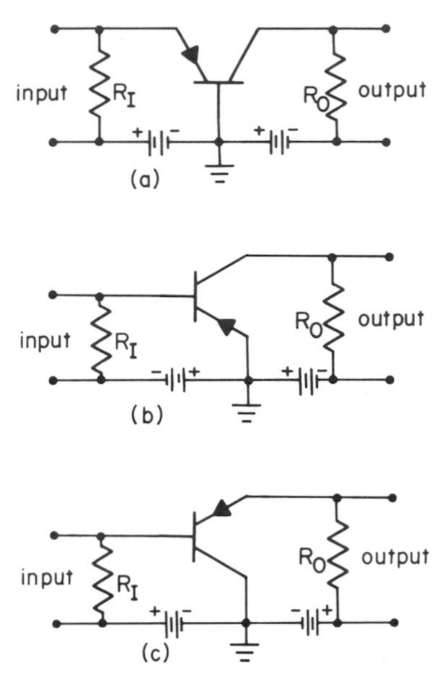


Fig. 33. Basic-transistor amplifier configurations: (a) common base; (b) common emitter; (c) common collector.

is in the base-emitter circuit, and the output is in the collector-emitter circuit. Input impedance for the CE amplifier is usually between 500 and 1500 Ω, while the output impedance is usually from 30 to 50 kΩ. The CE amplifier is analogous to the common triode amplifier which has a grounded or common cathode.

The common collector amplifier (CC) shown in Fig. 33(c) uses the base-collector circuit as an input and the emitter-collector circuit as an output. Input impedance is typically between 2 and 500 kΩ; output impedance usually ranges from 50 to 1000 Ω. This amplifier configuration, which is analogous to the cathode-follower vacuum-tube circuit, is often called an emitter follower.

Other bias configurations which require only one battery or power supply are often used. The CB amplifier in Fig. 34(a) uses a voltage divider to establish the proper bias voltages. The IR drop across $R_3$ keeps the base negative with respect to the emitter and positive with respect to the collector. In the CE and CC configurations shown in Figs. 34(b) and 34(c) respectively, no external voltage divider is used because the two pn junctions serve as an internal voltage divider and establish proper bias voltage levels.

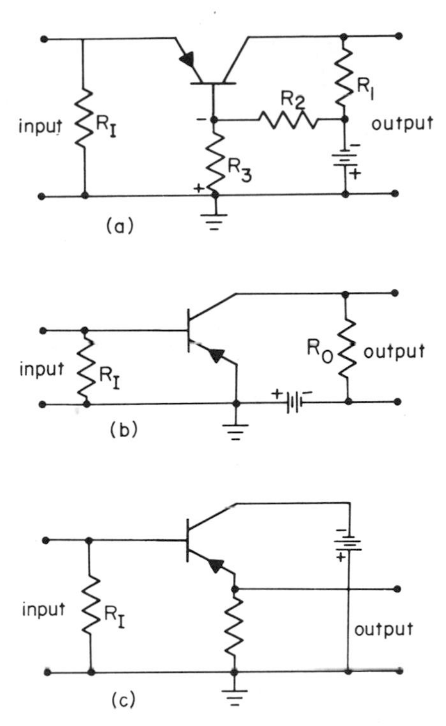


Fig. 34. Transistor amplifier configurations using one bias supply: (a) common base; (b) common emitter; (c) common collector.

## 1. Phase Relations and Current Flow

Consider the CE amplifier in Fig. 34(b). The phase relationship between the input and output voltages is easily deduced by observing the effect of the input signal on the forward-biased base-emitter junction. If the input signal is positive (negative for an npn transistor) the decreased forward bias on the base-emitter junction will decrease the emitter current. The resulting decrease in collector current will decrease the IR drop across $R_o$, thus causing the collector to swing more negative. Thus the output is inverted; as the input becomes more positive, the output becomes more negative. Similar arguments will show that the CB and CC configurations do not invert the input signal.

## 2. Graphical Analysis of Transistor Circuits

Since the common-emitter amplifier is the one most usually associated with digital circuitry, we will consider this case in some detail. The object of a graphical analysis is to select the operating or quiescent values of the base current $I_B$, collector current $I_C$, and collector-to-emitter voltage drop

$V_{CE}$ that will allow the maximum symmetric input-voltage swing with minimal output distortion. Quiescent values are so called because they are steady-state values in the absence of an input signal. Once selected, the quiescent values are obtained by the use of dc power supplies, voltage dividers, and current-limiting resistors. It should be noted that the quiescent values discussed in this section are used in transistor applications requiring linear amplification. When the transistor is used as a two-state current or voltage switch, the operating parameters are entirely different. This will be discussed later.

a. Biasing and Quiescent Values. The common-emitter output characteristics are used in obtaining a graphical analysis of a simple CE amplifier. These are plots of collector current $I_C$ vs. collector-to-emitter voltage drop $V_{CE}$ for a family of values of base current $I_B$. This family of curves, an example of which is shown in Fig. 35, appears very similar in shape to the plate characteristics of a vacuum pentode.

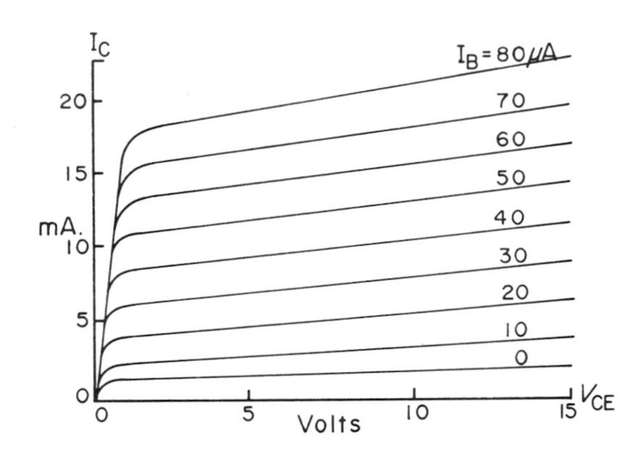


Fig. 35. Common-emitter output characteristics.

A typical CE amplifier circuit using an npn transistor is shown in Fig. 36. Resistors $R_1$ and $R_2$ along with bias supply $V_{CC}$ are chosen to produce the proper bias voltages over the anticipated range of input signal amplitude. Analysis of the circuit can be simplified by finding the Thévenin equivalent circuit for the divider network formed by $V_{CC}$, $R_1$, and $R_2$. This voltage divider and its Thévenin equivalent circuit are shown in Fig. 37. The values of the Thévenin equivalent voltage $V_T$ and resistance $R_T$ are given by

$$V_T = \left(\frac{R_1}{R_1 + R_2}\right)V_{CC}, \qquad R_T = \frac{R_1 R_2}{R_1 + R_2}.$$

These equations can be solved for $R_1$ and $R_2$:

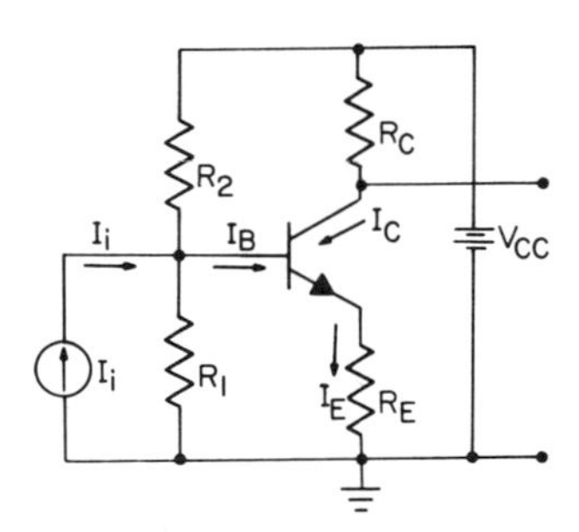


Fig. 36. Common-emitter amplifier circuit for graphical analysis.

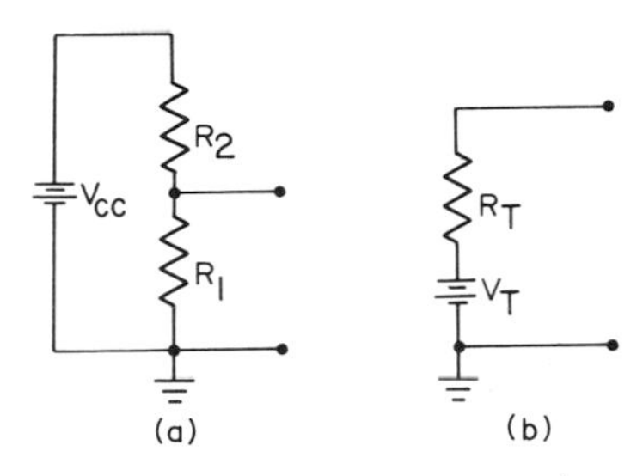


Fig. 37. Voltage-divider network (a), and its Thévenin equivalent circuit (b); used in biasing the transistor of Fig. 36.

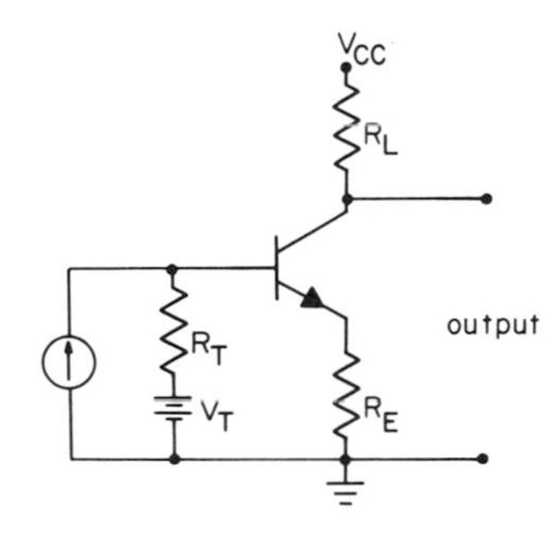


Fig. 38. Simplified common-emitter amplifier circuit showing Thévenin equivalent biasing.

$$R_1 = \frac{R_T}{1 - V_T/V_{CC}}, \qquad R_2 = R_T \frac{V_{CC}}{V_T}.$$

The simplified equivalent circuit is shown in Fig. 38.

Kirchhoff's voltage law is now applied to the input and output loops. For the output loop, we obtain

$$V_{CC} = V_{CE} + I_C R_L + I_E R_E.$$

Since $I_C$ is nearly equal to $I_E$, the output loop equation can be rewritten to give

$$V_{CC} = V_{CE} + I_C(R_L + R_E).$$

This is the equation of the circuit load line. It is conveniently plotted on the common-emitter output characteristic curves as shown in Fig. 39. The equation is usually evaluated at the intercepts where $I_C = 0$ and $V_{CE} = 0$. The equations for the intercept points are given by

$$V_{CE} = V_{CC} (I_C = 0)$$

$$I_C = \frac{V_{CC}}{R_L + R_E} (V_{CE} = 0).$$

It should be noted that the load line is independent of the type and properties of the transistor. It depends only on the values of the external circuit elements, that is, on $V_{CC}$, $R_L$, and $R_E$. The transistor must always operate along the load line, and values of $I_C$ and $V_{CE}$ for any value of $I_B$ are found from the intersection of the load line with the common-emitter characteristic curve representing that particular value of base current.

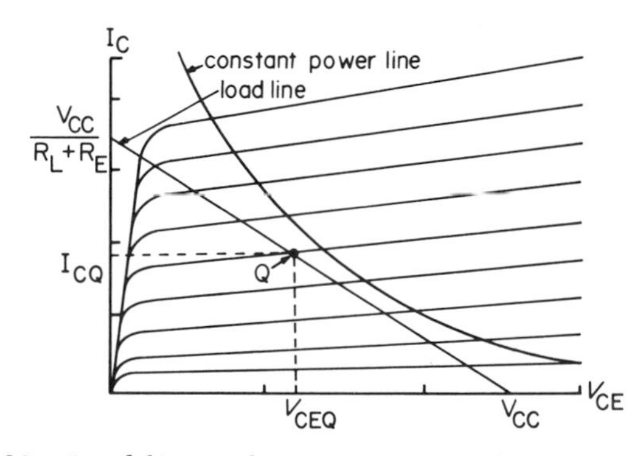


Fig. 39. Load line and constant power dissipation analysis.

The Kirchhoff voltage equation for the input loop is given by

$$V_T - I_B R_T = V_{BE} + I_E R_E$$

where $V_{BE}$ is the voltage drop across the forward-biased base-emitter junction and is about 0.7 V for silicon transistors. The input loop equation can be put in a more useful form by recalling that $I_B = I_E(1 - \alpha_f)$. After rearrangement, this gives

$$I_E = \frac{V_T - V_{BE}}{R_E + (1 - \alpha_f)R_T} .$$

Since the quantity $1 - \alpha_f$ may vary considerably, even for different transistors of the same type, in addition to being temperature dependent, $R_E$ and $R_T$ must be chosen such that $R_E \gg (1 - \alpha_f)R_T$. This will stabilize the emitter current against changes in $1 - \alpha_f$. This is one of the principal reasons for placing $R_E$ in the circuit. If this condition is met, the input loop equation reduces to the following very simple form:

$$I_E = \frac{V_T - V_{BE}}{R_E} .$$

This is the value of the quiescent emitter current $I_{EQ}$ and is approximately equal to the quiescent collector current $I_{CQ}$. When this equation is substituted for $I_C$ in the load line equation, the quiescent value of $V_{CE}$ can be found:

$$V_{CEQ} = V_{CC} - (V_T - V_{BE})\left(1 + \frac{R_L}{R_E}\right).$$

Quiescent values are usually chosen to give the maximum symmetric collector current swing. Since the collector current can vary from 0 to $V_{CC}/(R_L + R_E)$, half this maximum value or $V_{CC}/2(R_L + R_E)$ is usually chosen for $I_{CQ}$.

When constructing the load line, the power dissipation of the collector should be considered. This is accomplished by constructing a constant power-dissipation curve. The maximum collector dissipation is usually found in manufacturer's data. Since power dissipation is given by the product of collector current and voltage, maximum values of collector current can be computed for arbitrarily assigned values of voltage. Plotting corresponding values of current and voltage will produce a constant power-dissipation curve. The curve is usually drawn on the common-emitter characteristics. An example of this is shown in Fig. 39. When choosing values for $V_{CC}$, $R_E$, and $R_L$, the resulting load line should not cross the constant power-dissipation curve.

When designing an amplifier circuit, the maximum value of $V_{CC}$ usually is specified in manufacturer's data. Resistors $R_E$ and $R_L$ are then chosen to give the proper collector current without exceeding the power limitations, and $R_T$ is then chosen to satisfy the condition that $R_E \gg (1 - \alpha_f)R_T$. It is often more convenient to use a transistor parameter other than $\alpha_f$ in making these calculations. A parameter generally stated by the manufacturer is the ratio of collector to base current. This is called the current amplification factor and is symbolized by $\beta$.

$$\beta = \frac{I_C}{I_B} = \frac{\alpha_f I_E}{(1 - \alpha_f)I_E}$$

so

$$\beta = \frac{\alpha_f}{1 - \alpha_f} \quad \text{or} \quad \beta + 1 = \frac{1}{1 - \alpha_f}.$$

The inequality needed to stabilize the emitter current now can be described in terms of $\beta$,

$$R_T \ll (1 + \beta)R_E.$$

For given values of $R_E$ and $\beta$, $R_T$ is arbitrarily chosen to satisfy this inequality.

All that remains is to determine $V_T$, and then $R_1$ and $R_2$ can be calculated in the original amplifier circuit. Voltage $V_T$ is readily found from inspection of the previously described Kirchhoff voltage law expression for the amplifier input loop,

$$V_T = V_{BE} + I_B R_T + I_E R_E.$$

Since $I_B \ll I_E$ and $I_E \simeq I_C$, this equation can be rewritten to give

$$V_T = V_{BE} + I_{CQ} R_E$$

where $I_{CQ}$ is read from the quiescent point value on the common-emitter characteristic curves. Voltage $V_{BE}$ is about 0.7 V for silicon transistors. Now that $V_T$, $R_L$, $V_{CC}$, and $R_E$ have been specified, $R_1$ and $R_2$ can be computed.

b. Dynamic Transfer Characteristics. In an ideal transistor amplifier, the shape of the output waveform is exactly the same as that of the input waveform, differing only in amplitude and phase. This ideal situation is approximately realized only for small input signals. For larger input signals, the output waveform may be considerably distorted.

The degree of distortion and the shape of the output waveform for an arbitrary input signal can be obtained graphically by constructing the dynamic transfer characteristics for the amplifier. This is a plot of $I_C$ vs. $I_B$ evaluated along the load line. Construction of the dynamic transfer characteristic curve is simplified by using the projection technique shown in Fig. 40. The portion of the curve that is nearly linear gives the maximum and minimum values of base current $I_B$ for which the signal shape is preserved during amplification.

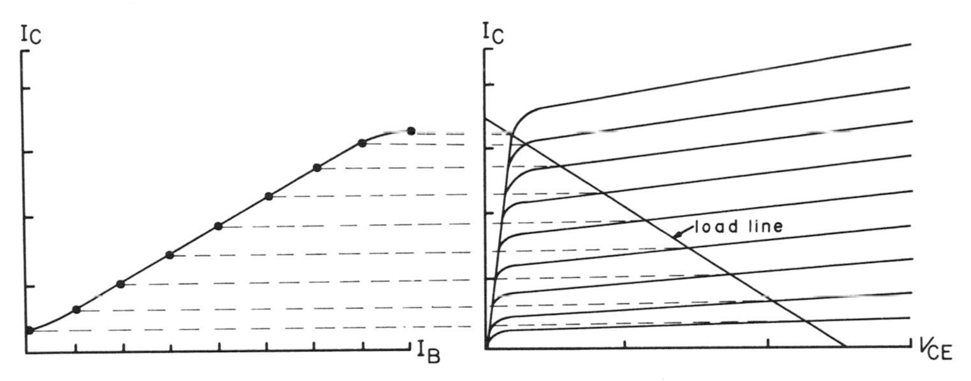


Fig. 40. Projection technique for constructing the dynamic transfer characteristics of a CE amplifier.

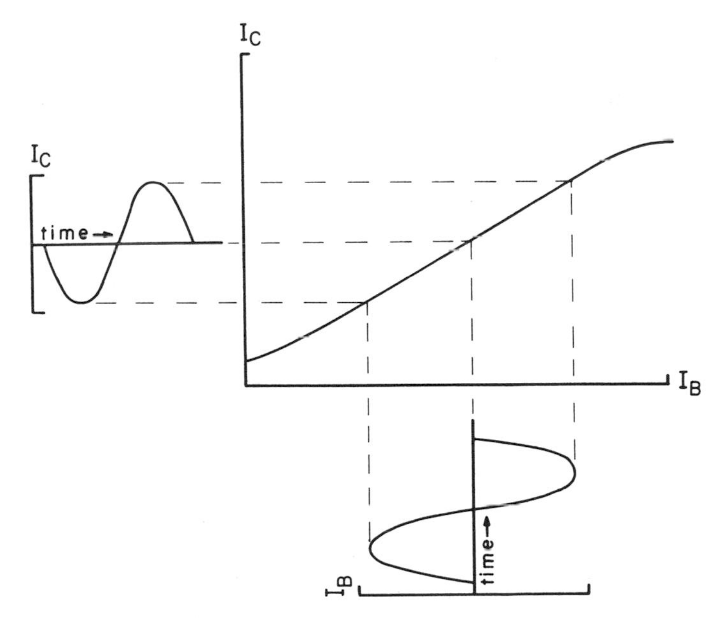


Fig. 41. Projection technique for obtaining the output signal waveform from the dynamic transfer characteristics.

The exact shape of the output signal can be found by projection as shown in Fig. 41. Note that the base and collector currents are always in phase. That is, an increase in base current produces an increase in collector current.

## F. Small-Signal Analysis and Equivalent Circuits

After the quiescent operating parameters of an amplifier have been determined, it usually is necessary to evaluate the input and output impedances as well as the current and voltage gains. While these can be evaluated graphically, it is usually more convenient to develop a linear equivalent circuit and study its response to small input signal changes. Since the transistor is a nonlinear device, small-signal, linear approximations will be used.

The CE amplifier is the most commonly used transistor amplifier, and we will develop its equivalent circuit in some detail. The reason for the popularity of the CE amplifier is the fact that it can develop large power, current, and voltage gains. The CB amplifier, which always has a current gain of less than 1, usually is used to drive a high-impedance load from a low-impedance source. The CC amplifier, which always has a voltage gain of less than 1, usually is used to drive a low-impedance load from a high-impedance source.

To derive the equivalent circuits and small-signal parameters, the transistor is considered a block box with two input and two output terminals. This is shown in Fig. 42. We will be concerned only with the terminal response and not with the internal transistor action. Currents and voltages will be indicated with lower case letters to show that they represent small-signal changes from the quiescent values. Upper case notation indicates total signal values or the algebraic sum of the quiescent and small-signal values.

For the CE amplifier, we have

$$i_1 = i_b, \quad i_2 = i_c, \quad v_1 = v_{be}, \quad v_2 = v_{ce}.$$

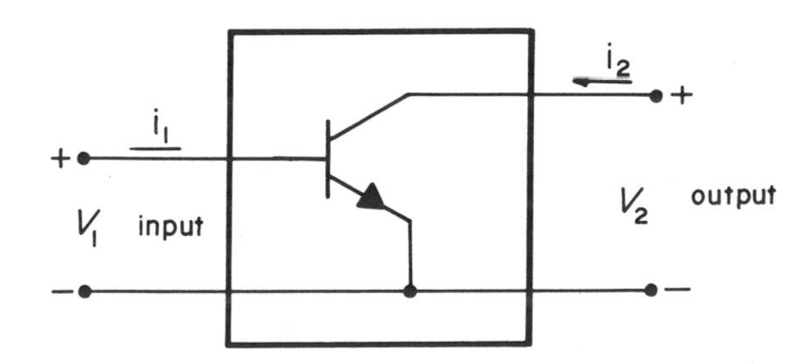


Fig. 42. Transistor model for small-signal analysis.

Any two of the four currents and voltages can be used as independent variables, and the other two are expressed as functions of these. Usually, the input current $i_1$ and the output voltage $v_2$ are used as independent variables because they are the easiest to measure and control. This gives the following small-signal equations:

$$v_1 = \left(\frac{\partial v_1}{\partial i_1}\right)_{v_2} di_1 + \left(\frac{\partial v_1}{\partial v_2}\right)_{i_1} dv_2 = h_{11}i_1 + h_{12}v_2$$

$$i_2 = \left(\frac{\partial i_2}{\partial i_1}\right)_{v_2} di_1 + \left(\frac{\partial i_2}{\partial v_2}\right)_{i_1} dv_2 = h_{21}i_1 + h_{22}v_2.$$

The coefficients of $i_1$ and $v_2$ are approximated by the slopes of the appropriate characteristic curves evaluated at the quiescent point. The values of $di_1$ and $dv_2$ are replaced by their small-signal equivalents, $i_1$ and $v_2$. The h parameters are called hybrid parameters because they use a voltage and a current as independent variables. As might be expected, the resulting equivalent circuits contain a Thévenin equivalent voltage source and a Norton equivalent current source.

The numerical subscripts on the h parameters are replaced with letters which indicate the nature of the parameters,

$$v_1 = h_i i_1 + h_r v_2$$

$$i_2 = h_f i_1 + h_o v_2$$

where

$$h_i = \left(\frac{v_1}{i_1}\right)_{v_2=0} = \text{short-circuit \underline{i}nput impedance}$$

$$h_r = \left(\frac{v_1}{v_2}\right)_{i_1=0} = \text{open-circuit \underline{r}everse voltage gain}$$

$$h_f = \left(\frac{i_2}{i_1}\right)_{v_2=0} = \text{short-circuit \underline{f}orward current gain}$$

$$h_o = \left(\frac{i_2}{v_2}\right)_{i_1=0} = \text{open-circuit \underline{o}utput admittance.}$$

Following this letter subscript, the letters c, b, or e are used to indicate a common-collector, common-base, or common-emitter configuration. Thus $h_{oe}$ is the open-circuit output admittance for a common-emitter amplifier.

## 1. The CE Amplifier

When graphical methods were discussed, the forward-biased base-emitter voltage drop was considered constant at 0.7 V for silicon transistors. This was valid because large signals were being considered, and small changes in $V_{BE}$ could be neglected. Here, small signals are

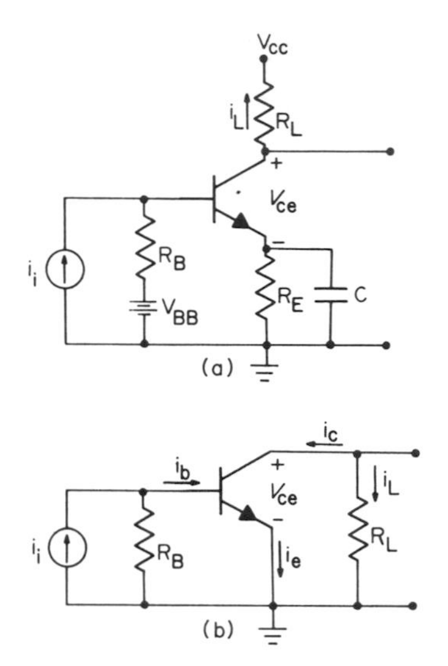


Fig. 43. Common-emitter amplifier circuit for small-signal analysis: (a) complete circuit; (b) simplified circuit for ac signals.

being considered; thus changes in $V_{BE}$ cannot be neglected. The basic CE circuit to be analyzed is shown in Fig. 43(a). In the figure, the emitter bypass capacitor C is used to restore the gain lost from including the emitter resistor $R_E$ in the circuit. The capacitor serves as a low-impedance shunt for ac signals. It has no effect on dc quiescent values, but it does shift the load line for ac signals. Since we are only interested in small ac fluctuations about the quiescent point, $V_{BB}$, $V_{CC}$, and C can be replaced with short circuits. This is equivalent to treating separately the ac and dc signal components. The simplified circuit is shown in Fig. 43(b).

Now, the four hybrid parameters must be evaluated. The common-emitter output characteristics are used to find $h_{oe}$. Small-signal values are replaced by incremental values evaluated at the quiescent point.

$$h_{oe} = \left.\frac{\Delta i_C}{\Delta v_{CE}}\right|_{\text{Q-point.}}$$

This is shown in Fig. 44. Figure 45 shows how $h_{fe}$ is evaluated from the same family of curves.

$$h_{fe} = \left.\frac{\Delta i_C}{\Delta i_B}\right|_{\text{Q-point.}}$$

Since $h_{oe}$ is usually less than $10^{-4}$ mho for silicon transistors, it can be

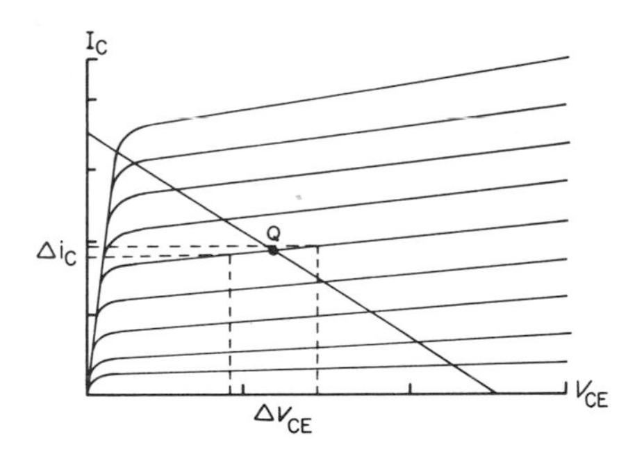


Fig. 44. Evaluation of $h_{oe}$ from the common-emitter output characteristics.

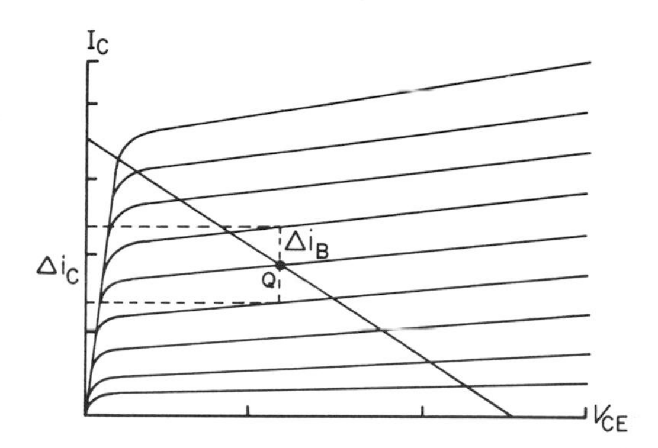


Fig. 45. Evaluation of $h_{fe}$ from the common-emitter output characteristics.

neglected if $R_L$ is less than about 2000 Ω.

An empirical expression is used to evaluate $h_{ie}$,

$$h_{ie} = \frac{0.035\, h_{fe}}{I_{CQ}}$$

where $I_{CQ}$ is the quiescent collector current expressed in amperes. The last parameter, $h_{re}$, is usually very small and neglected.

The equivalent linear circuit for the CE amplifier is shown in Fig. 46(a). The input is a voltage source $h_{re}v_{ce}$ in series with an input impedance $h_{ie}$. The output is a current source $h_{fe}i_b$ in parallel with an admittance $h_{oe}$. Since $h_{re}$ is usually negligible and $h_{oe}$ is short circuited by the emitter capacitor C, the simplified equivalent circuit of Fig. 46(b) is usually used.

When the external circuit resistors $R_B$ and $R_L$ along with the input

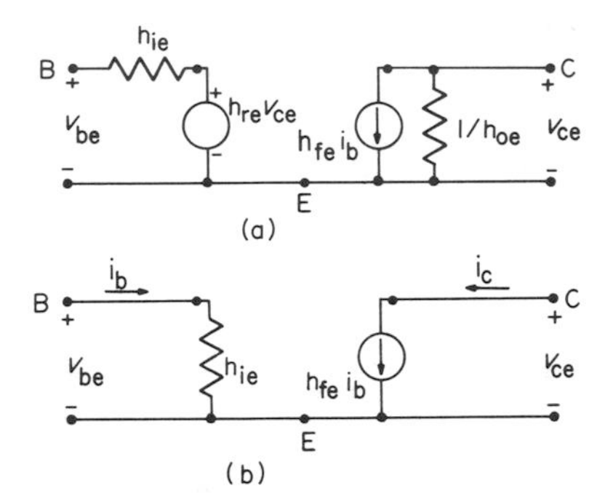


Fig. 46. Small-signal, linear equivalent circuit for the CE amplifier: (a) complete circuit; (b) simplified circuit neglecting $h_{re}$ and $h_{oe}$.

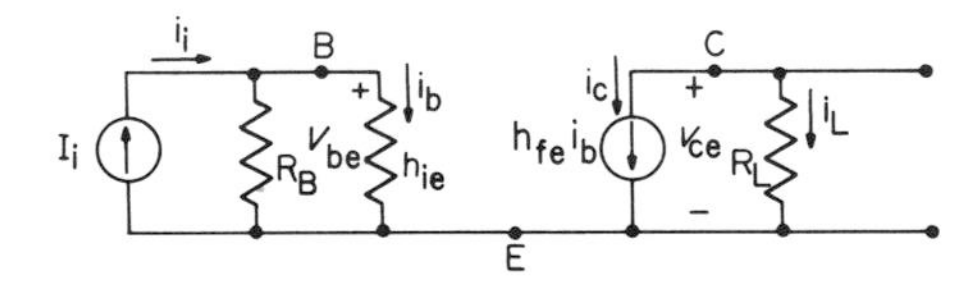


Fig. 47. Small-signal, linear equivalent circuit for the CE amplifier showing the external circuit resistors $R_B$ and $R_L$.

current source $I_i$ are considered, the circuit in Fig. 47 is obtained. From this circuit, the current gain $A_i$ is easily calculated.

$$i_b = i_i \left(\frac{R_B}{R_B + h_{ie}}\right) = i_i \left(\frac{1}{1 + h_{ie}/R_B}\right)$$

and

$$i_L = -i_c = -h_{fe} i_b$$

$$A_i = \frac{i_L}{i_i} = \left(\frac{i_L}{i_b}\right)\left(\frac{i_b}{i_i}\right)$$

so

$$A_i = \frac{-h_{fe}}{1 + h_{ie}/R_B} = \frac{-h_{fe}}{1 + 0.035\ h_{fe}/(I_{CQ}R_B)} .$$

The input impedance $Z_i$ is just the parallel combination of $R_B$ and $h_{ie}$ which reduces to $h_{ie}$ if $R_B \gg h_{ie}$:

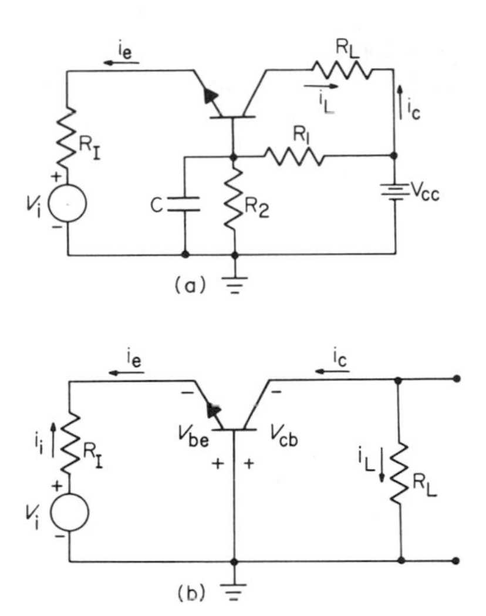


Fig. 48. Common-base amplifier circuit for small-signal analysis: (a) complete circuit; (b) simplified circuit for ac signals.

$$Z_i = \frac{R_B h_{ie}}{R_B + h_{ie}} \simeq h_{ie} \quad \text{if} \quad R_B \gg h_{ie}.$$

The output impedance $Z_o$ is equal to $1/h_{oe}$. If $h_{oe}$ is neglected, $Z_o$ is infinite.

## 2. The CB Amplifier

The basic CB amplifier is shown in Fig. 48(a). For ac signals, $V_{CC}$, $R_1$, $R_2$, and C are short circuited giving the simplified circuit shown in Fig. 48(b). If $v_{be}$ and $i_c$ are treated as the dependent variables, the following hybrid equations are obtained:

$$v_{be} = h_{ib}(-i_e) + h_{rb}v_{cb}$$

$$i_c = h_{fb}i_e + h_{ob}v_{cb}.$$

The forward current-amplification $h_{fb}$ is the ratio of collector to emitter currents.

$$h_{fb} = \left.\frac{i_c}{i_e}\right|_{v_{cb}=0}.$$

Since $i_c \simeq i_e$, then $h_{fb}$ is approximately equal to 1. The reverse voltage gain $h_{rb}$ is usually about $10^{-4}$ and is often neglected. The output admittance is given by the ratio $i_c$ to $v_{cb}$. Since $h_{ob}$ is typically about $10^{-6}$ mho, it is neglected. The input impedance is usually evaluated from the CE hybrid parameters:

$$h_{ib} = \frac{h_{ie}}{1 + h_{fe}}.$$

All CB hybrid parameters can be found by dividing the corresponding CE parameters by $1 + h_{fe}$:

$$h_{fb} = \frac{h_{fe}}{1 + h_{fe}}, \quad h_{ob} = \frac{h_{oe}}{1 + h_{fe}}, \quad h_{rb} = \frac{h_{re}}{1 + h_{fe}}.$$

The complete small-signal, linear equivalent circuit is shown in Fig. 49(a). The simplified circuit neglecting $h_{ob}$ and $h_{rb}$ and assuming that $h_{fb} = 1$ is shown in Fig. 49(b).

## 3. The CC Amplifier

The complete common-collector circuit is shown in Fig. 50(a). For ac signals, $V_{CC}$ is short circuited producing the circuit shown in Fig. 50(b). The small-signal, linear equivalent circuit for the CC amplifier is shown in Fig. 51. This circuit is developed by applying Kirchhoff's voltage law to the input and output portions of the circuit of Fig. 50(b).

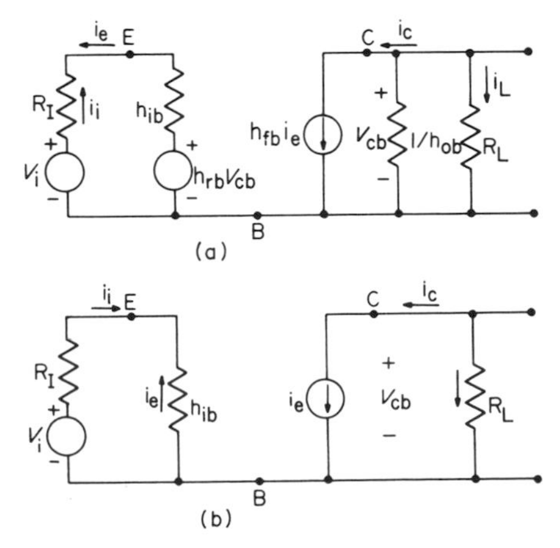


Fig. 49. Small-signal, linear equivalent circuit for the CB amplifier: (a) complete circuit; (b) simplified circuit neglecting $h_{ob}$ and $h_{rb}$.

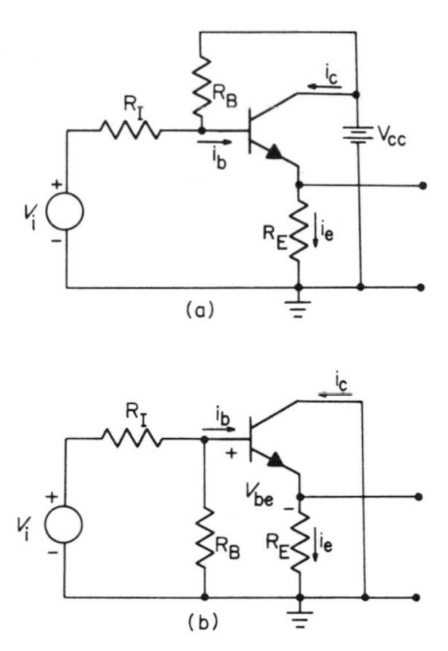


Fig. 50. Common-collector amplifier circuit for small-signal analysis: (a) complete circuit; (b) simplified circuit for ac signals.

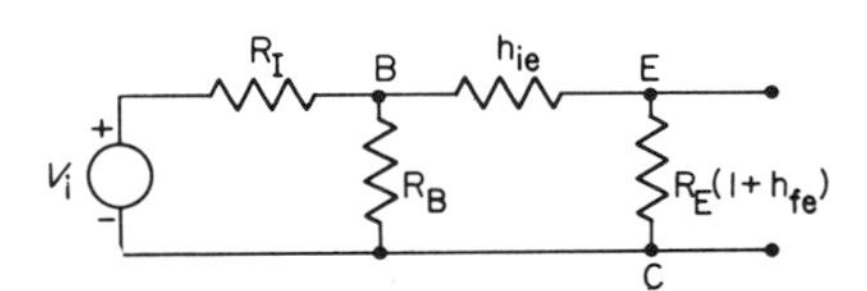


Fig. 51. Small-signal, linear equivalent circuit for the CC amplifier.

The common-collector amplifier has a voltage gain of about 1 and an input impedance of

$$Z_I = h_{ie} + (h_{fe} + 1)R_E.$$

The output impedance is given by

$$Z_o = h_{ib} + \frac{R_I R_B/(R_I + R_B)}{h_{fe} + 1}.$$

A more detailed description of small-signal transistor analysis and linear equivalent circuits may be found in the references listed at the end of this chapter.

## G. Electronic Logic Gates

Ideally the electronic circuits used for logic-gating operations should have two and only two well-defined and well-separated output voltage levels. These two states are arbitrarily assigned the logic values of Boolean 1 and Boolean 0. The actual voltage levels used to represent the two Boolean states are determined by the electronic properties of the particular circuits used in the logic implementation.

Four types of Boolean representations are commonly used in logic circuits. In positive logic, the voltage level assigned to Boolean 1 is more positive than that assigned to Boolean 0. In bipolar logic, the two Boolean states have the same voltage magnitude, but the 1 state is represented by a positive voltage while the 0 state is represented by a negative voltage. In negative logic, the Boolean 1 state is assigned a more negative voltage than the 0 state. Inverted logic is similar to bipolar except that the Boolean 0 state is assigned the positive voltage; while the Boolean 1 state is assigned the negative voltage.

The simplest electronic circuit having two well-defined output states is the mechanical switch in series with a load and a voltage source. This is shown in Fig. 52. When the switch is in a position 1, an open-circuit

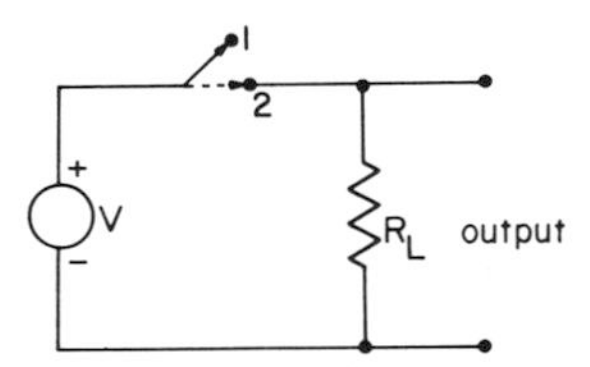


Fig. 52. The mechanical switch for logic implementation.

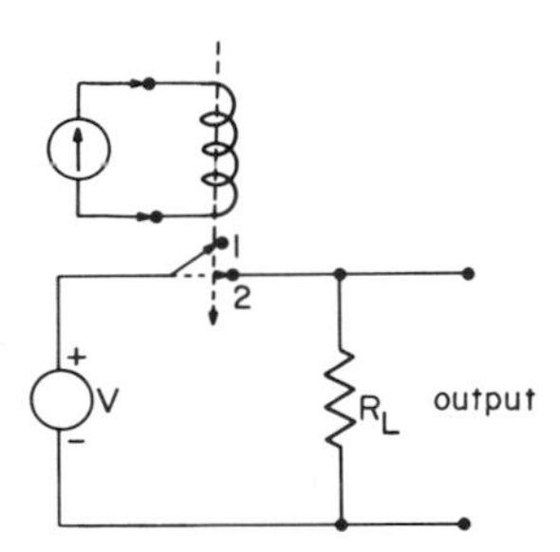


Fig. 53. The mechanical relay for logic implementation.

condition exists, and the output is 0 volts. When the switch is in position 2, the closed-circuit condition exists, and the output is V volts. It is clear that these are the only two possible output states.

A more interesting example is the mechanical relay shown in Fig. 53. Here the mechanical switch is operated by a magnet which is activated by an external input signal. It should be noted that while an infinite number of activating signal states are possible, there are only two values for the output signal, either 0 or V volts. If the activating signal is large enough to close the relay, the output is V volts. Increasing the activating signal beyond this point will have no effect on the output level.

## 1. Passive Logic Circuits

If a logic circuit contains amplification elements to maintain a certain logic-level output, the circuit is called active logic. If no such amplification elements are present, the circuit is called passive logic. In passive logic networks, an auxiliary logic-level restoration circuit may be required to restore signal losses which occur when a logic signal is processed by a gate. The simple transistor and the Schmitt trigger for level restoration will be discussed later.

### a. Passive Resistor Logic (RL).

Gating operations can be performed using resistor or diode networks. Consider the simple resistor network in Fig. 54. We will arbitrarily assign 0 volts to the 0 Boolean state and 4 volts to the 1 Boolean state for the three inputs A, B, and C. It should be noted that a Boolean 0 on any input corresponds to grounding that input. Table 24 gives the voltages developed at the output terminal D for all possible combinations of inputs. It is clear from this table that the number of possible output states is four rather than the two states required for logic implementation. In a simple resistive network, the number of output levels will increase with an increase in the number of inputs.

It is possible to place an auxiliary circuit in the output of the resistor logic network that will allow the network to function as an AND gate or an

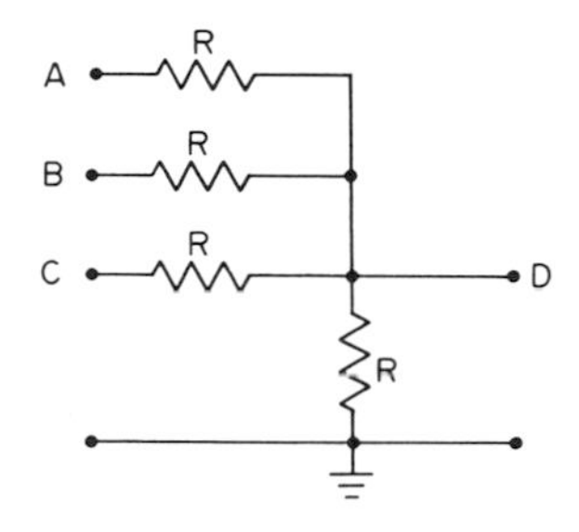


Fig. 54. Passive resistor logic network.

TABLE 24

Combinations for Resistor Logic

| A | B | C | D |
|---|---|---|---|
| 0 | 0 | 0 | 0 |
| 4 | 0 | 0 | 1 |
| 0 | 4 | 0 | 1 |
| 0 | 0 | 4 | 1 |
| 4 | 4 | 0 | 2 |
| 4 | 0 | 4 | 2 |
| 0 | 4 | 4 | 2 |
| 4 | 4 | 4 | 3 |

OR gate. One such auxiliary circuit is the Schmitt trigger. The output of the trigger has only two possible values. The voltage state of the output depends on the voltage at the input. Suppose a Schmitt trigger was designed such that the output was a Boolean 1 if the input voltage was greater than 0.5 volts. An OR gate will result if this trigger is placed in the output of the resistor network of Fig. 54. If any one or more of the inputs was at Boolean 1, the voltage at terminal D would be greater than 0.5 volts; thus the output of the trigger would be at Boolean 1.

If the Schmitt trigger was designed so that an input voltage of greater than 2.5 volts was required to produce a Boolean 1 output, the resulting circuit would be an AND gate. It is clear from the combinations in Table 24 that the input voltage to the Schmitt trigger is sufficient to produce a Boolean 1 output only if A AND B AND C are at logic 1. The need for logic-level restoration circuitry and the problem of input-signal interaction at the network summing point has resulted in only limited application of resistor logic in digital control and computing systems.

b. Passive Diode Logic (DL). The multiple output levels of the linear resistive network in RL can be eliminated by using nonlinear resistive devices such as diodes. In a high-quality switching diode, the ratio of backward to forward resistance may be greater than $10^6$. Examples of diode OR and AND gates are shown in Fig. 55. Again we will assume input voltages of 0 and 4 volts for Boolean 0 and 1, respectively. The bias voltages on the summing resistors are used to speed up switching by rapidly charging stray circuit capacitance.

If any one or more of the OR gate inputs is at Boolean 1, the output voltage at D will be just the input signal level reduced by the voltage drop across the forward-biased pn junction. If a 4-volt input is used with silicon

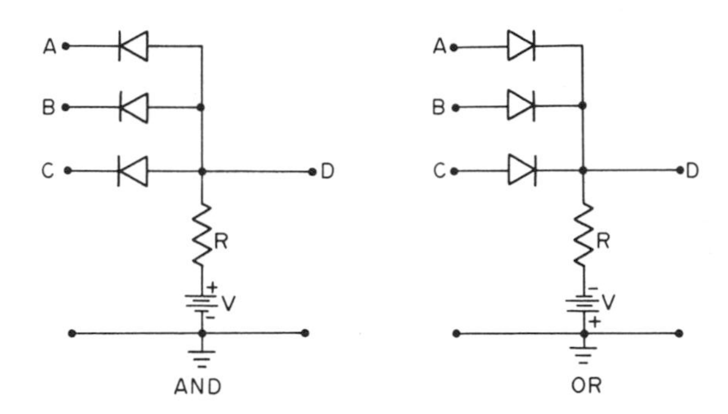


Fig. 55. Passive diode logic networks.

TABLE 25

Combinations for a Diode OR Gate

| A | B | C | D |
|---|---|---|---|
| 0 | 0 | 0 | -0.7 |
| 4 | 0 | 0 | +3.3 |
| 0 | 4 | 0 | +3.3 |
| 0 | 0 | 4 | +3.3 |
| 4 | 4 | 0 | +3.3 |
| 4 | 0 | 4 | +3.3 |
| 0 | 4 | 4 | +3.3 |
| 4 | 4 | 4 | +3.3 |

diodes, the output will be 4 - 0.7 = 3.3 volts. If all three inputs are grounded, corresponding to Boolean 0's at A, B, and C, the output will be just the voltage drop across the forward-biased junction or -0.7 volt for silicon diodes. The combinations for the diode OR gate are given in Table 25.

In the diode AND gate, all inputs must be at Boolean 1 to produce a Boolean 1 at the output. If any one or more of the inputs is at Boolean 0 (0 volts), the output will be the forward-bias voltage drop across the pn junction or +0.7 volt for silicon diodes. If all three inputs are at Boolean 1 (+4 volts), the output will be 4 volts plus the 0.7-volt drop across the junction or 4.7 volts. A table of combinations for the diode AND gate is given in Table 26.

Although the diode logic gate has only two well-defined output levels, it should be noted that the logic voltage levels are changed when logic signals are processed by diode gates. This change is the result of the voltage drop

TABLE 26

Combinations for a Diode AND Gate

| A | B | C | D |
|---|---|---|---|
| 0 | 0 | 0 | +0.7 |
| 4 | 0 | 0 | +0.7 |
| 0 | 4 | 0 | +0.7 |
| 0 | 0 | 4 | +0.7 |
| 4 | 4 | 0 | +0.7 |
| 4 | 0 | 4 | +0.7 |
| 0 | 4 | 4 | +0.7 |
| 4 | 4 | 4 | +4.7 |

across the forward-biased pn junction. After being processed by a series of gates, the logic levels may be changed to the point where level ambiguity results. When alternate AND and OR gates can be used, this problem is partially eliminated since the forward voltage drops will cancel for each pair of connected AND and OR gates. Since this arrangement is not usually possible in most logic circuit applications, level restoration usually is required.

### 2. Active Logic Circuits

We previously discussed how the transistor could be used as a simple linear amplifier. The operating or quiescent point was chosen to give the maximum symmetric collector current swing with minimum distortion in the output signal. Referring to the dynamic transfer characteristics in Fig. 41, the linear operating range can be considered that portion of the curve where the slope $dI_C/dI_B$ is nearly constant. It should be noted that the slope decreases rapidly at both the top and bottom of the curve. While these regions of the transfer characteristics are of little use in applications requiring linear amplification, they are well suited for current and voltage switching. This is because the collector current and hence output voltage change much more slowly with changing base current in these operating regions.

a. The Transistor Switch. A simplified CE transistor switching circuit is shown in Fig. 56. In Fig. 57, the CE output characteristics have been divided into three regions. The active region is used for linear amplification. In switching applications, the cutoff and saturation regions are used. In the cutoff region, the base current is too low to forward bias the base-emitter junction. The transistor is in a nonconducting state, and the

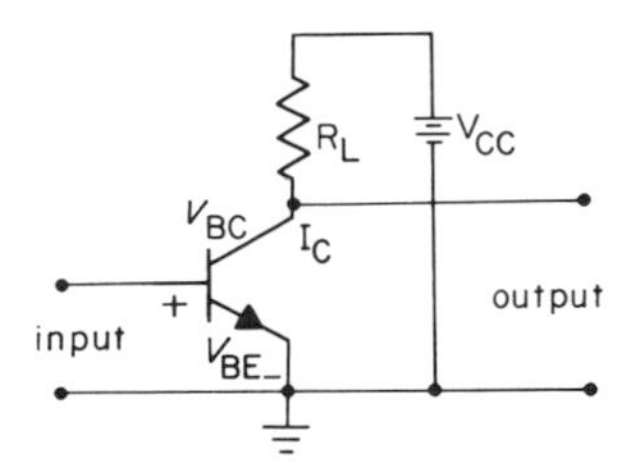


Fig. 56. Common-emitter transistor switching circuit.

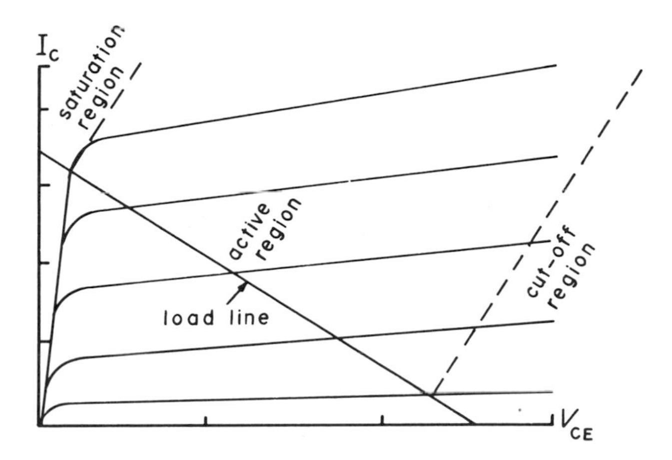


Fig. 57. Common-emitter output characteristics showing the cut-off and saturation regions.

voltage drop across the transistor is nearly equal to the supply voltage $V_{CC}$. If a positive voltage is applied to the base (negative voltage in a pnp transistor), base current will begin to flow if the voltage is great enough to forward bias the base-emitter junction. When the junction is forward biased, the transistor will start to conduct, and the voltage drop across it will decrease. If the base current is sufficiently large, the current through the transistor will nearly reach its limiting value of $V_{CC}/R_L$, and the voltage drop across it will decrease to a minimum value of about 0.2 volt. The transistor is then operating in the saturation region which physically corresponds to forward biasing the base-collector junction. This will cause the collector to inject carriers into the base as well as receive carriers from the emitter. This will produce a net accumulation of carriers in the base and hence saturate the base-collector junction. If the base is driven more positive, the transistor will go deeper into saturation but with little change in collector current or emitter-collector voltage drop.

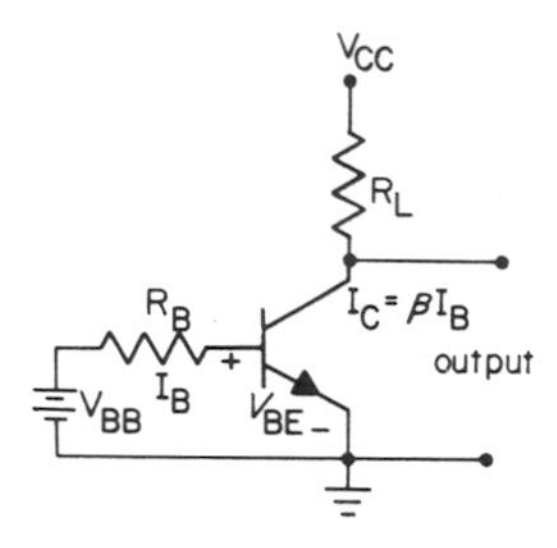


Fig. 58. Transistor switching circuit used to evaluate the base current needed to saturate the transistor.

The fact that the base-collector junction becomes forward biased during saturation is understood by considering the Kirchhoff voltage expression for the transistor output loop,

$$V_{CB} = V_{CC} - R_L I_C - V_{BE}.$$

Since $I_C$ is nearly equal to $I_E$, if $I_E$ is made sufficiently large, then

$$R_L I_C > V_{CC} - V_{BE}.$$

This will make the collector negative with respect to the base, thus forward biasing the base-collector junction. The base current needed to drive the transistor into saturation can be found by referring to Fig. 58. The saturation value of the collector current is given by

$$I_{C(sat)} = \frac{V_{CC} - V_{CE(sat)}}{R_L}.$$

To ensure saturation, it is essential that $\beta I_B > I_{C(sat)}$ where $I_B = (V_{BB} - V_{BE})/R_B$.

If the transistor is to be used as a current switch, the load is placed in series with the collector. A positive signal on the base will then produce a load current of nearly $V_{CC}/R_L$. If the transistor is to be used as a voltage switch, the load is placed in parallel with the transistor. When the transistor is switched on, the voltage across the load drops from $V_{CC}$ to about 0.2 volt. It is important to note that the transistor switch functions as a simple inverting gate. If positive representation is used, a Boolean 1 on the transistor base will result in a Boolean 0 at the output.

b. Switching Times and Nonsaturating Gates. One of the most important considerations in the design of high-speed transistor switching circuits is the switching or turn-on and turn-off times. The turn-on time is the sum of the delay time and the rise time, where the delay time is the time required

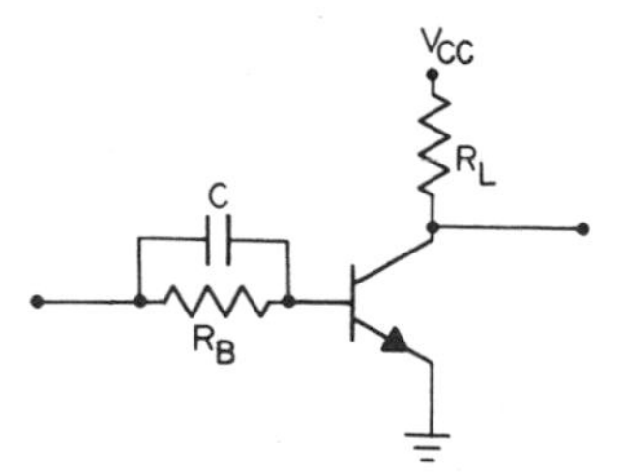


Fig. 59. Transistor switch with speed-up capacitor.

to charge the base-emitter capacitance and bring the collector current up to 10% of its saturation value, and the rise time is the time required to bring the collector current from 10 to 90% of its saturation value. The turn-off time is the sum of the storage and fall times, where the storage time is the time required to dissipate the saturation charge on the base-collector junction and reduce the collector current to 90% of its saturation value, and the fall time is the time required to reduce the collector current from 90 to 10% of its saturation value. Switching times for commercially available high-speed logic gates are typically about 10 nsec. In general, turn-off times are considerably longer than turn-on times.

Switching times can be reduced by placing a capacitor in parallel with the base resistor $R_B$. This is shown in Fig. 59. During turn-on, the capacitor supplies extra current which speeds up the saturation process. During turn-off, the saturation charge can rapidly dissipate through the capacitor thus decreasing the storage time. This improvement in switching time, however, is obtained at the expense of an increase in switching noise and ground currents. In addition, the circuit supplying the input logic signal is loaded more than in the absence of the speed-up capacitor. This requires a low output impedance in the previous logic stage.

Another way of decreasing switching time is to prevent saturation of the transistor. In Fig. 60 two diodes have been placed in the base circuit.

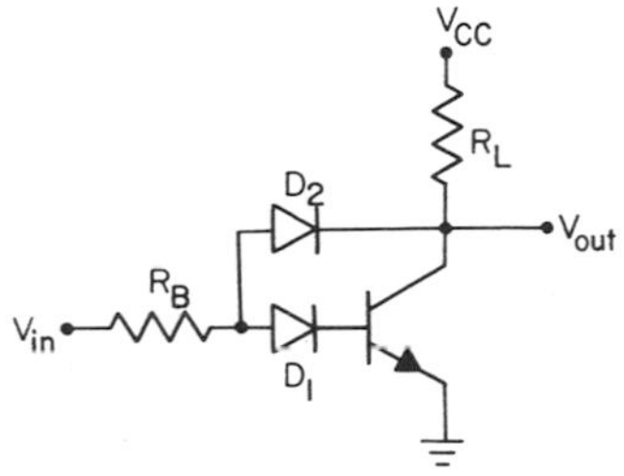

Fig. 60. Nonsaturating transistor switch using current-shunting diodes.

As the transistor starts to leave the active region, $D_2$ becomes forward biased and conducts surplus base current away from the base. Thus the transistor is kept just out of saturation. The other diode $D_1$ is required to compensate for the forward-biased voltage drop across $D_2$. If $D_1$ were not present, the transistor would enter the saturation region well before $D_2$ became forward biased. The transistor can be kept even further out of the saturation region by using a germanium diode for $D_2$ and a silicon diode for $D_1$. This is because the forward voltage drop across a germanium diode is several tenths of a volt less than across a silicon diode.

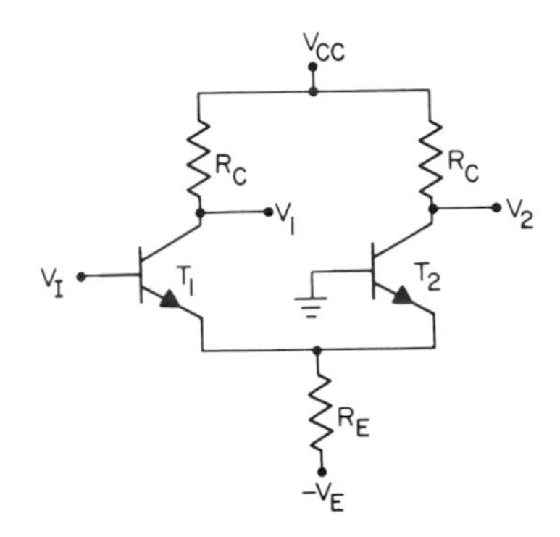


Fig. 61. Nonsaturating current-steering logic circuit.

Another method, known as current steering, can be used to clamp the transistor output to any point in the active region. Figure 61 shows a current-steering circuit. Two transistors are used with a common emitter resistor. This resistor is chosen such that $I_E R_C \leqslant V_{CC}$. This will ensure operation in the active region. The total current through the emitter resistor $R_E$ is just $V_E/R_E$. Thus current can be steered to either transistor by controlling the input voltage on the base of $T_1$.

When the input voltage $V_I$ is negative, $T_1$ is cut off and $T_2$ carries the total emitter current since its base is referred directly to ground. If $V_I$ is positive, $T_1$ is conducting and the voltage drop across $R_E$ reverse biases the base-emitter junction of $T_2$ thus cutting it off.

In a current-steering gate, the input signal does not affect the current amplitude but merely steers a constant current to one transistor or the other. Since currents are not rapidly switched off and on, the noise associated with the switching action is much lower in current-steering gates. This is an attractive feature in some switching applications. It should be noted that two complementary outputs are obtained in the current-steering gate. The output at $V_1$ is inverted while the output at $V_2$ is in phase with the input signal. While very short switching times can be obtained with current-steering gates, the change in logic representation between input and output represents a serious limitation. The gate input requires bipolar representation while the output is either positive or negative representation depending on whether the inverting or noninverting output is used.

c. The Schmitt Trigger. It was shown in Section IV.G.1.a that resistor networks are of limited use in logic processing applications because of the several possible output voltage levels. Multiple-input resistor OR and AND gates are practical only if a suitable voltage-level discriminating circuit is placed in the gate output. The Schmitt trigger circuit shown in Fig. 62 can be used here. The circuit is similar to the current-steering gate except that the common emitter resistor is referenced directly to ground, and the collector of $T_1$ is resistively coupled to the base of $T_2$.

In the absence of an input signal, $T_1$ is cut off. The high collector voltage of $T_1$ is applied to the base of $T_2$ through the voltage divider network R and $R_B$, thus saturating $T_2$. The saturation current from $T_2$ develops a voltage drop across $R_E$ which maintains $T_1$ in the cut-off state. An input signal large enough to forward bias the base-emitter junction of $T_1$ will cause current to flow through $T_1$, thus decreasing its collector voltage. This will decrease the base current into $T_2$, thus bringing it out of saturation and decreasing the current through $R_E$. As the voltage drop across $R_E$ decreases, $T_1$ is driven further into conduction thus decreasing its collector voltage. This process is regenerative, and very quickly $T_1$ is driven into saturation, and $T_2$ is cut off. Thus a certain critical input voltage $V_{on}$ will change the trigger output from Boolean 0 to Boolean 1. The value of $V_{on}$ depends on the voltage drop developed across $R_E$ which in turn depends on

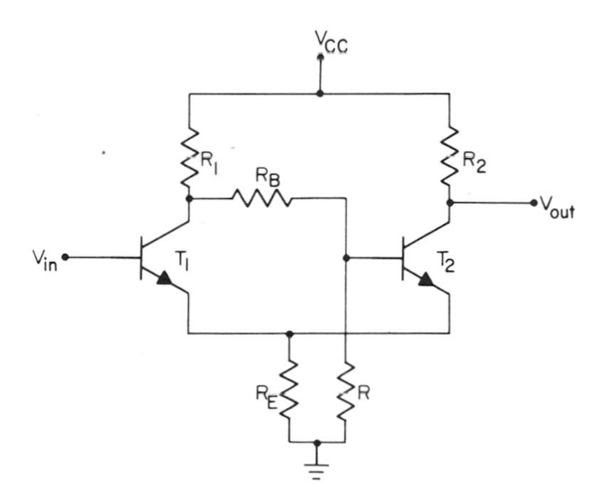


Fig. 62. The Schmitt trigger circuit.

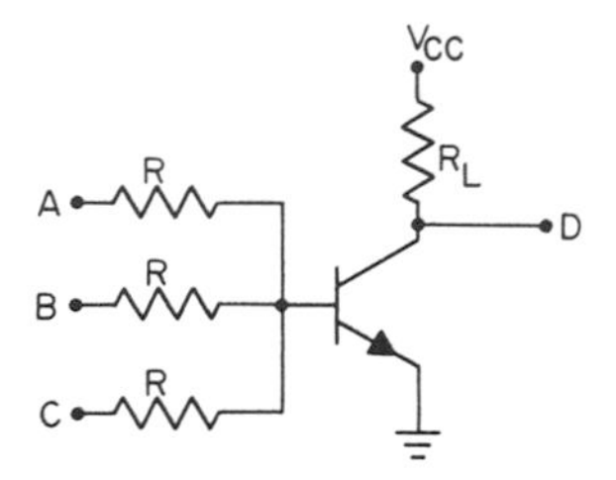


Fig. 63. Resistor-transistor logic NOR gate.

the collector current of $T_2$. Thus, changing either $R_2$ or $R_E$ will change the value of $V_{on}$.

If $V_{on}$ is set at about 0.5 volt, the resistive network of Fig. 54 will function as an OR gate when the network output is applied to the trigger input. If $V_{on}$ is set at about 2.5 volts, an AND gate will result.

In the Schmitt trigger, it is essential that $R_2$ is less than $R_1$. This will ensure that the voltage drop across $R_E$ decreases as $T_1$ turns on and $T_2$ turns off. Only in this way is the necessary regenerative switching obtained. Since $R_2$ is less than $R_1$, the saturation emitter current for $T_1$ is less than for $T_2$. Thus, after the circuit has switched, a lower input voltage will maintain the output at Boolean 1. This means that the turn-on voltage is greater than the turn-off voltage. This can be troublesome in certain logic applications. The difference between $V_{on}$ and $V_{off}$ is called the hysteresis or dead-zone voltage. The hysteresis is reduced by decreasing the difference between $R_1$ and $R_2$. This, however, increases the switching time. If $R_1$ is equal to $R_2$, the circuit functions as a linear amplifier.

In addition to its use as a voltage-level discriminating circuit, the Schmitt trigger is frequently used to produce sharp logic signals from irregular or slowly changing input waveforms.

d. Resistor-Transistor Logic (RTL). The restrictions placed on resistor logic with respect to output-level multiplicity and input-signal interactions largely can be overcome by placing a transistor switch in the gate output. The resulting circuit is called resistor-transistor logic (RTL). Consider the circuit of Fig. 63. Assume that all three inputs are at ground potential. Under this condition, no base current will flow and the transistor will be cut off. The output will then be $V_{CC}$ or Boolean 1 (positive representation with an npn transistor). If a positive signal corresponding to Boolean 1 is applied to any one or more inputs, the transistor will start to conduct and the output will fall to $V_{CC} - I_C R_L$. If the input signal is large enough, the transistor will be driven into saturation, and the output will

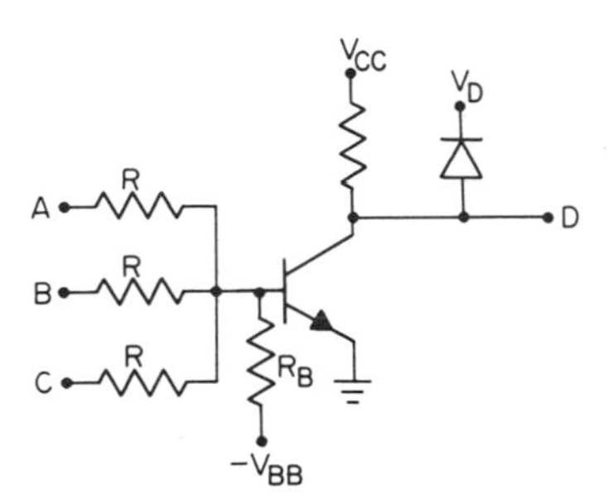


Fig. 64. Improved RTL NOR gate using a diode and bias supply to maintain a constant output voltage in the logic 1 state.

drop to $V_{CE(sat)}$ or Boolean 0. If positive signals are applied to any of the other inputs, the transistor will be driven deeper into saturation with only minor changes in the output voltage. Thus a Boolean 1 on any one or more inputs will produce a Boolean 0 at the output. This corresponds to a NOR operation.

One problem with RTL is that the output voltage depends on load current. If the gate is driving another circuit requiring a current of $I_D$, the actual gate output voltage is $V_{CC} - (I_C + I_D)R_L$. An improved RTL NOR gate is shown in Fig. 64. Here, a diode and bias supply are used to clamp the Boolean 1 output voltage at $V_D$ independent of the load current. The $V_{BB}$ bias supply ensures that the transistor is completely cut off when all inputs are at Boolean 0.

e. Diode-Transistor Logic (DTL). A natural extension of RTL and DL is the diode-transistor logic gate. A transistor is placed in the diode gate output to isolate input and output signals and to prevent changes in logic level caused by the diode forward voltage drop. Since the transistor output is inverted, the DL AND and OR gates become DTL NAND and NOR gates, respectively. Examples of DTL NAND and NOR gates are shown in Fig. 65.

In the NAND gate, $R_1$, $R_2$, and $R_B$ form a voltage divider that ensures transistor saturation when all inputs are at Boolean 1. If any one or more inputs is at Boolean 0, the corresponding diodes conduct and the transistor is cut off. This produces an output of $V_D$ or a Boolean 1. In the NOR gate, a positive signal corresponding to a Boolean 1 on any one or more of the inputs drives the transistor into saturation and produces a Boolean 0 at the output.

f. Direct Coupled Transistor Logic (DCTL). In DCTL the input signals are applied directly to the bases of the transistor switches. Because of the very simple circuitry involved, DCTL has become quite popular. The DCTL NOR and NAND gates are shown in Fig. 66. In the NOR gate, a

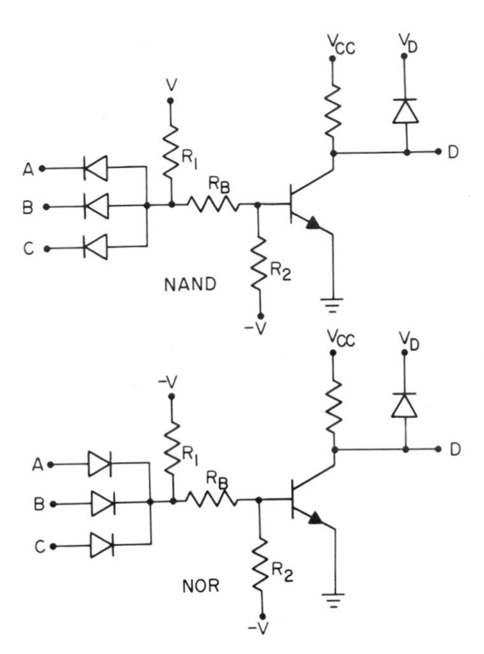


Fig. 65. Diode-transistor logic NAND and NOR gates.

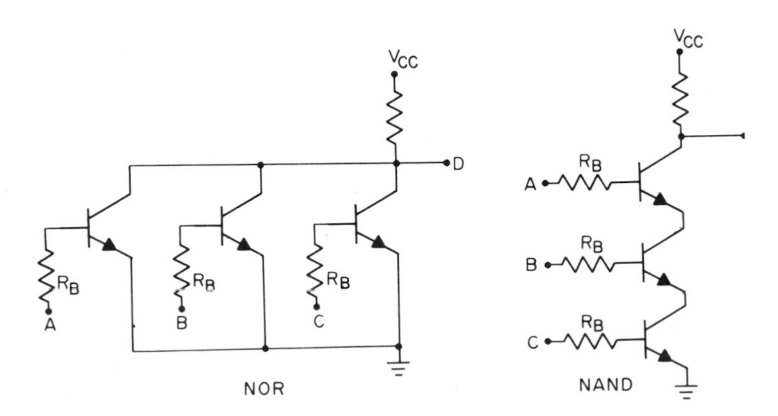


Fig. 66. Direct-coupled transistor logic NAND and NOR gates.

Boolean 1 on any one or more inputs drives the corresponding transistors into saturation producing a Boolean 0 at the output. If all inputs are at Boolean 0, all transistors are cut off and an output voltage of $V_{CC}$ or a Boolean 1 will result. There is a slight change in output voltage if more than one transistor is saturated, but the change is small and presents no real problem.

In the NAND gate, all inputs must be at Boolean 1 in order to draw enough load current to change the output from Boolean 1 to Boolean 0. The base resistors $R_B$ are used to stabilize the base current with respect to differences in the characteristics of the various transistors.

g. Current-Mode Logic (CML). Current-mode logic, sometimes called emitter-coupled logic (ECL), is the only commonly used logic circuit that operates in the active rather than in the saturation region. The principal advantage of operating in the active region is the improvement in switching time. With CML, switching times of 1 nsec are possible. Nonsaturating logic circuits are difficult to operate by controlling the base voltage. This is because of fairly large differences in transistor parameters. A CML gate is shown in Fig. 67.

The voltage source $-V_{EE}$ along with $R_E$ forms a constant current source which supplies a constant current I. This current is less than the saturation current of any of the gate transistors. If all inputs are at a negative voltage (Boolean 0), the constant current I will be steered through $T_0$ since its base is at ground potential. The output at $V_1$ will then be $V_{CC} - IR_E$ which corresponds to Boolean 0, and output $V_2$ will be at $V_{CC}$ or Boolean 1. A positive voltage applied to any one or more of the inputs will steer the current I to the corresponding transistors, thus changing the logic states at both $V_1$ and $V_2$. Thus we see that the output $V_1$ will implement

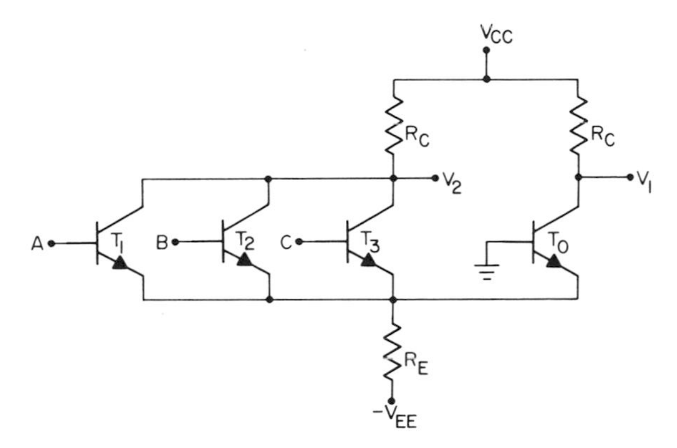


Fig. 67. Current-mode logic gate.

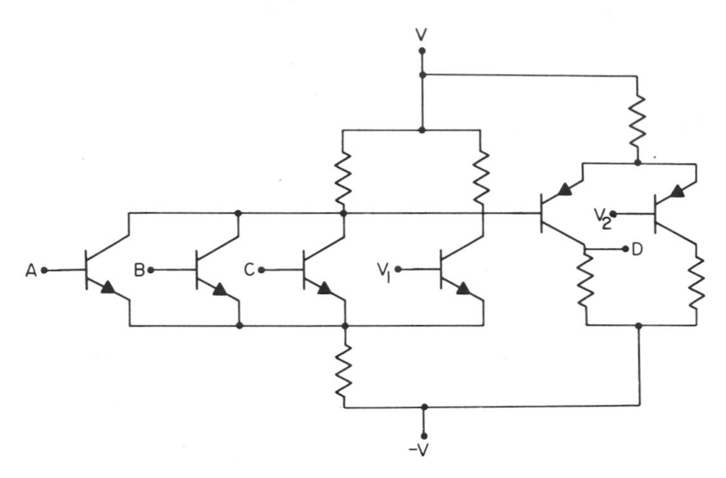


Fig. 68. Current-mode logic gate with logically compatible input and output voltage levels.

the OR operation, and the output $V_2$ will implement the NOR operation.

It should be noted that CML gates cannot be directly coupled because of the change in logic representation between input and output. Figure 68 shows a method for coupling CML gates. This circuit uses the complementary properties of pnp and npn transistors. If the reference voltages $V_1$ and $V_2$ are properly chosen, the output at D is logically compatible with the inputs A, B, and C. Thus the output D can be used as an input to another CML gate.

h. Transistor-Transistor Logic (TTL). All logic circuits discussed so far have used the CE configuration. Transistor-transistor logic uses the CB configuration. A basic TTL NAND gate is shown in Fig. 69. If all

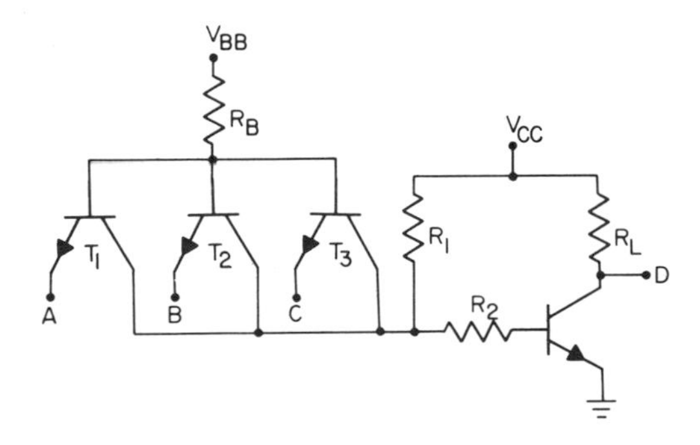


Fig. 69. Transistor-transistor logic NAND gate.

inputs are positive corresponding to Boolean 1, then $T_1$, $T_2$, and $T_3$ are cut off. Then $V_{CC}$ can supply sufficient current through $R_1$ and $R_2$ to saturate $T_0$, thus producing a Boolean 0 at the output. If any one or more of the inputs is grounded corresponding to Boolean 0, then $V_{BB}$ will supply sufficient base current to drive the corresponding transistors into saturation. The current from $V_{CC}$ will then be shunted away from the base of $T_0$ thus turning it off. This will result in a Boolean 1 at the output.

The modified TTL gate in Fig. 70 is used in integrated microcircuits. Here, all inputs are common to one transistor. This greatly simplifies the circuitry without changing the basic operation of the gate. If all inputs are at Boolean 1, $T_1$ is cut off. This turns on $T_2$ which draws sufficient current to saturate $T_4$, thus producing an output of Boolean 0. If any input is at Boolean 0, $T_1$ will conduct, thus turning off $T_2$. This will reduce the base current of $T_4$ to the point where it turns off, thus producing an output of Boolean 1. When $T_2$ is off, its high collector voltage turns on $T_3$ which allows considerable output current to flow. A principal advantage of this circuit is that a large output current is possible with either 0 or 1 output states. This greatly increases the number of other gates which may be driven from the gate output. The diode in the circuit is used to limit the leakage current through $T_3$ when the output is at Boolean 0.

Figure 71 shows a TTL NOR gate. If both inputs are at Boolean 0, $T_1$ and $T_2$ are conducting. This turns off $T_3$ and $T_4$, thus turning off $T_6$ and producing a Boolean 1 at output C. When $T_3$ and $T_4$ are off, $T_5$ is on, thus supplying considerable output current. If either or both inputs are at Boolean 1, the corresponding transistors are turned off. This will turn on $T_3$ or $T_4$ or both, thus supplying base current to $T_6$. This will drive $T_6$ into saturation, thus producing a Boolean 0 output.

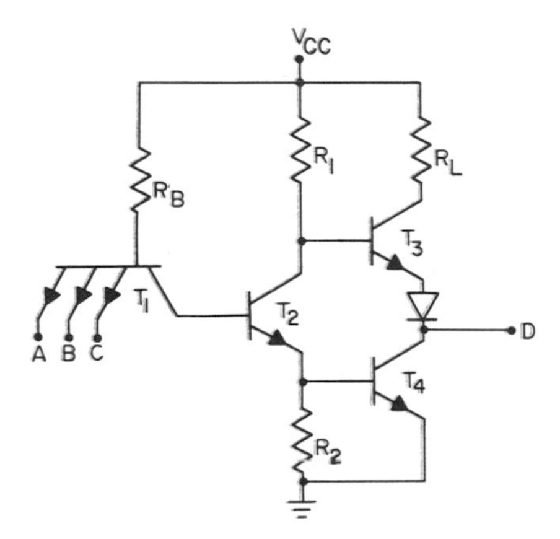


Fig. 70. Transistor-transistor logic NAND gate used in integrated microcircuits.

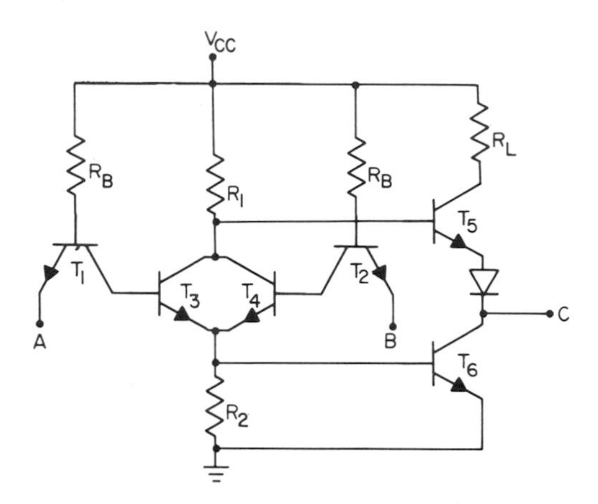


Fig. 71. Transistor-transistor logic NOR gate.

## 3. Summary of Electronic Gate Properties

The choice of logic type for a particular application may be governed by several factors. These include cost, switching speed, noise immunity, fan-in and fan-out, and logic operation. Fan-in is the maximum number of inputs that a logic gate may have, and fan-out is the maximum number of similar gates that can be driven by a gate output. Type DTL, DCTL, CML, and TTL are the most frequently used logic circuits in modern digital instruments.

Diode-transistor logic is low in cost and has good noise immunity. Switching times are rather poor, typically about 25 nsec. Fan-out is only moderate, usually between 5 and 10. Circuits can be designed for either NAND or NOR operation.

Direct-coupled transistor logic is very low in cost. This is its principal advantage. Its switching time is about the same as DTL, but its noise immunity is much poorer. Fan-out is also poorer than DTL, seldom being greater than 5. DCTL circuits are usually designed for NAND or NOR operation.

Current-mode logic is used where very high switching speeds are required. Total switching time can be on the order of 1 nsec. Noise immunity is poor, and cost is quite high. Fan-out properties are exceptionally good. A fan-out of 20 or more is not uncommon. The most serious limitation of CML is the logic representation mismatch between input and output. This problem largely has been overcome, and integrated-circuit CML gates with matched input and output representation are now available. This should greatly increase the popularity of CML circuitry.

Transistor-transistor logic is presently very popular. Its noise immunity is excellent, and fan-out may be greater than 10. Switching speed is good, usually between 10 and 15 nsec. Cost is moderate, and a wide variety of integrated-circuit packages are available. Both NAND and NOR gates are used with TTL, but the variety of NOR gate packages is quite limited. One type of TTL circuit is available which has an open or floating collector output. Several of these gates can be connected to perform the OR operation.

## V. MEMORY DEVICES AND APPLICATIONS

The logic-gate circuits discussed in Section IV.G immediately respond to any changes in the logic state of the inputs. For example, if positive representation is used with an AND gate, the output is at logic 1 only if all inputs are at logic 1. If any one or more inputs change to logic 0, the output changes to logic 0. Another very important class of logic circuits possesses hysteresis or memory properties. That is, the output logic state depends not only on the present logic state of the inputs but also on their previous logic state. These two-state memory devices have found numerous applications in computer memory, arithmetic processing units, and digital counters. Examples of these circuits and their applications will be considered in this section.

### A. Basic Bistable Memory Circuits

The Schmitt trigger circuit discussed in Section IV.G.2.c is an example of a regenerative switching circuit. The collector of one transistor was resistively coupled to the base of the other transistor. This resulted in a positive feedback loop which rapidly changed the output states of both transistors when the input voltage exceeded a certain critical value. If the collector of each transistor is resistively coupled to the base of the other transistor, a bistable memory circuit will result.

Consider the basic bistable circuit shown in Fig. 72. Assume that transistor $T_2$ is saturated. The collector of $T_2$ is at $V_{CE(sat)}$ or about 0.2 volt. This is coupled to the base of $T_1$ through the voltage divider $R_2$ and $R_{B1}$. The resulting voltage at the base of $T_1$ is too low to forward bias its base-emitter junction. Thus $T_1$ is cut off and has a collector voltage of about $V_{CC}$. Since the collector of $T_1$ is resistively coupled to the base of $T_2$ through the voltage divider $R_1$ and $R_{B2}$, $T_2$ will be kept in saturation if the divider ratio and $V_{CC}$ are properly chosen. Thus the circuit is stable with $T_1$ cut off and $T_2$ saturated. This corresponds to a logic 0 (positive representation) at V and a logic 1 at $\overline{V}$. As the circuit is completely symmetrical, it also should be stable with $T_1$ saturated and $T_2$ cut off. This corresponds to a logic 1 at V and a logic 0 at $\overline{V}$. Since the circuit is stable in two different logic states, it is called a bistable multivibrator or simply a binary.

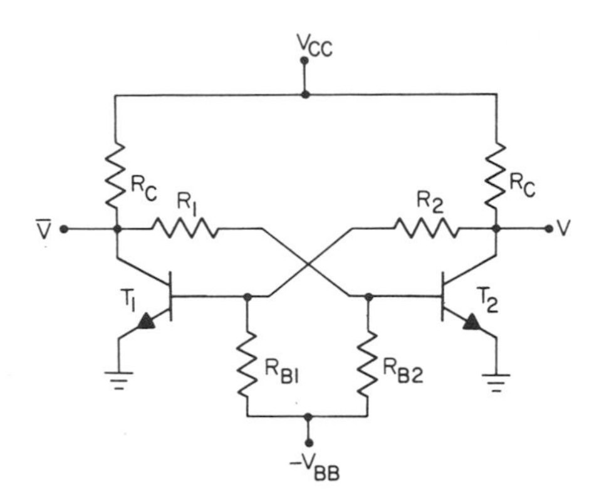


Fig. 72. Basic bistable memory circuit.

The resistive coupling between the transistors results in a positive feedback loop. Regenerative switching occurs only when the loop gain is greater than 1. Because the CE transistor configuration has appreciable voltage gain, a loop gain of greater than 1 is readily obtained if the transistors are operating in their active regions. As the transistors enter cutoff or saturation, the collector voltage changes much more slowly with changing base voltage so the loop gain decreases and falls below unity. The regenerative switching action is then terminated, and a stable output state is obtained. The base resistors $R_{B1}$ and $R_{B2}$ are returned to a negative bias to ensure that the nonconducting transistor is firmly clamped in the cut-off region.

As with simple transistor switches, loading often can be a problem in bistable circuits. To prevent loading by the circuit itself, $R_1$ and $R_2$ should be much larger than $R_C$. This will prevent the collector voltage of the cut-off transistor from falling much below $V_{CC}$. However, if $R_1$ and $R_2$ are made too large, the active-region loop gain during transitions between stable states may fall below unity. To ensure an active-region loop gain of greater than unity, $h_{fe}R_C$ must be greater than $R_1$ or $R_2$. In addition to loading by the circuit itself, the external load being driven by the bistable may draw sufficient current to appreciably reduce the collector voltage of the cut-off transistor. In extreme cases, the collector voltage of the cut-off transistor may be reduced to the point where the base current into the conducting transistor is insufficient to keep it in saturation. If the conducting transistor enters the active region, regenerative switching will cause the bistable to change states. This problem may be particularly severe in computing applications where the load may be highly variable.

This problem can be reduced by clamping the collector of the cut-off transistor to some voltage $V_D$ which is less than $V_{CC}$. This is accomplished using catching diodes as shown in Fig. 73. The base bias $-V_{BB}$ and the

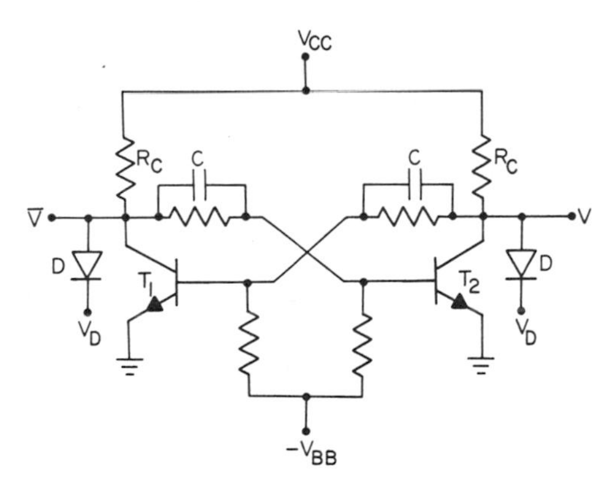


Fig. 73. Bistable circuit using catching diodes to maintain a constant output voltage in the logic 1 state.

feedback voltage dividers must be chosen to ensure that the conducting transistor is saturated when the cut-off transistor has a collector voltage of $V_D$ rather than $V_{CC}$. With this circuit, considerable load current can be drawn in the logic 1 output state with little change in output voltage. However, the output voltage swing between stable states has been reduced.

In computing applications, it frequently is necessary for a bistable to supply appreciable current in both the logic 0 and the logic 1 output states. This can be accomplished using the binary shown in Fig. 74. Here an extra output stage has been added which can supply considerable load current in either output logic state. Consider the right side of the figure; if $T_2$ is cut

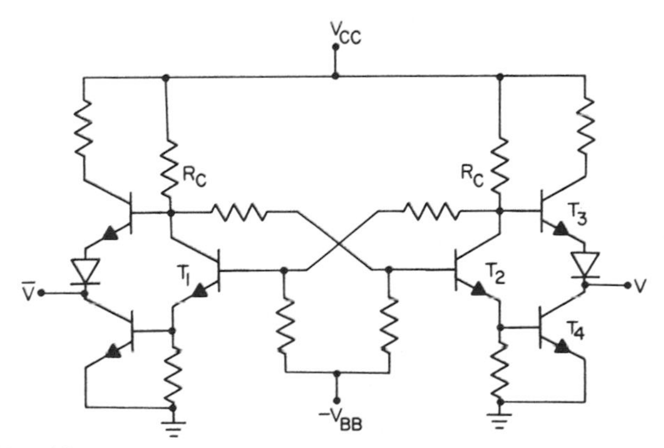


Fig. 74. Bistable circuit using emitter-follower output stages to supply large load currents in either logic state.

off, its high collector voltage turns on $T_3$ while its low emitter current turns off $T_4$. This produces an output of logic 1 across $T_4$, but the emitter follower configuration of $T_3$ can supply a large current to an external load. In other words, $T_4$ serves as a dynamic emitter resistor for the $T_3$ emitter follower output stage. If $T_2$ is saturated, its low collector voltage turns off $T_3$ while its high emitter current turns on $T_4$. This produces a logic 0 output across $T_4$ and a considerable current can be supplied to an external load through $T_4$.

In addition to supplying load current in both logic states, the bistable of Fig. 74 has excellent switching-time characteristics. During transitions from logic 0 to logic 1, transistor $T_3$ rapidly pulls up the output voltage as it goes into conduction.

### 1. Methods for Improving Switching Time

The capacitors C in Fig. 73 are called speed-up or commutating capacitors. They are used to improve the binary's switching time. When a transistor is leaving saturation, the commutating capacitor provides a low impedance path for the dissipation of the saturation storage charge in the base of the conducting transistor. When a transistor is leaving the cut-off region, the capacitor supplies extra current which decreases the time needed to forward bias the base-emitter junction.

This improvement in switching time is obtained only at the expense of a considerable increase in switching noise and ground currents. In addition, considerable time is required for the entire circuit to settle into a quiescent state, since this will not occur until the capacitors are completely discharged. For this reason, commutating capacitors usually are quite small, seldom being greater than about 100 pF.

Binary switching times can be improved even more by preventing the transistors from saturating. This can be accomplished by placing a diode-current shunt between the base and collector of each transistor as shown in Fig. 75. As a transistor leaves the active region, the corresponding diode $D_1$ or $D_2$ becomes forward biased thus shunting surplus current away from the transistor base. A more complete description of this method for preventing transistor saturation is given in Section IV.G.2.b.

### 2. Bistable Triggering

Thus far we have discussed how the bistable circuit has two stable logic output states. To be useful as a digital logic device, the bistable circuit must be able to change logic states on command. This is accomplished by bringing one or both transistors into the active region. Once this occurs, the loop gain will exceed unity, and the regenerative switching action will complete the switching process. Since the bistable circuit will change output states on command, it is frequently called a flip-flop.

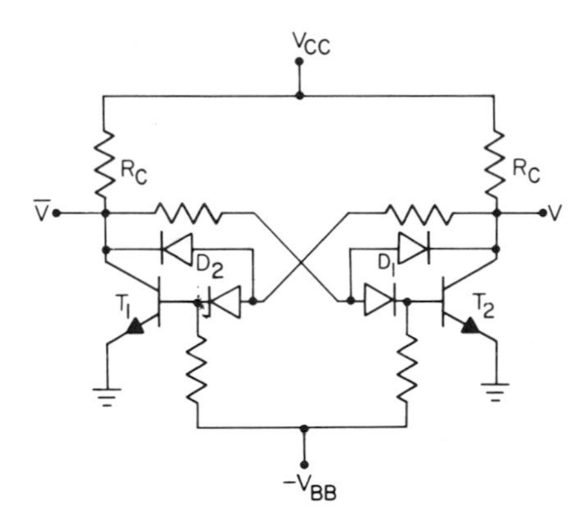


Fig. 75. Bistable circuit using diode current shunts to prevent transistor saturation.

a. Dc Triggering. Consider the bistable shown in Fig. 76. Assume $T_1$ is cut off and $T_2$ is saturated, corresponding to a logic 0 at V and a logic 1 at $\overline{V}$. If a positive voltage of sufficient amplitude is applied to the base of $T_1$, it will start to conduct and enter the active region. Regenerative switching will then quickly drive $T_1$ into saturation and $T_2$ into the cut-off region, thus changing the output states at V and $\overline{V}$. The circuit is now stable in this new state, and removal of the switching signal at the base of $T_1$ will have no effect.

The bistable can be returned to its original state by applying a positive voltage to the clear input. It should be noted that a negative voltage applied to the collector of $T_1$ will produce the same effect as a positive voltage

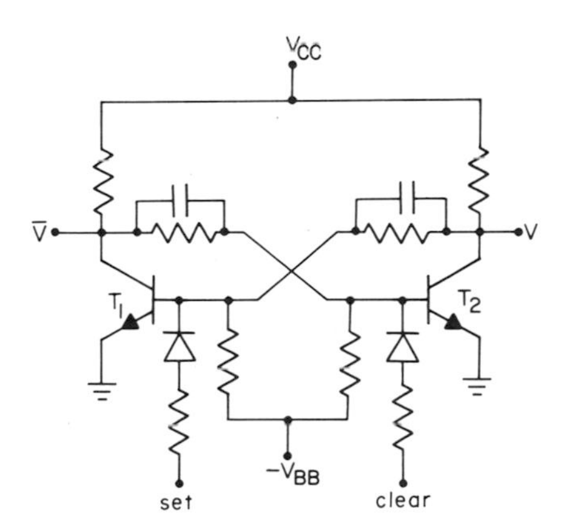


Fig. 76. Bistable circuit with dc triggering inputs at the transistor bases.

applied to its base. Alternatively, if $T_1$ is cut off and $T_2$ saturated, the diode polarities can be reversed and a negative voltage applied to the base of $T_2$. This again will cause the bistable to change output states. In general, the bistable is more sensitive to an input signal that will turn off the conducting transistor than to one that will turn on the nonconducting transistor.

b. AC Triggering. In many applications, it is necessary that the bistable changes states during a change in the input or trigger voltage, that is, during the rising or falling portion of a trigger pulse. This is called ac triggering and is accomplished by differentiating the trigger signal. A bistable with inputs for both ac and dc triggering is shown in Fig. 77. A trigger pulse applied to the set or clear ac input will be differentiated by the input coupling capacitor $C_I$ producing a positive and a negative spike. Only the negative spike can pass the ac input diode and be effective in triggering the bistable. Thus the circuit will trigger on the leading edge of a negative-going trigger pulse.

The bistable shown in Fig. 78 uses ac triggering at the transistor bases. Here a trigger pulse is delivered to the clock input T. After differentiation by the capacitors C, the positive spike reverse biases diodes $D_1$ and $D_2$. The negative spike may forward bias $D_1$ and $D_2$ depending on the voltage levels applied to the set (S) and reset (R) inputs. A positive voltage corresponding to a logic 1 applied to either the S or R input will prevent the negative trigger spike from forward biasing the corresponding diode. Thus the logic levels at S and R can steer the clock pulse to the base of either or both transistors.

For example, if R is at logic 1 and S is at logic 0, then the negative spike from the trigger pulse will be steered through diode $D_2$, thus turning

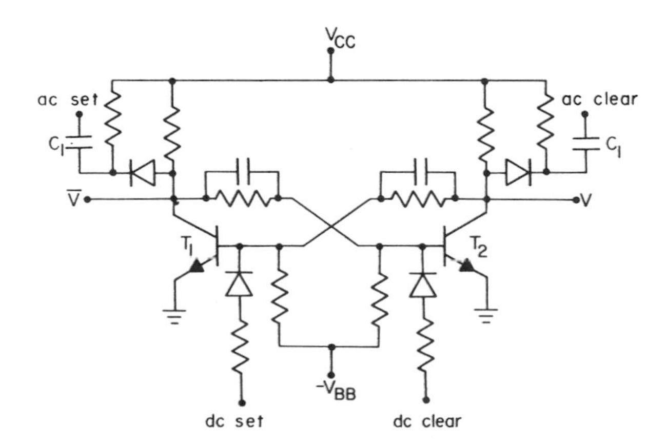


Fig. 77. Bistable circuit with dc triggering inputs at the transistor bases and ac triggering inputs at the transistor collectors.

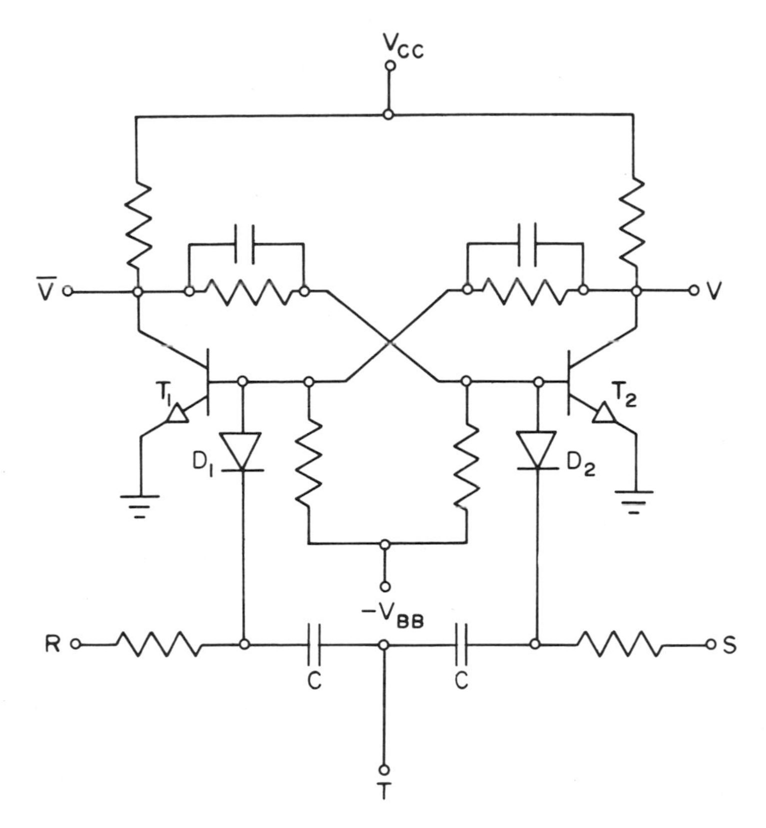


Fig. 78. Clocked RS bistable circuit using ac triggering.

off $T_2$ and producing an output of logic 1 at V and logic 0 at $\overline{V}$. If R is at logic 0 and S is at logic 1, the negative trigger spike will be steered to the base of $T_1$, thus turning it off and producing an output of logic 0 at V and logic 1 at $\overline{V}$. If both R and S are at logic 1, the output states will be the same after the clock pulse as before it. If both R and S are at logic 0, the negative trigger spike will be applied to the bases of both transistors, and the output will be indeterminate. These results are summarized in Table 27. Here $R_n$ and $S_n$ are the logic levels at the R and S inputs before the arrival of the n-th negative trigger spike, and $V_{n+1}$ is the output at V after the n-th negative trigger spike. This circuit represents the basic clocked RS bistable multivibrator.

3. The T Binary

In most counting applications, the bistable circuit is designed to change output states after every clock pulse. The clocked RS circuit of Fig. 78 can

TABLE 27

Combinations for a Clocked RS Flip-Flop

| $S_n$ | $R_n$ | $V_{n+1}$ |
|---|---|---|
| 0 | 0 | Indeterminate |
| 1 | 0 | 0 |
| 0 | 1 | 1 |
| 1 | 1 | $V_n$ |

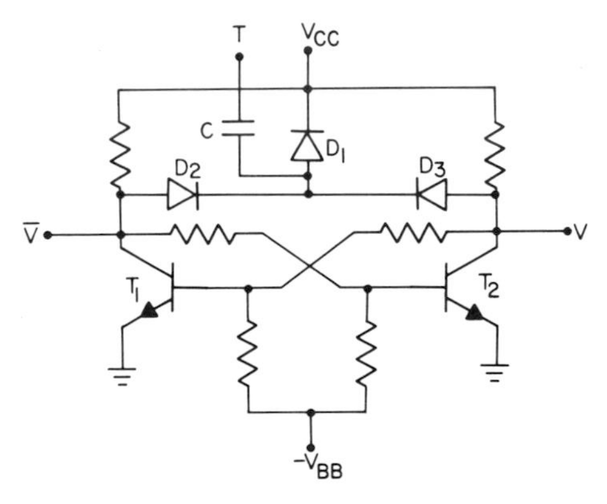


Fig. 79. The T binary using ac triggering at the transistor collectors.

be made to function in this way by connecting the V output to S and the $\bar{V}$ output to R. After a clock pulse, the outputs at V and $\bar{V}$ will change logic states as will the steering inputs R and S. This prepares the circuit for another transition during the next clock pulse. The resulting circuit is called a T binary or T flip-flop, the T indicating the toggle-switchlike action of the circuit.

The T binary shown in Fig. 79 also will change output states after every clock pulse. Assume $T_2$ is saturated and $T_1$ is cut off. The voltage drop across the collector resistor of $T_2$ reverse biases $D_3$. Since $T_1$ is cut off, the voltage drop across the collector resistor of $T_1$ is nearly zero; thus $D_2$ is at zero bias. The negative spike from the differentiation of the trigger pulse by C and $D_1$ will forward bias $D_2$ and appear at the base of $T_2$, thus turning it off. Once $T_2$ is cut off, $D_2$ will be reverse biased while $D_3$ will be at zero bias. Thus the next trigger pulse will be steered through $D_3$ to the base of $T_1$. This will result in another change in the output logic states.

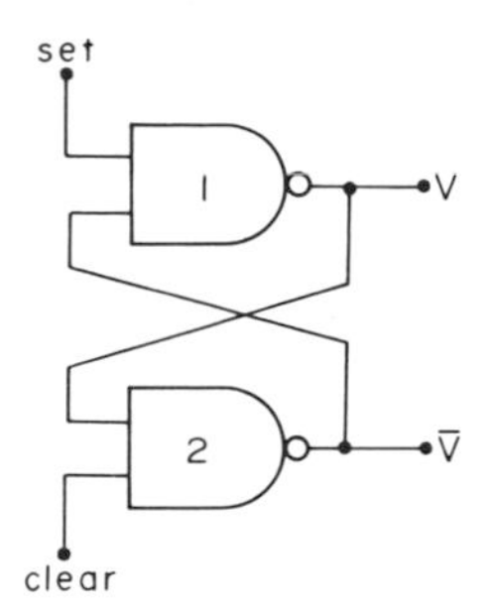


Fig. 80. Set-clear dc bistable made by cross coupling NAND gates.

## 4. Logic-Gate Bistable Circuits

Many of the discrete component bistable circuits thus far discussed can be constructed using cross-coupled inverting logic gates. Figure 80 shows how two 2-input NAND gates can be cross coupled to make a simple dc set-clear bistable. Assume that both the set and clear inputs are at logic 1. Also assume that output V is at logic 0. Since the output of V is used as an input to gate 2, the logic 0 output at V guarantees a logic 1 output at $\overline{V}$. This means that both inputs to gate 1 are at logic 1 thus maintaining an output of logic 0 at V. This corresponds to one of the stable states of the circuit. Since the circuit is completely symmetrical, it must also be stable with outputs of logic 1 at V and logic 0 at $\overline{V}$.

Consider the stable state with V at logic 0 and $\overline{V}$ at logic 1. If the set input is momentarily changed to logic 0, the output at V will go to logic 1. This will place both inputs to gate 2 at logic 1 thus producing a logic 0 at output $\overline{V}$. Since this is coupled back to the gate 1 input, V will stay at logic 1 even after the set input returns to logic 1. In this way, the cross-coupled NAND gate circuit can be made to change states. Table 28 shows the output logic states at V and $\overline{V}$ for the various combinations of set and clear inputs.

TABLE 28

Combinations for a Cross-Coupled NAND Gate Bistable

| Set | Clear | V | $\overline{V}$ |
|---|---|---|---|
| 0 | 0 | 1 | 1 |
| 1 | 0 | 0 | 1 |
| 0 | 1 | 1 | 0 |
| 1 | 1 | Indeterminate | |

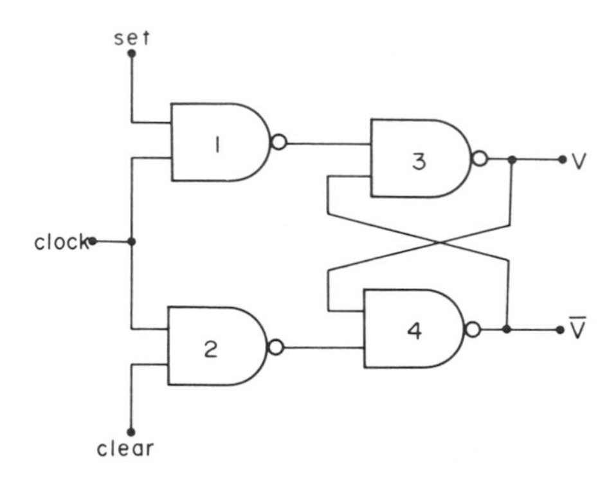


Fig. 81. Clocked set-clear logic gate bistable using dc triggering.

In Fig. 81, a dc gating or clock input has been added to the cross-coupled NAND bistable circuit. Only when the clock input is at logic 1 will the logic states of the set and clear inputs be transmitted to the bistable circuit. Thus the clock input serves as a means of controlling when and if the bistable will change states. Since the set and clear logic signals are inverted by gates 1 and 2, the table of combinations for this circuit shown in Table 29 is different from the table for the basic cross-coupled logic gate bistable.

TABLE 29

Combinations for a Gated Cross-Coupled Bistable

| $S_n$ | $C_n$ | $V_{n+1}$ |
|---|---|---|
| 0 | 0 | $V_n$ |
| 1 | 0 | 1 |
| 0 | 1 | 0 |
| 1 | 1 | Indeterminate |

The dc bistable of Fig. 81 cannot be used as a T binary unless the clock pulses are very short (less than the propagation time through the circuit); otherwise, the output logic states may change many times during a single clock pulse. This is because the circuit responds to the actual logic state at the clock input rather than to changes in the logic state of the clock input. Such a circuit could not be used in counting applications.

This problem is overcome by using ac coupling as shown in Fig. 82. Since this circuit responds only to the falling edge of the clock pulse

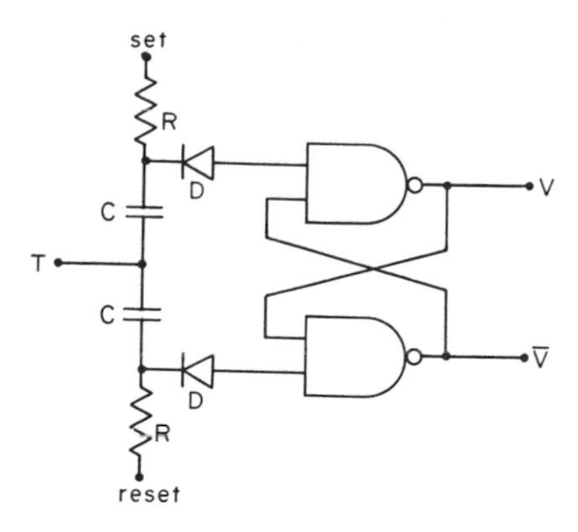


Fig. 82. Clocked RS logic gate bistable using ac triggering.

(transitions from logic 1 to logic 0), only one change in output states can occur during a clock pulse. If the V output of Fig. 82 is connected to the set input and the $\bar{V}$ output connected to the reset input, the output logic states will change after every clock pulse.

## 5. Master-Slave Bistable Circuits

It would be very useful to have a dc-coupled bistable that could be used in counting applications. While the ac-coupled circuit of Fig. 82 can be used for digital counting, the falling edge of the clock pulse must be sufficiently steep to produce a negative spike large enough to trigger the circuit. The dc-coupled circuit of Fig. 81 will respond to any shape of clock pulse, but the output states may change many times during a clock pulse if the circuit is used as a T binary.

The RS master-slave bistable shown in Fig. 83 uses dc coupling and can readily be converted for use in digital counting applications. Two

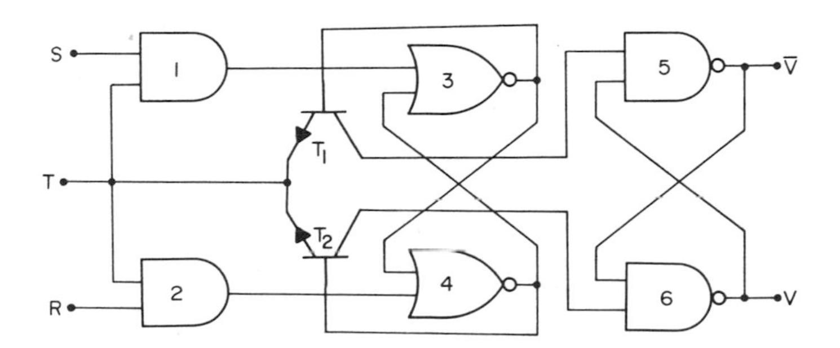


Fig. 83. The RS master-slave bistable circuit.

cross-coupled bistable circuits are used here; one uses NAND gates, and the other uses NOR gates. During a clock pulse, four distinct switching steps are performed. The coupling transistors $T_1$ and $T_2$ respond to a lower clock voltage than do the input gates 1 and 2. At the beginning of the clock pulse, $T_1$ and $T_2$ cut off producing logic 1's at the inputs to gates 5 and 6. This isolates the master bistable from the slave bistable. At a later time, the leading edge of the clock pulse reaches sufficient amplitude to activate input gates 1 and 2. The information at the S and R inputs is then transferred to the master output gates 3 and 4. It should be noted that the master is still isolated from the slave. During the decreasing voltage at the end of the clock pulse, gates 1 and 2 are deactivated before the coupling transistors $T_1$ and $T_2$. As the clock pulse voltage continues to decrease, $T_1$ and $T_2$ are activated. The logic levels at the master outputs are then inverted by $T_1$ and $T_2$ and transferred to the slave inputs. It should be noted that the output states of the master may change many times during a clock pulse, depending on the logic states at S and R. However, only the final output of the master at the termination of the clock pulse will be transmitted to the slave. The table of combinations for the RS master-slave circuit of Fig. 83 is given in Table 30.

TABLE 30

Combinations for the RS Master-Slave Bistable

| $S_n$ | $R_n$ | $V_{n+1}$ |
|---|---|---|
| 0 | 0 | $V_n$ |
| 1 | 0 | 1 |
| 0 | 1 | 0 |
| 1 | 1 | Indeterminate |

If the V output of the slave is connected to an additional R input of the master, and the $\overline{V}$ output of the slave is connected to an additional S input of the master, a JK master-slave bistable will result. This circuit, which is the basic digital counting unit, uses dc coupling but will undergo only one change in output states per clock pulse. The dc coupling allows the use of clock pulses of almost any shape. Frequently, several auxiliary J and K inputs are used in JK master-slave circuits. Where these occur, all J input must be at logic 1 for the circuit to respond to a J = 1 input. The same is true for the auxiliary K inputs. The table of combinations for the JK master-slave bistable is given in Table 31.

TABLE 31

Combinations for the JK Master-Slave Bistable

| $J_n$ | $K_n$ | $V_{n+1}$ |
|---|---|---|
| 0 | 0 | $V_n$ |
| 1 | 0 | 1 |
| 0 | 1 | 0 |
| 1 | 1 | $\overline{V}_n$ |

## B. Registers and Applications

The bistable memory circuit can be used to store one bit of binary information. Information is entered into the S and R or the J and K inputs, and a clock pulse will result in the information being stored in the V and $\overline{V}$ outputs of the bistable. Several RS or JK bistable circuits can be connected to store several bits of binary information. Such an array of bistables is called a memory or storage register.

Consider the 4-bit memory register shown in Fig. 84. Here each RS bistable is shown as a rectangle with R, S, and T inputs and V and $\overline{V}$ outputs. Each bit of the 4-bit binary number ABCD is presented to the input of the corresponding bistable. The inverter gates are used to present the complements of A, B, C, and D to the R inputs of the bistables. Clock pulses simultaneously are delivered to all clock inputs. When a clock pulse is delivered, the binary information at the four inputs is entered into the memory register and appears at the V outputs of the corresponding bistables. The complements of A, B, C, and D are stored at the $\overline{V}$ outputs. When new information is to be entered into the memory, it is presented to the four

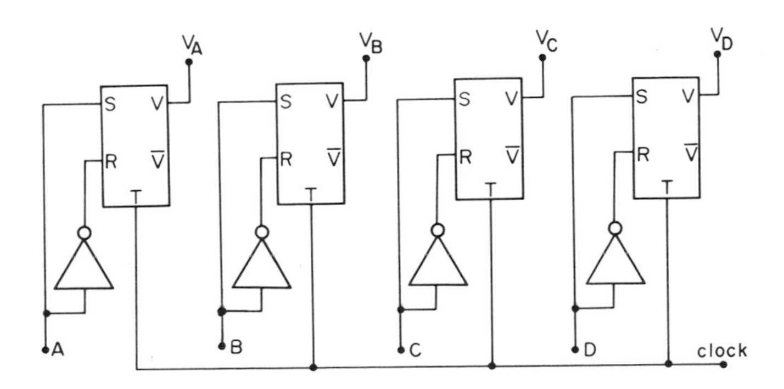


Fig. 84. Basic 4-bit memory register.

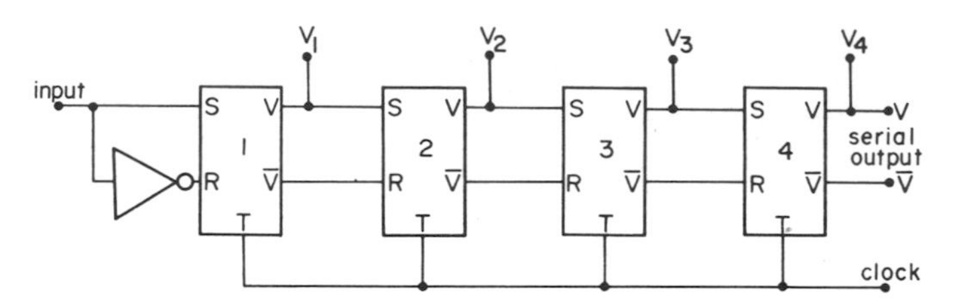


Fig. 85. Basic 4-bit shift register.

inputs, and another clock pulse will store it in the bistables.

A more useful register can be made by connecting the V output of each bistable to the S input of the next bistable and connecting the $\overline{V}$ output of each one to the R input of the next. The resulting network is called a shift register. An example of a 4-bit shift register is shown in Fig. 85. In this network, each clock pulse shifts the contents of each bistable to the next bistable on the right.

Assume the binary number 1010 is to be entered in the register. The lowest-order bit is presented to the input. A clock pulse will enter this 0 bit in bistable number 1. The next-lowest-order bit is then presented to the input and a clock pulse delivered. This will shift the contents of bistable 1 into bistable 2 and enter the second bit in bistable 1. This procedure is repeated twice more. The V outputs 1, 2, 3, and 4 then will contain the binary bits 1, 0, 1, and 0, respectively.

It may be necessary to deliver the contents of the register sequentially to some other unit in the computer. In the 4-bit shift register of Fig. 85, a

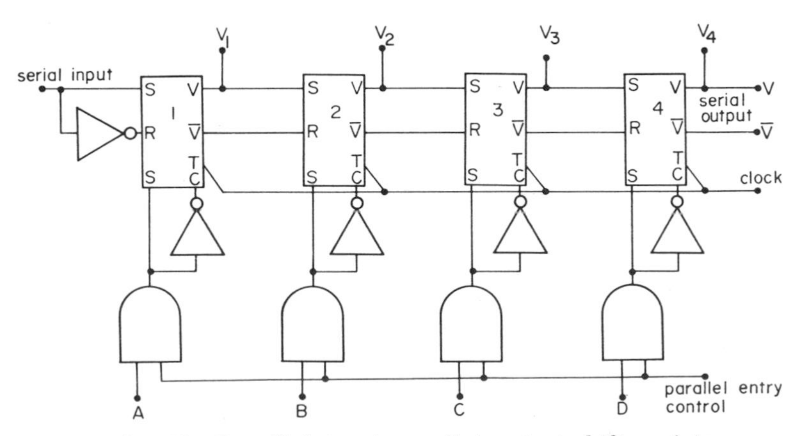


Fig. 86. Parallel-input-parallel-output shift register.

series of four clock pulses will deliver the four bits of binary information stored in the register to the serial output. Each successive clock pulse will shift one binary bit out of the register.

In many arithmetic operations, it is necessary to enter information into a shift register in parallel fashion; that is, all bits are entered simultaneously. A 4-bit parallel-entry shift register is shown in Fig. 86. Here each RS bistable has an additional set of inputs for direct parallel entry. A logic 1 pulse applied to the parallel-input control will enter the information presented to the four parallel-entry inputs. Clock pulses then can be used to shift the register contents. If necessary, a series of clock pulses will present the contents of the register sequentially at the serial output. A serial-entry input also has been provided. Thus binary information can be entered and retrieved in both serial and parallel fashion. Applications of this basic parallel-input-parallel-output shift register will be discussed.

In the shift registers thus far discussed, information is lost when it is shifted out of the register. In a circulating shift register, the V and $\overline{V}$ outputs of the last bistable are connected to the R and S inputs of the first bistable. As each binary bit is shifted out of the register it reappears in the first bistable. In this way, any number of shifts can be performed without losing the register contents. A 4-bit circulating-shift register is shown in Fig. 87. When the control input is at logic 0, clock pulses will circulate the register contents. When the control is at logic 1, new information is entered at each clock pulse, and the previous register contents are shifted out.

## 1. Binary Addition Using Shift Registers

Numerous circuits have been developed for the addition of binary numbers. Most of these use some combination of shift registers and the binary full- and half-adding circuits discussed in Section III. F. 3.a. Fig. 88

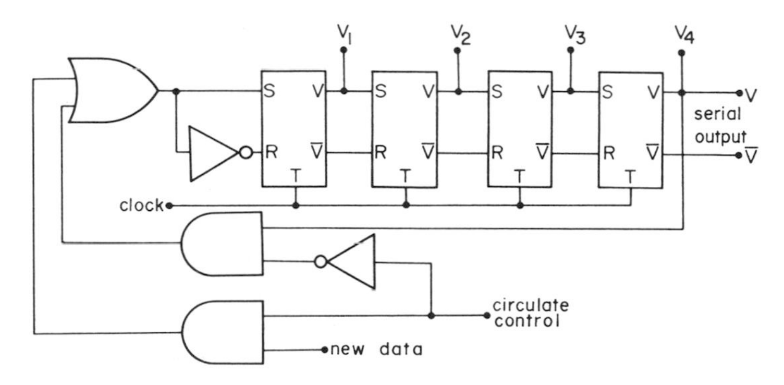


Fig. 87. Circulating shift register.

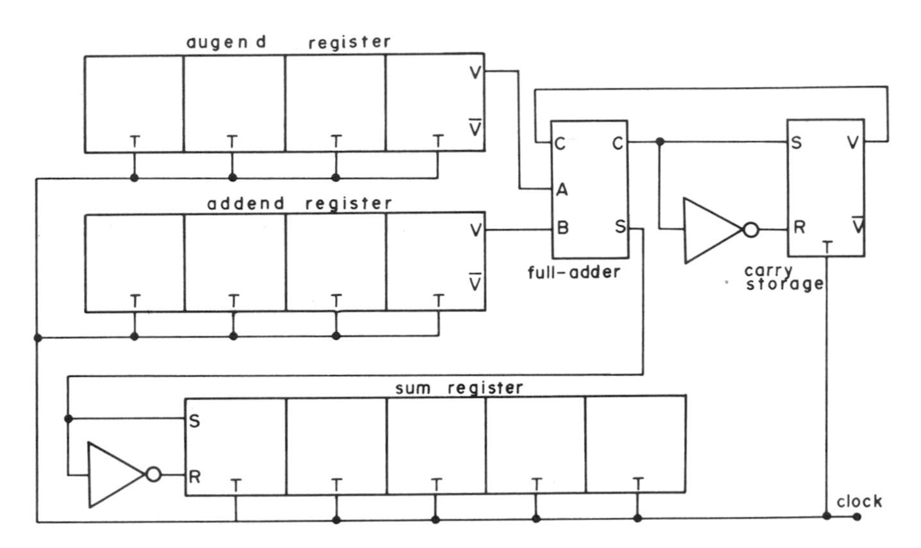


Fig. 88. The 4-bit serial adder.

shows a diagram of a serial adder. The augend and addend are stored in shift registers as is the resulting sum. A 1-bit register (single bistable) is used for temporary storage of the carry output from the full adder. Assume the augend and addend registers have been loaded with binary numbers A and B, respectively, and the 1-bit temporary storage register has been cleared producing a logic 0 at its V output. The lowest-order bits of A and B immediately are presented to the adder. The first clock pulse will load the first (lowest-order) sum bit into the sum register and the first carry bit into the temporary storage register. In addition, the second-lowest-order bits of A and B will be shifted into the adder. If a carry was present from the addition of the lowest-order bits, it also will be presented to the adder since it was stored in the V output of the temporary storage register. The second clock pulse will shift the second sum bit into the sum register and supply the adder with the next-lowest-order bits of A, B, and carry C. This process is continued until all bits of A and B have been added. The sum register then contains the final sum.

Since each augend and addend bit is added sequentially, the serial adder is rather slow. The parallel adder shown in Fig. 89 uses a separate adder for each binary bit. Although more costly, the parallel adder is considerably faster than the serial adder. In the parallel adder, each corresponding bit of A and B is presented to a separate adder. One clock pulse will gate all sum bits into the sum register. The carry output from each adder is connected to the carry input of the next-higher-order adder. The time required to perform parallel addition is usually limited by the time

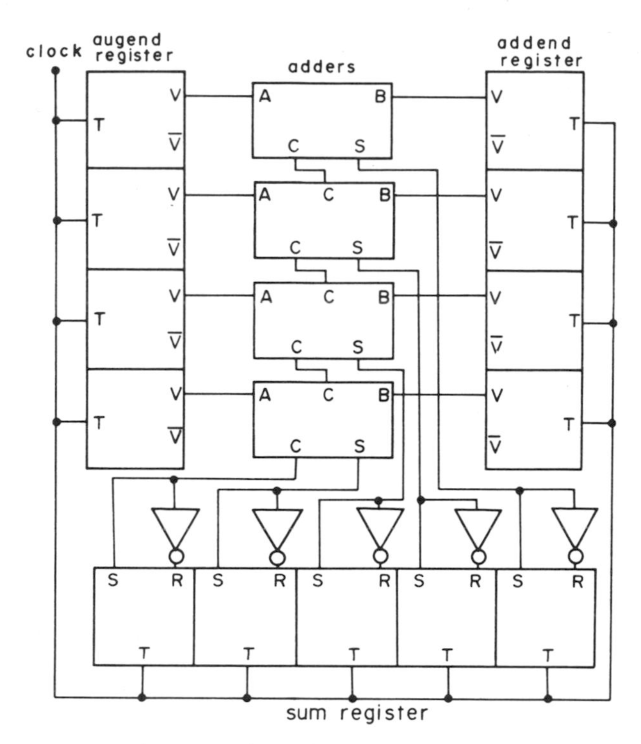


Fig. 89. The 4-bit parallel adder.

needed for the carry outputs to ripple through the string of adders.

If three or more numbers are to be added using the serial or parallel adders of Figs. 88 and 89, respectively, the first two numbers are added, and the contents of the sum register is transferred to the augend register. The third number is then loaded into the addend register and a second addition performed. These steps must be repeated for every additional number to be added. This requires considerable time and circuitry.

A more satisfactory method for adding a series of numbers is by accumulation. Several types of accumulators have been designed, and one using adding circuits is shown in Fig. 90. The first number to be added is loaded into the augend register. The first clock pulse will gate this number through the adders and into the accumulator register. This register now contains the sum of one number. The V output from each stage of the accumulator register is fed back into the B or addend input of the corresponding adder. A second number then is loaded into the augend register. The sum of the two numbers then is gated into the accumulator register by the second clock pulse. This sum immediately appears at the B inputs to the adders. A

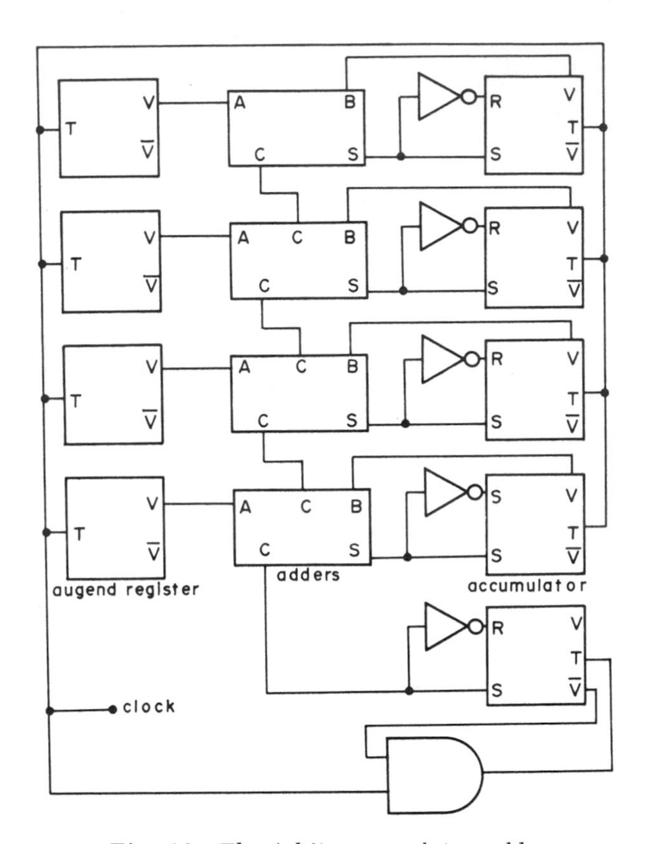


Fig. 90. The 4-bit accumulator adder.

third number then can be loaded into the augend register and added to the sum of the first two. A third clock pulse will gate the sum of the three numbers into the accumulator register. This process is repeated as many times as necessary to add a series of binary numbers. The AND gate in the circuit is used to prevent the loss of the highest-order carry if this appears before the final addition.

## 2. Binary Subtraction Using Shift Registers

Slight modification of the serial and parallel adding circuits of Figs. 88 and 89, respectively, will allow them to be used for 1's complement subtraction. Consider the parallel subtractor shown in Fig. 91. Here the V outputs of the minuend register are connected to the A inputs of the full adders, and the $\overline{V}$ outputs of the subtrahend register are connected to the B inputs of the adders. This circuit will perform 1's complement binary subtraction. Notice that the carry output of the highest-order adder is connected

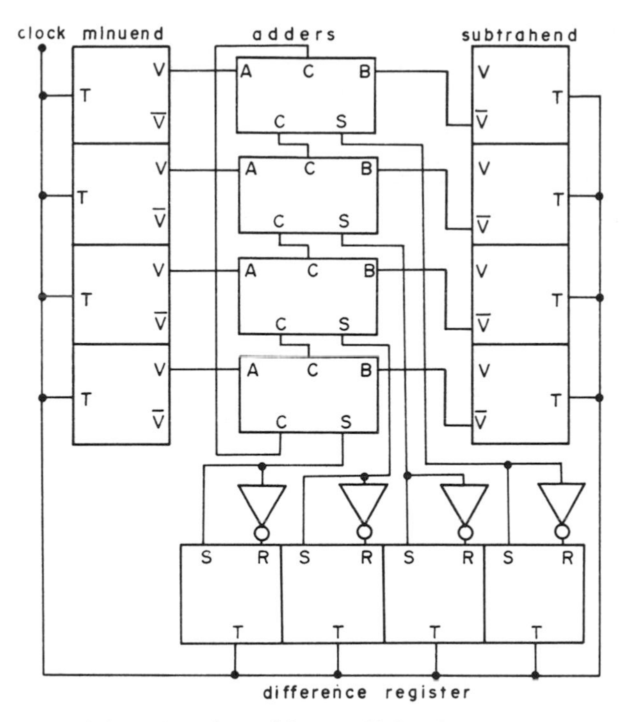


Fig. 91. The 4-bit parallel subtractor.

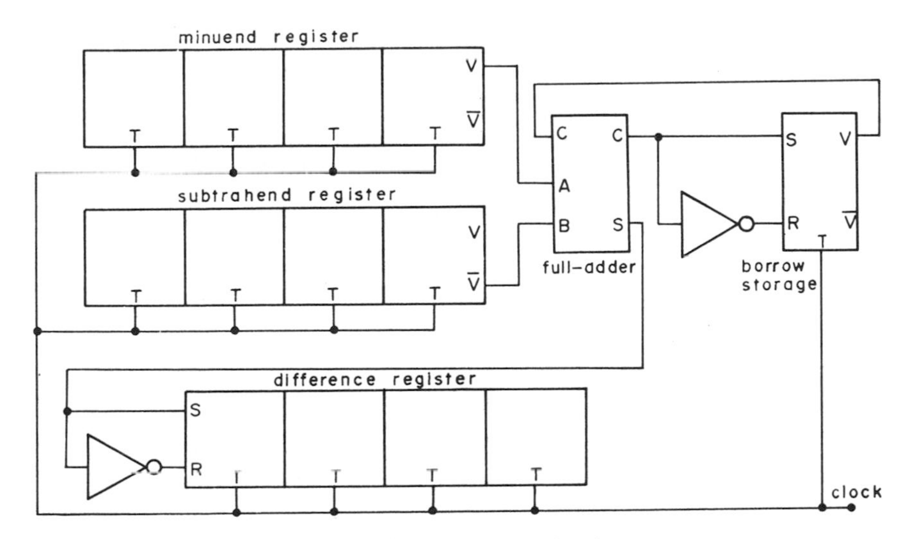


Fig. 92. The 4-bit serial subtractor.

to the carry input of the lowest-order adder. This provides for the end-around carry to be added to the lowest-order binary place. As in parallel addition, subtraction time usually is limited by the time required for the carry outputs to ripple through the string of adders. However, subtraction time usually is greater than addition time since the end-around carry must ripple through the network in addition to the regular carry.

A 1's complement serial subtractor is shown in Fig. 92. Here the V output of the minuend register is sequentially added to the $\overline{V}$ output of the subtrahend register. The end-around carry is obtained by initially setting the V output of the 1-bit carry register at logic 1.

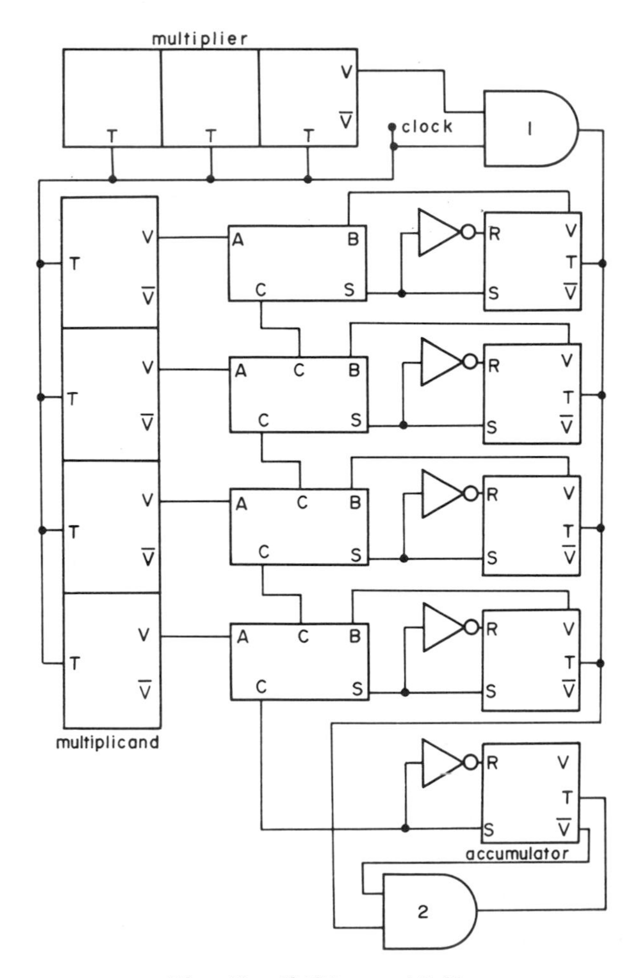


Fig. 93. Shifting multiplier.

## 3. Multiplication by Shifting

The accumulator circuit of Fig. 90 can be modified for use as a shifting multiplier. The complete multiplier is shown in Fig. 93. The multiplicand register is designed to shift left while the multiplier register shifts right. Each clock pulse will shift the multiplicand one place to the left and the multiplier one place to the right. In addition, the contents of the multiplicand register will be added to the previous contents of the accumulator register when the lowest-order multiplier place contains a 1 bit. This is accomplished by gating the clock and the lowest-order place in the multiplier register through AND gate 1. If, after any shift, the lowest-order place in the multiplier is a 0 bit, the next clock pulse will not be delivered to the accumulator register. Sufficient time must be allowed between clock pulses for the carry output to ripple through the string of adders.

## 4. Comparison of Binary Numbers

In many arithmetic and control operations it is necessary to determine the relative magnitude of two or more binary numbers. For example, in 1's complement subtraction if the subtrahend is larger than the minuend, no end-around carry is generated, and the contents of the difference register must be complemented to obtain the magnitude of the final answer. Since the computational procedures are somewhat different for positive and negative differences, the computer must determine the relative magnitudes of the minuend and subtrahend.

Figure 94 shows a simple method for detecting equality between two binary numbers A and B. The output bistable initially is set to give a logic 1 at the V output. This logic 1 activates the AND gate and will allow clock pulses to be delivered to the output bistable. Each clock pulse will shift a corresponding pair of bits from registers A and B into the exclusive OR network. If any corresponding pair of bits of A and B are not equal, the

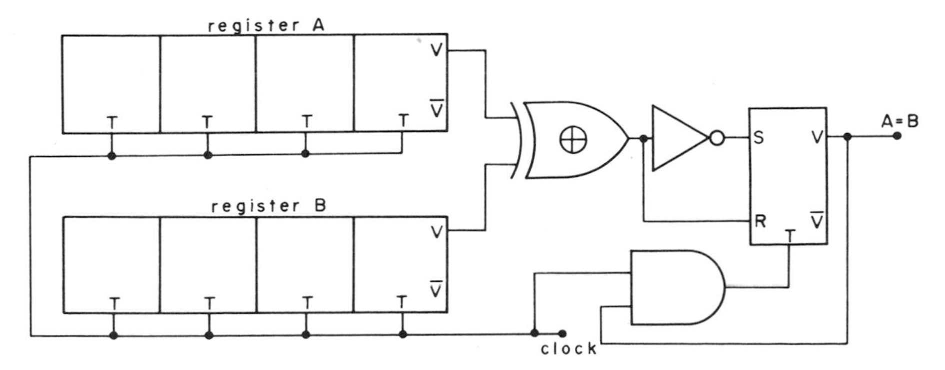


Fig. 94. Serial equality detector.

logic 1 output from the exclusive OR will change the V output of the output bistable to logic 0. This logic 0 will deactivate the AND gate, thus preventing additional clock pulses from being delivered to the output bistable. This will keep the V output at logic 0, indicating an inequality.

The relative magnitude of two binary numbers can be determined using the network shown in Fig. 95. The two numbers are stored in left-shift registers A and B. Corresponding pairs of bits from A and B are shifted out of the register and into the gating network, starting with the highest-order bits. Assume that the highest-order bits of A and B are 1 and 0, respectively. This means the V output of register A AND the $\overline{V}$ output of register B will be at logic 1, producing a logic 1 output at AND gate 1 indicating that A is larger than B. If the highest-order bits of A and B are 0 and 1, respectively, the output at AND gate 2 will be at logic 1, indicating that B is larger than A. In either case, the output of OR gate 3 will be at logic 1, deactivating AND gate 4. This will prevent further shifting, thus holding the answer at the gate 1 or 2 output. In addition, a logic 1 output at gate 3 will produce a logic 0 at the V output of the equality-detecting bistable, thus indicating an inequality.

If the highest-order bits of A and B are equal, the outputs at gates 1 and 2 will be logic 0, as will the output of gate 3. This will keep gate 4 activated, and another clock pulse can be delivered to the shift registers. The process is repeated until an inequality is detected. If the outputs at gates 1 and 2 are both at logic 0 after all bits of the two numbers are shifted

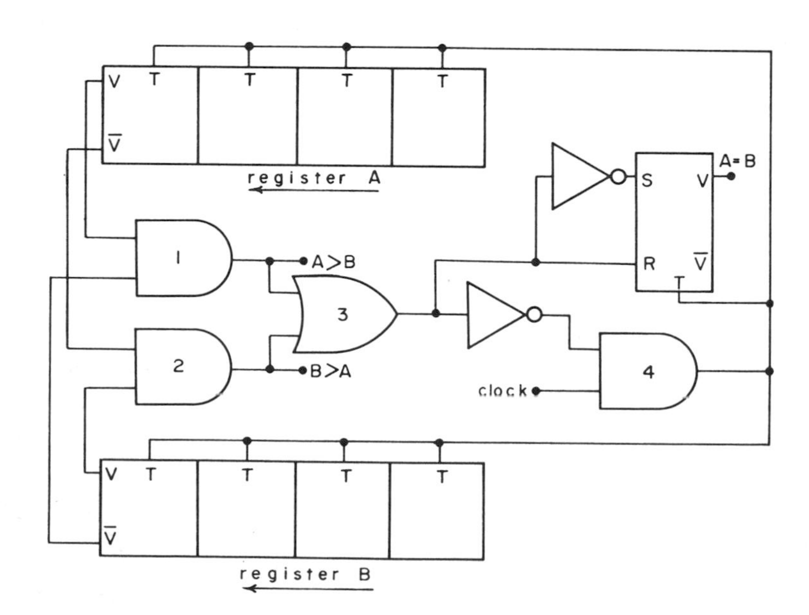


Fig. 95. Serial relative-magnitude determination network.

out of the registers, the V output of the equality-detecting bistable will be at logic 1, indicating that the numbers are equal.

Only a few of the most common applications of registers were discussed. Numerous variations have been developed for each of the arithmetic processing networks mentioned in this section, and the interested reader is referred to the references at the end of this chapter.

## C. Digital Counting Circuits

In any number system, counting is performed by sequentially increasing the lowest-order place by one unit for each count starting with 0. The procedure continues until the lowest-order place has a magnitude of one less than the base or radix of the number system. The next count returns the lowest-order place to 0 and carries one unit to the next-lowest-order position.

Since the binary number system has a base of 2, the lowest-order place returns to 0 after every other count. Thus every count will complement the lowest-order position. This repeated complementation is easily accomplished using a JK master-slave bistable. Every clock pulse applied to the clock input will complement both the V and $\overline{V}$ outputs. This produces an output logic waveform that is half the frequency of the input or clock waveform.

Assume a JK master-slave bistable initially has been cleared, producing a logic 0 at the V output. The falling edge of the first clock pulse will change the V output from logic 0 to logic 1. The falling edge of the second clock pulse will change the V output back to logic 0. This constitutes one complete cycle of the V output. Assume now that the V output of the bistable is connected to the clock input of a second bistable. After the second clock pulse is delivered to the first bistable, its V output will change from logic 1 to logic 0. This will produce a change from logic 0 to logic 1 at the V output of the second bistable. A third clock pulse will change the V output of the first bistable from logic 0 to logic 1. This will have no effect on the output of the second bistable since it will change states only during a 1 to 0 transition of its clock input. The fourth clock pulse will produce a 1 to 0 transition at the V output of the first bistable which will cause a 1 to 0 transition at the V output of the second bistable. Thus four complete clock pulses are required at the input of the first bistable to produce one complete cycle of the V output of the second bistable.

### 1. Binary Up Counters

If a series of bistables is connected such that the V output of each one is used as a clock input to the next one, the frequency of the V output from each bistable is just one-half the frequency of the V output of the previous bistable. Thus a series of bistables connected in this way can be used as a binary

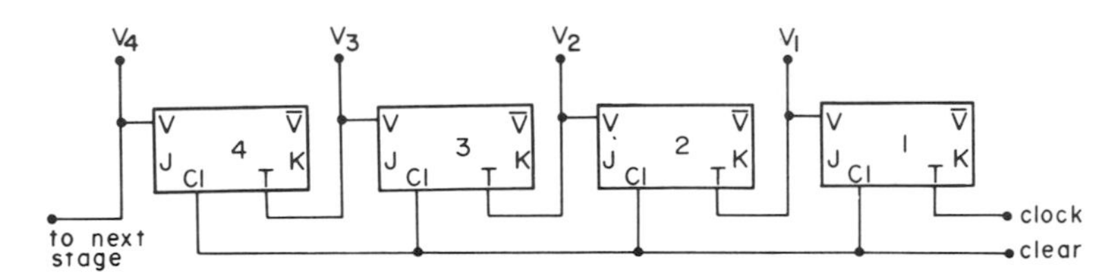


Fig. 96. Basic 4-bit binary up counter.

frequency divider.

In Fig. 96, four bistables have been connected. Assume all four have been cleared, producing logic 0's at the four V outputs. If these four outputs are considered the four places in a 4-bit binary number, the resulting number is binary 0000 or decimal 0. The first clock pulse will change the output of bistable 1 to logic 1. This will produce the binary number 0001 at the four outputs. The second clock pulse will complement the output of bistable 1 producing a logic 1 to logic 0 transition. This will cause the output of bistable 2 to complement, producing a logic 1 at its output. The four outputs then contain the binary number 0010 or decimal 2. The third clock pulse will produce a 0-to-1 transition at the output of bistable 1, but this will have no effect on bistable 2. The four outputs will then contain the binary number 0011 or decimal 3. The fourth clock pulse will produce a 1-to-0 transition at $V_1$ which will cause a 1-to-0 transition at $V_2$. This in turn will complement $V_3$ producing a logic 1. The four outputs then contain the binary number 0100 or decimal 4. Thus the four outputs contain the binary number equal to the number of clock pulses delivered to the clock input of bistable 1. This forms the basis of digital counting circuits.

The 4-bit counter of Fig. 96 can count up to binary 1111 or decimal 15. In general, a counter using n bistables can count to $2^n - 1$. When the 16th clock pulse is delivered to the 4-bit counter, a 1-to-0 transition will occur at $V_1$. This will cause a 1-to-0 transition at $V_2$ which in turn will produce a 1-to-0 transition at $V_3$. Thus these 1-to-0 transitions ripple through the string of bistables producing an output of binary 0000. This resets the counter. In addition, the counter can be reset after any number of counts by applying a momentary pulse to the clear input.

## 2. Binary Down Counters

In many counting applications it is required that the counter output decreases by one unit for every clock pulse. Down counting is readily accomplished by connecting the $\overline{V}$ output rather than the V output of each bistable to the clock input of the next one. A 4-bit down counter is shown in Fig. 97. Assume that the counter is initially cleared, producing the binary number 0000 at the four V outputs. The first clock pulse will produce a

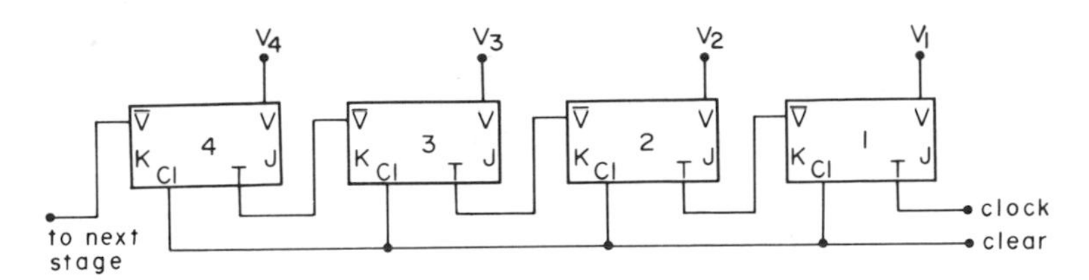


Fig. 97. The 4-bit binary down counter.

0-to-1 transition at $V_1$ and a 1-to-0 transition at the $\overline{V}$ output of bistable 1. Since the $\overline{V}$ output of bistable 1 is connected to the clock input of bistable 2, $V_2$ also will go from logic 0 to logic 1. The $\overline{V}$ output of bistable 2 will thus undergo a 1-to-0 transition which will cause $V_3$ to go to logic 1. This switching action proceeds along the entire string of bistables until all V outputs are at logic 1. Thus the first clock pulse fills the counter to capacity or binary 1111 for the 4-bit counter.

The second clock pulse will produce a 1-to-0 transition at $V_1$, thus forming the binary number 1110. The third clock pulse will cause a 0-to-1 transition at $V_1$ and a 1-to-0 transition at $V_2$, thus forming the binary number 1101. In the same way each successive clock pulse will reduce the binary number at the outputs by one unit. After 16 clock pulses, the counter will have counted down to binary 0000, and the 17th clock pulse will reset the counter to binary 1111.

## 3. Binary Up-Down Counters

Figure 98 shows a counter circuit that can count either up or down on command. If the up input is at logic 1 and the down input is at logic 0, the V output from each bistable will be gated to the clock input of the next bistable. This will result in up counting. If the up input is at logic 0 and the down input at logic 1, the $\overline{V}$ output of each bistable will be gated to the clock input of the next one thus causing the circuit to count down.

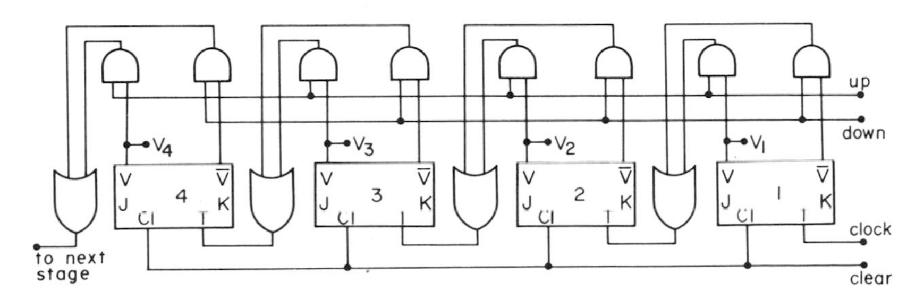


Fig. 98. The 4-bit binary up-down counter.

## 4. Binary-Coded Decimal Counters

In many counting applications the counter output is to be displayed on a decimal output device such as a typewriter or series of decimal display tubes. In such cases, it often is more convenient to do the counting directly in binary-coded decimal. This eliminates the need for binary-to-decimal or binary-to-BCD conversion. The outputs from a 4-bit BCD counter can be connected directly to a 4-bit decoder. The outputs from the decoder then are used to drive the decimal display unit.

In a 4-bit binary counter, the counter counts up to binary 1111 or decimal 15 and then recycles on the 16th clock pulse. For BCD counting, the counter must count to binary 1001 (8421 BCD) or decimal 9 and recycle on the tenth clock pulse. An example of an 8421 BCD counter is shown in Fig. 99. The $\overline{V}$ output of bistable 4 and the V output of bistable 1 are connected to the clock input of bistable 2 through AND gate 1. For the first seven clock pulses, the $\overline{V}$ output of bistable 4 is at logic 1, thus keeping AND gate 1 active. This is equivalent to directly connecting $V_1$ to the clock input of bistable 2. Thus the network functions as a regular binary counter for the first seven counts.

After the eighth clock pulse, the counter will contain binary 1000, resulting in a logic 0 at the $\overline{V}$ output of bistable 4, thus deactivating AND gate 1. This will inhibit further changes in the $V_2$ output. At the same time, AND gate 2 is activated by the logic 1 at $V_4$. The next 1-to-0 transition at $V_1$ will then be delivered to bistable 4. The ninth clock pulse will produce an output of binary 1001. On the tenth clock pulse, $V_1$ will go to logic 0. Since AND gate 1 is not active, $V_2$ will stay at logic 0. On the other hand, AND gate 2 is active, so the 1-to-0 transition at $V_1$ will deliver a clock pulse to bistable 4, thus changing $V_4$ from logic 1 to logic 0. This results in an output of binary 0000.

Figure 100 shows a 2421 BCD counter. This counter functions as a simple binary counter for the first four clock pulses. After the fourth clock pulse, the 0100 output activates the AND gate with the result that the fifth

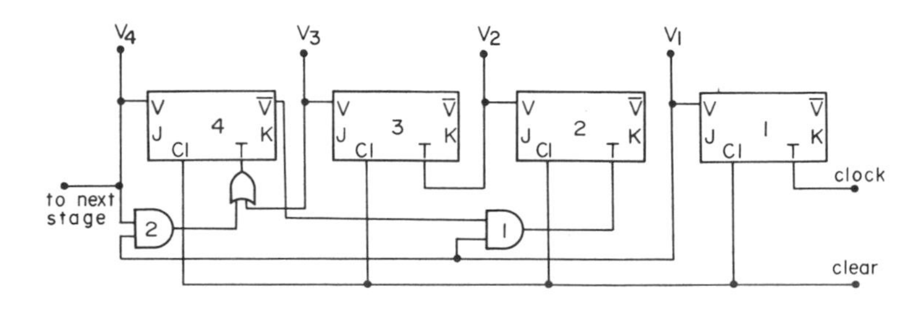


Fig. 99. The 8421 binary-coded decimal counter.

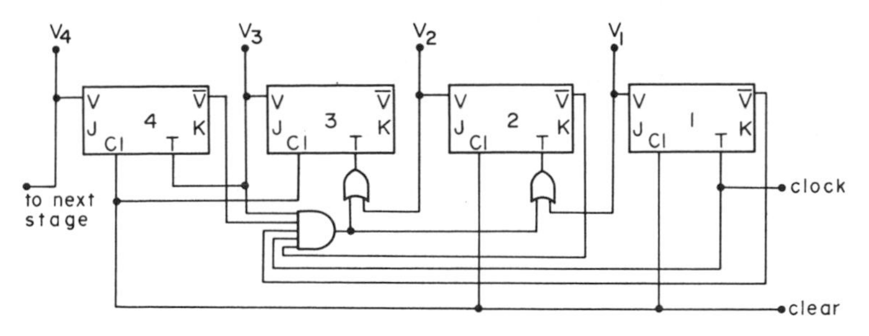


Fig. 100. The 2421 binary-coded decimal counter.

clock pulse is delivered directly to bistables 1, 2, and 3. Thus $V_1$, $V_2$, and $V_3$ all will complement on the fifth clock pulse. The 1-to-0 transition of $V_3$ will also result in complementation of $V_4$. This will produce the required 1011 after the fifth clock pulse. The logic 0 at $\overline{V}$ of bistable 4 then deactivates the AND gate, and the counter continues the normal binary count. After nine clock pulses the output will be 1111, and the tenth clock pulse will produce a 1-to-0 transition in all four bistables, thus resetting the counter.

## 5. Rippleless Binary Counters

In the counters thus far discussed, the clock pulse must ripple through the entire string of bistables before being applied to bistable 4. In high-speed counting applications, the time required for this ripple action can limit the maximum counting rate. This problem can be overcome by directly applying the input clock pulses to the clock inputs of all bistables in the counter. Several schemes have been developed for doing this, and one using cascaded AND gates is shown in Fig. 101. Each AND gate has inputs from the clock, the V output of the previous bistable, and the output from the previous AND gate.

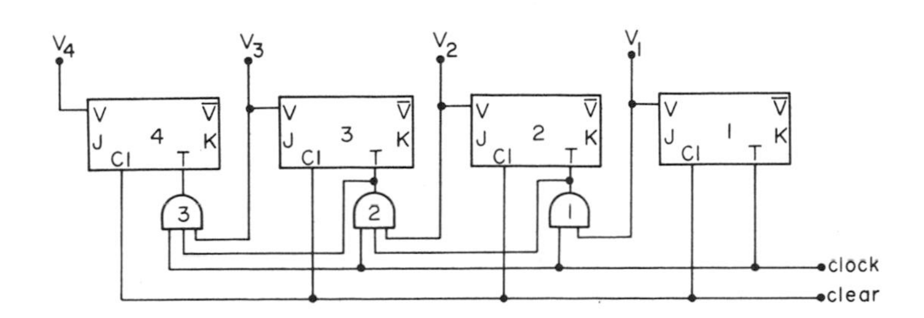


Fig. 101. Rippleless binary counter.

Assume the counter initially is cleared, producing an output of binary 0000. This deactivates all three AND gates, thus allowing the first clock pulse to be delivered only to bistable 1. The first clock pulse then produces an output of 0001. The logic 1 at $V_1$ then activates AND gate 1, and the second clock pulse will be delivered to bistables 1 and 2, producing an output of 0010. This deactivates AND gate 1, the output of which keeps AND gate 2 deactive. The third clock pulse then will be delivered only to bistable 1, producing an output of 0011. The logic 1 at $V_1$ will activate AND gate 1, the output of which, together with the logic 1 at $V_2$, will activate AND gate 2. Thus the fourth clock pulse will be applied to bistables 1, 2, and 3. This will produce an output of 0100. In similar fashion, the outputs of the AND gates are properly activated or deactivated, depending on the logic states of the V outputs and the states of the other AND gates. In this way, the input clock pulses simultaneously are applied to the clock inputs of all bistables which must undergo a transition during a given clock pulse. This allows the counter to perform at its maximum counting rate.

## REFERENCES

[1]. R. C. Baron and A. T. Piccirilli, Digital Logic and Computer Operations, McGraw-Hill, New York, 1967.

[2]. T. C. Bartee, Digital Computer Fundamentals, McGraw-Hill, New York, 1960.

[3]. R. Benrey, Understanding Digital Computers, John F. Rider, New York, 1964.

[4]. S. H. Caldwell, Switching Circuits and Logic Design, John Wiley, New York, 1959.

[5]. Y. Chu, Digital Computer Design Fundamentals, McGraw-Hill, New York, 1962.

[6]. L. Delhom, Design and Application of Transistor Switching Circuits, McGraw-Hill, 1968.

[7]. J. N. Harris, P. E. Gray, and C. L. Searle, Digital Transistor Circuits, John Wiley, New York, 1966.

[8]. P. M. Kintner, Electronic Digital Techniques, McGraw-Hill, New York, 1968.

[9]. H. V. Malmstadt and C. G. Enke, Digital Electronics for Scientists, W. A. Benjamin, New York, 1969.

[10]. J. Millman and H. Taub, Pulse, Digital, and Switching Waveforms, McGraw-Hill, New York, 1965.

[11]. M. Phister, Logical Design of Digital Computers, John Wiley, New York, 1958.

[12]. D. L. Schilling and C. Belove, Electronic Circuits: Discrete and Integrated, McGraw-Hill, New York, 1968.

[13]. R. K. Richards, Arithmetic Operations in Digital Computers, D. Van Nostrand, Princeton, N. J., 1957.

[14]. U. S. Department of the Army, Basic Theory and Application of Transistors, Dover, New York, 1963.

Chapter 4

# PROGRAMMING LANGUAGES

Clarence H. Thomas

Department of Chemistry
University of Cincinnati
Cincinnati, Ohio 45221



## I. INTRODUCTION

One of the most significant developments in the short history of digital computers is that of the programmable computer. A programmable computer can be used to perform an almost infinite variety of calculations. In fact the main limitation often lies in the programmer's ability to creatively use the options which a given computer possesses. Initially only the largest computers could be programmed. But recently even small computers have come on the market which are programmable. These computers may be used in a single laboratory or shared among several laboratories. Even in cases where the computer is used most or all of the time in conjunction with a single instrument, the capability of programming it gives the researcher a flexibility in designing and changing experiments which he otherwise would not have. A drawback to these small computers is that many of them cannot be programed in a "high level" language such as FORTRAN. The greater flexibility in designing experiments gained by the chemist will usually more than repay him for his efforts in learning the programming language.

In this chapter we will discuss both the machine-oriented languages and the programmer-oriented languages. The machine-oriented languages will be discussed first; even for those who plan to program only in the "high level" languages some introduction to the machine language is very useful. In order to program and "debug" (or find errors in a program) efficiently, a programmer must have some knowledge of how the computer will translate his program into machine language. This knowledge is also necessary if he is to make best use of the information which the computer gives in debugging his program.

## A. Types of Languages

In considering the languages in which a computer is programmed we may divide them into three levels. At the lowest level is the program as it appears in the machine. In this form the program is coded in the binary system, with one instruction in each word of memory. In this form the instructions in memory cannot be distinguished from the numbers which are stored in the memory except by their location. At the second level are assembly languages; in these languages there is a one-to-one correspondence between the instructions of the language and the machine's binary-coded instructions. The assembly language will be easier for a person to program in than the binary, since each instruction will be represented by a mnemonic which can be more easily recognized by a person than a string of binary digits. Nevertheless, programming in an assembly language is a fairly laborious procedure, since it requires the writing of an instruction for each simple operation which the machine is to perform, such as a data transfer, addition, and jumping to an instruction word. At the highest level are the user-oriented languages, such as FORTRAN and ALGOL. These languages are not translated into the machine language on a one-to-one basis. Each statement in the language may represent many statements in the machine language. These languages have been designed to be not too "machine-dependent." That is, the same program may be run on any machine available for which a translating program has been written. Thus a program written in Japan may be run on computers in the United States or Europe. These high-level languages are truly international languages, and cross the barriers between different languages and cultures as well as those between different manufacturers of computers. Of course, long and complicated programs will ordinarily contain a few statements which will be machine dependent, but changing a few statements is much less work than rewriting a whole program. Therefore, when given a choice, the computer user will generally prefer to write his programs in one of these languages. The initial work of programming is less, and the eventual possible use of the program is greater.

## B. Flow-Charting a Program

Before one sets out to actually program a problem, one must have a very clear idea of exactly what the problem is and of the steps which are necessary in solving it. No one has any doubts about the necessity of having building plans before starting the construction of a house. A plan of attack is also necessary before one starts the programming of a problem on a computer.

In emphasizing the importance of having a plan we should not become so rigid as to demand that all programs be planned in exactly the same fashion. For an experienced programmer all that may be necessary will be a few brief notes which will indicate what input data will be used, the calculations to be done with this data, and what output will be needed. In fact, for the really top-notch experienced programmer all of these notes may be carried mentally without even writing down a few notes. But the more successful a programmer is the more we may be sure that he has planned his programs carefully before starting to write them.

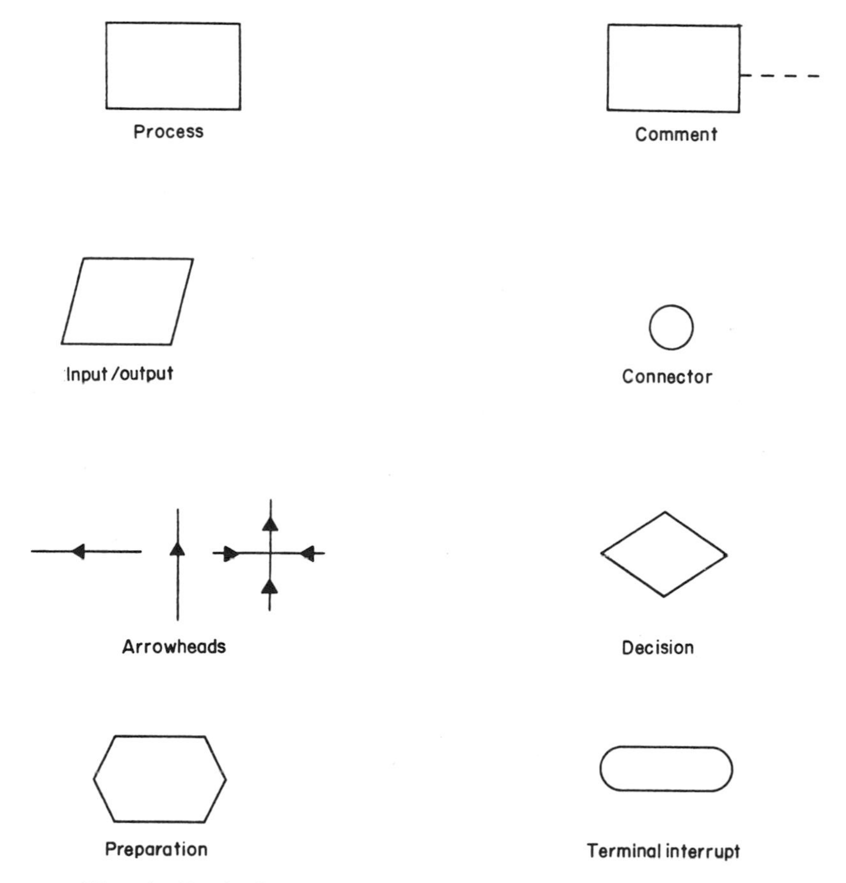


Fig. 1. Basic American National Standard flow chart symbols.

The beginning programmer may find it helpful to use the American National Standard Flow Chart Symbols. These are given for reference in Fig. 1. The rectangular box is used for a process which is carried out in the program, such as addition and transferring a number from one memory location to another. The rectangular box may also enclose a comment which is not part of the program, but which is included to explain a step in the program. A dotted line is drawn from the comment box to the appropriate box in the flow chart. Input/output functions are written in a trapezoidal box; this includes information read in from cards or tape and the processed data to be printed, punched on cards, written on tape, etc. A small circle is used to indicate a connection to another part of the flow chart in cases where it is inconvenient to draw a line. Arrowheads are used to indicate the direction of flow of the flow chart when this is not left-to-right or top-to-bottom. A diamond-shaped box is used to indicate a decision point; the program may go any of several directions from this box, depending on the value of a variable tested at this point. The rectangular box with diamond-shaped ends is used for a step which is necessary to prepare for the rest of the calculation, such as setting counters or switches. Finally, the rectangular box with rounded ends is used to indicate the beginning or end of a program or subroutine, or a halt or delay in the program, such as a halt to wait for more data to be entered.

In flow-charting a program with this set of standard symbols several conventions should be followed. First of all, flow of the chart is assumed to be from left to right and from top to bottom. In these cases no arrows need be drawn on the lines connecting boxes. Otherwise, arrows should be drawn on the lines connecting boxes to indicate the direction of program flow. A most important point is that no box should have more than one line coming into it, and except for the decision box, no more than one line going out of it. Figure 2 shows a decision box in which the variable I is being compared with IBIG. The two variables being compared are separated by a colon. There are three possible exits from the box for the cases I < IBIG, I = IBIG, and I > IBIG. For these cases the program will continue execution at the connecting points 5, 3, and 8, respectively.

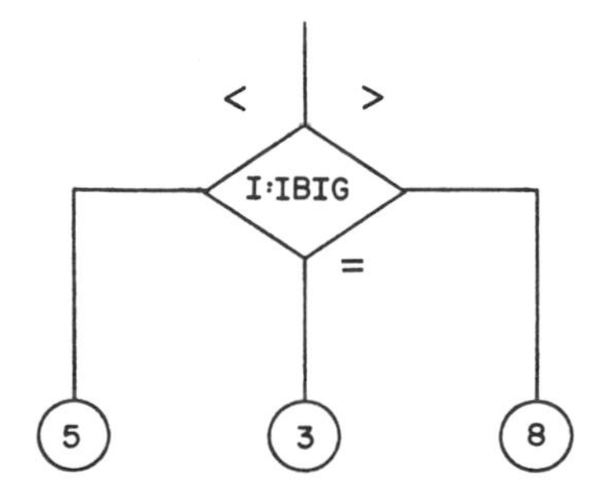


Fig. 2. An example of a decision box used in a flow chart.

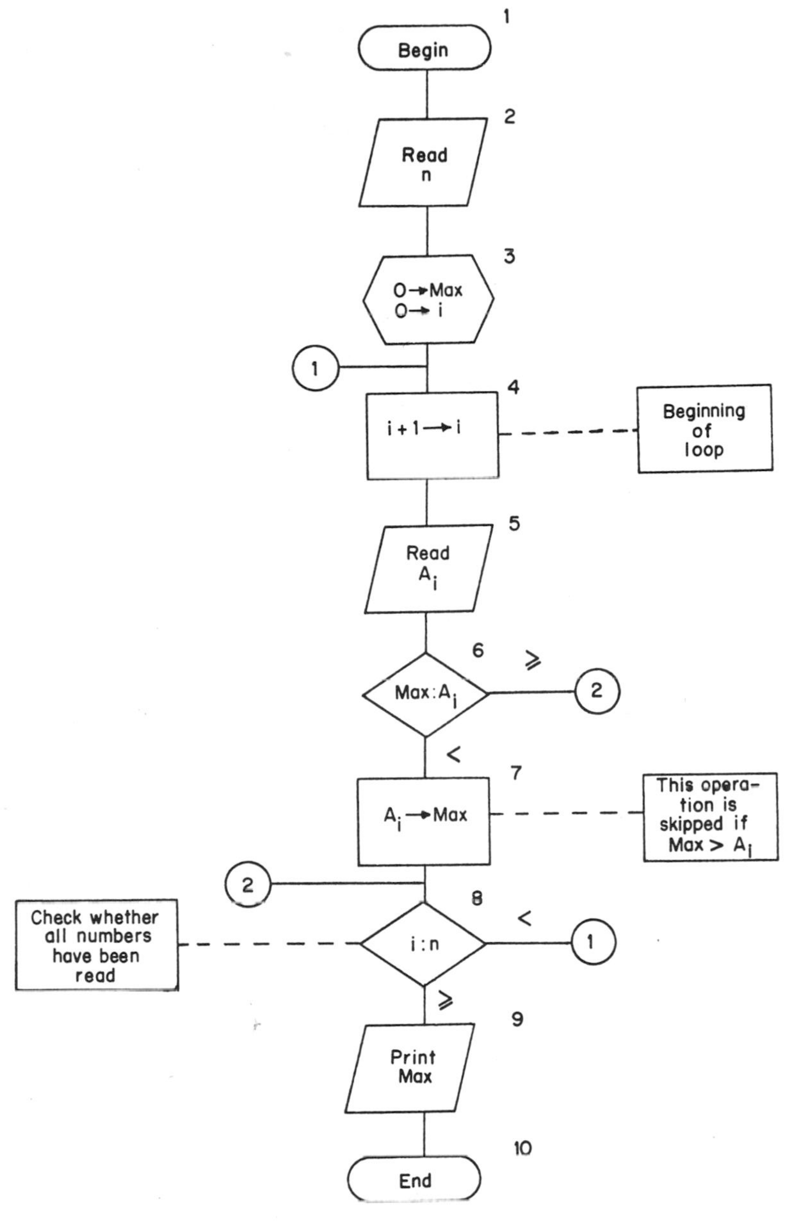


Fig. 3. An example showing the flow chart for a simple program.

Figure 3 illustrates the flow chart for a very simple problem. The problem here is to read in n positive numbers, find the largest number in the group, and then print it out. Although this is a very simple program, the flow chart illustrates the basic principles involved in flow-charting any problem. The first input is the number of values to be read in, n. The next step, in box 3, is to set the value of the maximum (called Max here) to zero, as well as the counter, i. In box 4 the counter is incremented by 1, and then the ith value is read in. Box 4 is the beginning of a loop which continues to box 8. The program will loop through these five boxes one time for each number read in. In the sixth box the number just read in is compared with the current value of Max. If $A_i$ is larger than Max the value of Max is changed to that of $A_i$, in box 7. Otherwise box 7 is skipped, and the test for whether all n numbers have been read in yet is made. If not, the program goes back to the beginning of the loop, to box 4. Otherwise it jumps out of the loop, and prints the value of Max, the largest number in the set read in.

A very detailed flow chart, such as that illustrated in Fig. 3, is most useful when the program is being written in a machine-oriented language. In preparing a flow chart for programming in a high-level language, such as FORTRAN, fewer details need to be included, and the individual steps can be stated in the language in which the program is to be written. For example, the loop from boxes four through eight in Fig. 3 could be indicated as a DO loop, and the whole loop placed in one box. In short, the important thing to remember about flow-charting is not some particular method, but simply that some plan of attack on the problem must be worked out before any program can be written.

### C. Debugging a Program

Once a program has been written the programmer is anxious to try it out on a computer; but when he gets his first results back, alas! The program has not worked. In all but the simplest of programs it is almost inevitable that errors in programming will be made, and more of the programmer's time will probably be spent in detecting and correcting these errors than were spent in writing the original program. The errors which occur in the running of a program are almost always due to the programming; in rare instances there may be an error in translating the program into machine language (the mistake of the programmer who wrote the translating program); in still rarer instances there may be an error in the electronics of the computer. Unfortunately, however, the programmer will have only himself to blame for almost all of the mistakes made by his programs. There is really no general way of debugging a program, but a few hints will be given here to help the novice programmer.

The first time a program is submitted for running it may contain typographical or syntactical errors — that is, parts of the program which do not conform to the rules of the language in which the program is written. These will usually be found by the translating program, or compiler. These are generally easy to correct, the main difficulty being in interpreting the manual which describes the language rules. Once these errors have been corrected the real work of debugging the program begins.

In order to know whether a program is doing what we want it to do we must compare the actual output of the program with what we expect the program to produce. And in order to make this comparison we must know what the output should be. It is wise then to choose a set of "test data" as input to the program which will result in simple and readily recognizable output. It will probably be a good idea to initially have the program give some output at each step in the execution, so that the step in which an error occurs may be more readily found. Particular care should be paid to the "decision points" in the program — is the program going through each loop the number of times it should, and not more or less; is it jumping from one part of the program to another when it should? It may be necessary to prepare several sets of trial data to test different sections of the program, since some parts may be skipped entirely with certain data. This will generally be better than trying to combine all possible cases in one set of data. When no more errors are found in the test cases the program should be tested with the actual data which will be processed by it. In this case the correct output is generally not known, and only by comparison with other experimental data, or by the self-consistency of several different calculations, can we be assured that the program is working correctly for the cases we are interested in. In the cases of extremely long and complicated programs there always exists the possibility that some particular combination of input data will not be correctly processed, and even on well-tested programs we should be on the lookout for this possibility.

In order to make the debugging process as painless as possible we should make use of all the debugging aids which the computer manufacturer makes available to us. The compiler ordinarily provides a "map" giving the location of the program variables in memory. We may also find useful a detailed listing of the program in the assembly language of the computer, which can be used to determine the exact location in the program at which an error has occurred which is referred to in a computer error message. A dump of all or part of the memory used by the program at the point at which the program "dies" or ceases to work as it should will also generally be useful. If the test data have been carefully chosen the values of the variables at the time of "death" and their expected values will often be most illuminating. Certain systems also have the capability of displaying the value of variables whenever they change, or the number of a statement whenever it is executed. If one uses these options for every variable and every statement one is likely to be overwhelmed with a lot of useless output, but for certain variables in a certain section of the program they make debugging a great deal simpler.

## II. MACHINE-ORIENTED LANGUAGES

When writing a program in one of the high-level languages, such as FORTRAN, it is not necessary to know exactly how the program is going to be translated by the computer in order to be able to program a problem. However the programmer who wants to utilize the computer most efficiently will need to understand something of what is happening in the computer during the execution of his program, so that the computer will not be entirely a "black box" for him. For the programmer who wishes to program a problem on one of the many mini-computers now available, such knowledge may be absolutely essential. The smaller versions of these computers, with only 4000 or so memory locations available, cannot translate the high-level languages into machine language because too many memory locations are required for the translating program, or compiler.

For the purposes of illustration we will imagine a small computer which has a 16-bit word and about 400 words in memory. We will consider only a few instructions; those which are common to all of the small computers. We will see how a few simple problems can be programmed on this imaginary computer.

On computers of this sort the hexadecimal number system, which uses 16 rather than 10 as its base, is often used in input and output. The reason

TABLE 1

Conversion Table for Binary and Hexadecimal

| Decimal | Binary | Hexadecimal |
|---|---|---|
| 0 | 0000 | 0 |
| 1 | 0001 | 1 |
| 2 | 0010 | 2 |
| 3 | 0011 | 3 |
| 4 | 0100 | 4 |
| 5 | 0101 | 5 |
| 6 | 0110 | 6 |
| 7 | 0111 | 7 |
| 8 | 1000 | 8 |
| 9 | 1001 | 9 |
| 10 | 1010 | A |
| 11 | 1011 | B |
| 12 | 1100 | C |
| 13 | 1101 | D |
| 14 | 1110 | E |
| 15 | 1111 | F |

for this is that conversion from binary to hexadecimal is very easy; four digits of binary correspond to a single hexadecimal digit. The hexadecimal is more easily read by a person than a binary; with some practice addition and subtraction in hexadecimal becomes almost as easy as the same operations in decimal. Table 1 gives the binary and hexadecimal equivalents of the decimal integers from 0 to 15. The first ten decimal digits are used for hexadecimal; for the six more digits required the first six letters of the alphabet are commonly used. To convert a number from binary to hexadecimal the first step is to separate the number into groups of four digits, starting from the right. Then each group of four is converted into a single hexadecimal digit, using the hexadecimal equivalent given in Table 1. An example of the conversion is given below:

0010 1011 0101 0111

| 0010 | 1011 | 0101 | 0111 |
|---|---|---|---|
| 2 | B | 5 | 7 |

Therefore the binary number 0010101101010111 has the hexadecimal equivalent 2B57.

Conversion from hexadecimal to decimal is somewhat more difficult, and conversion tables for this purpose are provided in the reference manuals of many computers. If such a table is not available conversion can be made in the following way:

$$\begin{aligned} 2B57 &= 2 \times (16)^3 + B \times (16)^2 + 5 \times (16)^1 + 7 \times (16)^0 \\ &= 2 \times 4096 + 11 \times 256 + 5 \times 16 + 7 \times 1 \\ &= 8196 + 3816 + 80 + 7 \\ &= 12095 \end{aligned}$$

## A. Form of Program in Machine

The program is stored in memory in the same way that numbers are stored; as a series of electrical "hots" and "colds" which may be represented as binary 1's and 0's. The only way that the computer can distinguish between the program and data is by their location in memory. If by some programming error the computer is instructed to start executing instructions at an address which contains data rather than instructions it will do just that — although the results will be far from what the programmer intended.

For the imaginary computer which we are considering we will specify a certain format for a word in memory which is to be an instruction. The 16 bits of an instruction word will be divided as follows:

| 0 | 3 4 | 5 6 | 15 |
|---|---|---|---|
| INS | X | I | MA |

INS — instruction code — first four bits.

X — indicates whether or not the contents of the index register are to be added to the memory address of the instruction word to form the final address. A 1 indicates that the index register contents are to be added, a 0 indicates that they are not to be added.

I — indicates whether the instruction addresses a memory location directly or indirectly. In a direct address, the actual location of the data to be operated on is contained in the memory word addressed. In indirect addressing, the memory word addressed contains the address of another memory word. The instruction will operate upon this indirectly addressed word. 1 indicates indirect addressing, 0 indicates direct addressing.

MA — contains the address of the data which is to be operated upon.

The INS field of the instruction contains four bits, so that we may have $2^4$ or 16 different instructions. For the simple examples we are considering we will not need even this many instructions; however most computers have a repertory of many more instructions than this, with hundreds available on very large computers. The part of the instruction word which contains the address of the data is 10 bits long, and thus we may directly address $2^{10}$ or 1024 different memory locations. If we wish to address more than this we must use indirect addressing.

We will now give a short repertoire of instruction which may be used on our fictitious computer. These instructions are typical of those found in most computers. The binary and hexadecimal code will be given for each instruction, as the hexadecimal is more readily recognized.

| Binary | Hexadecimal | Instruction |
|---|---|---|
| 0101 | 5 | The contents of the memory location addressed are transferred into the accumulator register |
| 1101 | D | The contents of the accumulator are stored in the memory location addressed |
| 1100 | C | The contents of the addressed memory location are added to the accumulator |
| 1110 | E | The contents of the addressed memory location are subtracted from the accumulator |

| Binary | Hexadecimal | Instruction |
|---|---|---|
| 0100 | 4 | Take the next instruction from the memory location addressed |
| 1001 | 9 | Compare the contents of the accumulator and the memory location addressed. If the contents of the accumulator are greater, skip the next instruction. Otherwise execute the next instruction |
| 1010 | A | Stop execution and wait |

The steps which are involved in machine execution of an instruction can be understood, in a very general way, by examining the schematic given in Fig. 4. Each cycle may be divided into two parts; the fetch and execute. At the beginning of the cycle the address of the instruction to be executed is

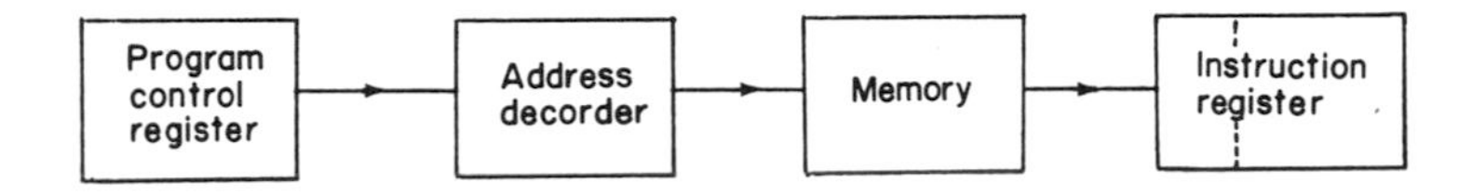


Fetch

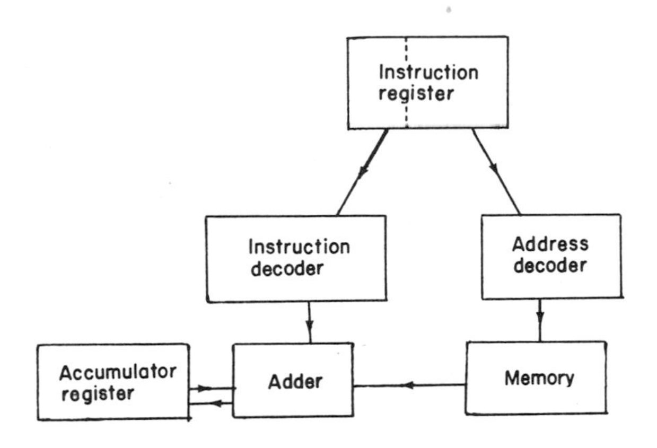


Execute

Fig. 4. Schematic of steps in one cycle of machine operation.

contained in the Program Control Register. This address is decoded by the Address Decoder which activates the necessary circuits to transfer the contents of this memory location to the Instruction Register. The dotted line across the instruction Register box indicates that the instruction word is divided into two parts — the first part giving the operation and the second part the address of the operand. In the execute part of the cycle the contents of the instruction register are decoded; the first part by the instruction decoder, which activates the necessary circuits to perform, for example, an addition. The address decoder forms the address of the operand (this may involve several steps which are not indicated in the schematic if

TABLE 2

A Machine Language Program

| Address | Contents | Comments |
|---|---|---|
| 0010 | 5022 | |
| 0011 | D023 | $0 \rightarrow$ Max |
| 0012 | 5025 | |
| 0013 | D024 | $1 \rightarrow i$ |
| 0014 | 5027 | Load $A_i$ into Accumulator |
| 0015 | 9023 | Compare $A_i$ and Max |
| 0016 | 4018 | Skipped if $A_i >$ Max |
| 0017 | D023 | Store $A_i$ in Max |
| 0018 | 5024 | |
| 0019 | C025 | |
| 001A | D024 | $i + 1 \rightarrow i$ |
| 001B | 9026 | Compare i and n |
| 001C | 401E | Skipped if $i > n$ |
| 001D | A000 | Stop — end of calculation |
| 001E | 5014 | |
| 0018 | C025 | Instruction 0014 is incremented |
| 0020 | D014 | |
| 0021 | 4014 | Jump back to beginning of loop |
| 0022 | 0000 | Zero |
| 0023 | 1FA2 | Max } May be anything initially |
| 0024 | 2A51 | i } May be anything initially |
| 0025 | 0001 | One — for incrementing |
| 0026 | 0013 | n |
| 0027 | 00AD | $A_1$ |
| 0028 | 0529 | $A_2$ |
| ⋮ | ⋮ | ⋮ |

indirect addressing or indexing are used) and activates the circuits to relay the contents of this memory position to the Adder. The results of the operation are transferred to the Accumulator Register, and the machine is ready to go to the next step. During the cycle at some point the contents of the Program Control Register have been incremented by one, so that the next instruction in sequence will be executed next. In the case of those statements involving a jump to another part of the program the PCR will be modified during the execute half-cycle to contain the location of the next instruction, wherever that is.

These basic ideas will perhaps be clearer after an example is studied. In Table 2 a machine language program (written out in hexadecimal, rather than binary, for easier reading) is given. The purpose of this program is similar to that of the program flow-charted in Fig. 3. In this example the program is in addresses $0010_{16}$ to $0021_{16}$. (In the remainder of this section 4-digit numbers will all be in hexadecimal.) The data are in addresses 0022 to 0039. No input/output operations are used in the program; it is assumed that some means of reading the program and data into and out of the memory is available.

The first instruction of the program loads a zero into the accumulator. This zero is then stored into the address reserved for Max. The instructions in addresses 0012 and 0013 set the counter i equal to 1. A loop begins with 0014 which goes to 0021, which is an instruction to jump back to 0014. In this loop the current $A_i$ is loaded into the accumulator (0014) and compared with the current maximum value (0015). If $A_i$ is greater than Max, 0016 is skipped and $A_i$ is stored in Max (0017). Otherwise no change in Max is made. The counter i is then incremented (0018-001A) and compared with the value of n (001A). If the new i is greater than n the program stops (001D). Otherwise 1 is added to the instruction 0014 — changing 5026 to 5027 in the first instance. The new instruction will load $A_2$ into the accumulator, and then $A_3$, $A_4$, etc., in each successive loop. Here an instruction is treated just like data — the computer cannot distinguish between the two.

This example shows some of the difficulties involved in writing a program in machine language. If we wanted to modify the program, say to keep a running total of the A's, we would have to insert this within the loop; in doing so we would change the numbering of some of the instructions, and therefore we would have to change also the addresses of these instructions in other instructions which referred to them. It would also be necessary to change the addresses of the data, so that we would have to change the addresses in the statements referring to data. In short, even a small change in the program would involve a great deal of work on the part of the programmer and would involve hundreds of possibilities of introducing errors. Therefore, although programming in machine language is possible, it is generally not advisable, except for making small changes in larger programs already written in another language, or for writing very short and simple routines for use with other programs. The professional programmer may have to use machine language if he is preparing routines for a machine for which no translater is as yet available, but the ordinary user generally does not have to program in machine language.

## B. Assembly Language

The programming of a problem in machine language requires the programmer to keep track of a great many mechanical details: where all of the variables are going to be stored, the addresses of any instructions referred to in another instruction. It is apparent that a computer itself should be able to be programmed to keep these details straight, and should be able to do this more easily than any human programmer. There is the additional problem of remembering what binary (or hexadecimal) digits correspond to a given instruction. A mnemonic device, such as LD representing the instruction "Load the contents of the operand address into the Accumulator Register" is much easier to remember, and fewer mistakes in writing the wrong instruction may be expected.

Assembly languages handle these details for the programmer. Variables are indicated by a name, and the computer assigns them to some storage location. Likewise instructions may be referred to by a name, rather than by their address; the assembler program will insert the proper address in each instruction in which the named instruction is referenced. If we assign the mnemonics given in Table 3 to each of the instructions, and the names given in Table 4 to the instructions and variables, the program of Table 2 will appear as in Table 5. This program is much easier to read than that given in Table 2. A more important advantage is that if we wish to make any changes in the program written in assembler language we do not have to make changes in the addresses of variables referenced in the instructions. For example, the instruction ST, MAX will refer to the address of MAX,

TABLE 3

Mnemonics for Instructions in Assembly Language

| Binary | Hexadecimal | Mnemonic |
|---|---|---|
| 0101 | 5 | LD |
| 1101 | D | ST |
| 1100 | C | ADD |
| 1110 | E | SUB |
| 0100 | 4 | JMP |
| 1001 | 9 | CMP |
| 1010 | A | STP |

TABLE 4

Assignment of Names to Memory Locations for Program in Table 2

| Hexadecimal location | Name |
|---|---|
| 0022 | ZERO |
| 0023 | MAX |
| 0024 | I |
| 0025 | ONE |
| 0026 | N |
| 0027-0039 | A |
| 001E | BILL |
| 0016 | JOE |
| 0014 | FRED |
| 0018 | TOM |

TABLE 5

The Program of Table 2, as It Would Appear in an Assembly Language

| | | | |
|---|---|---|---|
| | LD, ZERO | TOM | LD, I |
| | ST, MAX | | ADD, ONE |
| | LD, ONE | | ST, I |
| | ST, I | | CMP, N |
| FRED | LD, A | | JMP, BILL |
| JOE | CMP, MAX | | STP |
| | JMP, TOM | BILL | LD, FRED |
| | ST, MAX | | ADD, ONE |
| | | | ST, FRED |
| | | | JMP, FRED |

which will be assigned by the computer when the program is translated. Adding new variables and instructions will change the address of MAX in the machine, but the programmer does not have to be concerned with this problem.

In short, then, an assembly language has two important advantages over a machine language: (1) the use of mnemonics makes the likelihood of error in writing the instructions less likely and (2) assignment of memory locations is handled by the computer, relieving the programmer of much tedious book-keeping.

## C. Loading an Object Program

The assembler program accepts a program written in the assembler language and translates it into a binary-coded program. The program as written in assembler language is termed the "source program" and the translated program is called the "object program." The memory locations assigned in the object program are relative to the beginning of this program and not absolute addresses. A larger program may consist of several subprograms, each assembled separately. In addition certain standard routines may be supplied by the manufacturer — such as I/0 routines, error-handling routines, and standard math functions — square root, sine, etc. Before the object program may be run it is necessary that all of these various subprograms be properly concatenated, and absolute memory addresses assigned for each instruction and data reference. This operation is handled by a second program known as the "loader." The loader accepts all of the assembled subprograms (or looks them up in the computer's memory or from a tape if they are standard functions) and assigns absolute memory locations for them in such a way that no overlaps between subprograms will occur. The program is now ready to be accepted by the computer in its binary form, and to be executed. Thus, in the case of an assembly language program we find three distinct steps in going from the program as written in assembly language to the actual computations using the program: (1) assembly of the program to give an object program, (2) concatenation of the assembled subprograms to give a program with absolute addresses ready to be executed, and (3) execution of the program. In the case of user-oriented languages, such as FORTRAN, the same series of steps takes place, except that in the first step a more difficult translation must be made. The program for translating a high-level language into binary is called a "compiler." In this case the first step is therefore compilation of the source program to give an object program.

# III. USER-ORIENTED LANGUAGES

Whenever possible, most computer users will want to program their problems in one of the high-level languages, such as FORTRAN or ALGOL. The advantages of using these languages are manifold. First of all, the initial programming is greatly simplified. The languages use a vocabulary and grammar which are similar in many respects to spoken languages and the conventions of mathematics. Secondly, the languages are more or less independent of the machine on which they are to be used. Thus programs may be exchanged between different users who use machines made by different manufacturers, or the same user may run his programs on different machines. Finally, most computer manufacturers provide extensive debugging facilities for these languages, so that it is not only easier to write a program originally in one of these languages, but it is also easier to debug the program, an important consideration for a program of any complexity.

The rather short descriptions of the languages FORTRAN, ALGOL, and PL/I that follow should be read in conjunction with a reference manual for the computer on which the user wishes to run his program. Not all of the options of a given language are available on every computer, and certain special options, not general for the language, are available for some computers. The following descriptions of these languages will provide enough information for the user to decide which language is best suited to his needs, and to write simple routines in the language.

## A. FORTRAN

The FORTRAN (FORmula TRANslator) language was introduced by IBM in 1957 for its 704 computer. It was the first user-oriented language to be widely accepted, and has since been made available by almost all computer manufacturers for the larger models of their computers. FORTRAN has gone through several main versions; the most generally accepted version at present is FORTRAN IV, introduced by IBM in 1964 for the 7090 computer. FORTRAN IV has been adopted by other manufacturers with various extensions and restrictions; no attempt will be made here to discuss all of these. The "dialect" of FORTRAN discussed here will be American Standard FORTRAN; programs written in this version of FORTRAN may be run on the computers of almost all American manufacturers without change; if a program is expected to be run on different machines it is wise to use only these features of FORTRAN in writing it, and none of the extensions available for the machine on which the program is originally run and tested.

### 1. Basic Layout of Cards, Comments

The FORTRAN program is ordinarily prepared for submission to the computer by punching the program on cards. A rather strict setup of the cards is required. Each new FORTRAN statement must begin on a new card; only columns 7 through 72 may be used for the statement; if more columns are needed for the statement another card may be used which has anything in column 6; the symbol in column 6 is ignored, except for telling the compiler that the card is a continuation card. (This feature may lead to errors; if the programmer accidentally starts a statement in column 6 rather than column 7 the first symbol of the statement will be ignored and the rest interpreted as a continuation of the preceding statement.) Columns 73 through 80 may be used for numbering the cards in the deck, a useful feature, since the deck may be accidentally spilled. If the cards have been numbered a card sorter can be used to arrange them in the correct order again in a few minutes. Columns 1 through 5 may be used for a number identifying the statement.

Comments may be inserted anywhere in the program. A letter C in the first column indicates that the card is a comment and will be ignored by the compiler. It is generally a good idea to insert many comments in a program to remind oneself what is intended at each step. These comments will prove very helpful in debugging the program, and a month or a year later if one wishes to revise the program it will be much easier to see exactly what was intended at each step.

## 2. Symbols for FORTRAN

The symbols which may be used in FORTRAN may be divided into three classes: numeric, alphabetic, and special. The characters used in American Standard FORTRAN are as follows:

Numeric: 0 1 2 3 4 5 6 7 8 9
Alphabetic: A B C D E F G H I J K L M N O P Q R S T U V W X Y Z
Special: Blank = + - * / ( ) , . $

(The character $, which is a special character in American Standard FORTRAN, is defined as alphabetic in IBM FORTRAN IV.)

## 3. Constants and Variables

a. Constants. The types of constants permitted in FORTRAN and examples of each type are:

Integer constants: A signed or unsigned string of digits, with no decimal point. Examples: 25, -25738, 32.

Real constants: A signed or unsigned string of digits with a decimal point included. These may be written in two forms. The first form does not include an exponential factor, examples being 3.25, .0028954, 1059.00. The second form uses the letter E followed by a positive or negative one or two digit integer to indicate a power of ten. Examples of real constants written in the exponential format are .325E01, 2.55E-22, 35.2E11.

Double Precision constants: Where greater accuracy is required than is possible with real constants, double precision may be used. A double precision constant occupies two successive memory positions. In American Standard FORTRAN only the exponential form is permitted, with a D replacing the E. Examples: .2846949782314D03, 75.D-08, 345.D00.

Complex constants: A pair of ordered real constants enclosed in parentheses is recognized as a complex constant. Examples: (3.59, .25E-05), (25.29, 3.15).

Logical constants: Only two types of logical constants are possible: .TRUE. and .FALSE.. Note the periods before and after the symbol. These are part of the symbol and may not be omitted.

Hollerith constants: These are formed by prefixing a string of n characters with nH. These are most commonly used when it is desired to output the n characters as a message during the running of the program, to indicate whether certain steps have been performed, and to label output. Examples: 2HX=, 32HTHE ROOTS OF THE POLYNOMIAL ARE, 28H THE ITERATION HAS CONVERGED. Note that in counting the number of characters to determine n that blanks must be counted.

b. Variables. Variables are those quantities which can be changed during the course of the calculation. Variable names (identifiers) are used to designate the variables in a program. In FORTRAN the identifiers consist of 1 to 6 characters, the first of which must be a letter, and the rest may be letters or numbers. The convention is followed that identifiers beginning with the letters I, J, K, L, M, N are of type INTEGER and all other identifiers are of type REAL. The convention may be overridden by an explicit declaration of variable types of the beginning of the program. Examples:

```
REAL IMP, LEG, HAT
DOUBLE PRECISION MATRIX, CAT, ROOT
INTEGER R, S, T
COMPLEX Z, HAP, ABE
LOGICAL TEST1, TEST2, SWITCH
```

A Hollerith variable may have an identifier of any of the above types; however, the representation of a Hollerith variable in memory is very machine dependent, so that parts of a program using such variables may not be transferable from one machine to another.

## 4. Arrays

There are many cases in which it is convenient to designate a group of variables by the same name, and distinguish among them by the use of subscripts. For example, it might be convenient to indicate the coefficients of a polynomial by the identifiers A(1), A(2), A(3), etc. If the coefficients of several different polynomials were to be used, these might be designated by a doubly subscripted array, A(1, 1), A(1, 2), A(1, 3), etc., for the first polynomial, and A(2, 1), A(2, 2), etc., for the second. American Standard FORTRAN allows up to three subscripts for a variable; other versions may allow up to seven subscripts. The form which the subscripts may take is rather limited. If k and n are unsigned integer constants, and L is an integer variable name, the forms which are permitted are as follows:

```
k
L
L+k
L-k
n*L
n*L+k
n*L-k
```

The order of k, n, L in the above subscripts is important, for example k-L or L*n are not valid subscripts. It is necessary that a subscripted variable be declared as such. This may be done by a DIMENSION statement or in a type declaration statement. The size of the variable for each subscript is given in parentheses after the name of the variable. Examples:

```
DIMENSION VECTOR(100), A(10, 10)
REAL MATRIX (100, 100), JIV(3, 10, 20)
DOUBLE PRECISION A(200, 10)
```

A variable should be declared in only one place; if it is dimensioned in a type declaration statement, it should not be included again in a DIMENSION statement.

## 5. Expressions

Expressions are sequences of operands (variables and constants) and operators, which indicate the ways in which the operands are to be combined together to form new operands. There are two types of expressions in FORTRAN: arithmetic and logical (Boolean).

a. Arithmetic Expressions. There are five arithmetic operators in FORTRAN:

- \+ addition
- \- subtraction
- * multiplication
- / division
- ** exponentiation

The order in which these operations are carried out is: (1) exponentiation, (2) multiplication and division, (3) addition and subtraction. In order to carry out the operations in a different order the operands and operators may be grouped into units by parentheses — in this case the operations in the innermost parentheses are carried out first, following the same convention as in algebra. Examples:

| FORTRAN | Algebraic form |
|---|---|
| A/B*C - D**(A-B+C) | $\frac{A}{BC} - D^{(A-B+C)}$ |
| X*Y+3. *Z | $XY + 3Z$ |
| SUM+3. *(X-(Y-YMIN)**(V+W)) | $S + 3(X - (Y - Y_{min})^{(V+W)})$ |

In Table 6 the combinations of operand types which are allowed are indicated by a Y, those which are disallowed by an N.

TABLE 6

Combinations of Operand Types in FORTRAN for A op B
Y = allowed, N = not allowed

| A \ B | Operator = +, -, *, / | | | | Operator = ** | | | |
|---|---|---|---|---|---|---|---|---|
| | Integer | Real | Complex | Double precision | Integer | Real | Complex | Double precision |
| Integer | Y | N | N | N | Y | N | N | N |
| Real | N | Y | Y | Y | Y | Y | N | Y |
| Complex | N | Y | Y | N | Y | N | N | N |
| Double precision | N | Y | N | Y | Y | Y | N | N |

b. Boolean Expressions. There are three Boolean operators in FORTRAN, and six relational operators which are closely associated with the Boolean operators. The Boolean operators are .AND., .OR., .NOT.. (Note the periods at the beginning and end of the operator — these are an essential part of the operators.) The operator .NOT. changes a true value to false, and vice versa. The multiplication tables for .AND. and .OR. are given here for reference:

A .AND. B

| B \ A | T | F |
|---|---|---|
| T | T | F |
| F | F | F |

A .OR. B

| B \ A | T | F |
|---|---|---|
| T | T | T |
| F | T | F |

The relational operators are used to combine arithmetic expressions to form a logical expression. The relational operators are:

.EQ. equal to
.NE. not equal to
.GT. greater than
.GE. greater than or equal to
.LT. less than
.LE. less than or equal to

The order in which operations are performed in expressions involving logical operators is as follows:

(1) All arithmetic expressions
(2) The relational operators given above form logical expressions from pairs of arithmetic expressions

(3) .NOT.
(4) .AND.
(5) .OR.

Examples:

```
.NOT.((A-B).GT.(B+C).AND.(A.LT.C))
((A*B+C).GE.D).AND.(D.GT.A)
((SUM+X**I).LT. TEST) OR.(I.GE. ITEST)
```

In order to make the program easier to read and to remove any possible ambiguity for the programmer, it is often wise to insert parentheses even in places where the ordinary rules for ordering the operations would give the desired result. In the above examples, the meaning of the expression would not be changed by removing all the parentheses in the second and third examples.

## 6. The Assignment Statement

The assignment statement is used to assign the value of an expression to a variable. The general form of the assignment statement is V=e, where e is some expression. The meaning of the equality sign here is not the same as in algebra; it means rather that V is replaced by e. There are several rules regarding the types which V and e must belong to:

(a) If e is a logical expression, V must be a logical variable.
(b) If e is a complex expression, V must be a complex variable.
(c) If e is an arithmetic expression (integer, real, or double precision) V must be an arithmetic variable (integer, real, or double precision).

If V and e are not of the same type the combinations are handled as shown in the following:

| V | e | |
|---|---|---|
| Integer | Real<br>Double precision | Result of e rounded to next smaller integer |
| Real | Double precision | Least significant part of e is dropped |
| Real | Integer | e is converted to real form, adding zeros if necessary to least significant places |
| Double precision | Real<br>Integer | e is converted to double precision form, filling in zeros in least significant places |

Examples:

```
CAT=1.5
I=I+1
J=L*K*M-5
TEST=(A.GT.B).OR.(I .LT. IMAX)
```

7. Control Statements

With the assignment statement we are able to change the value of a variable. However, the order in which the values of variables are to be changed is just as important as the calculations themselves. In a typical FORTRAN program there are almost as many statements controlling the flow of the program as there are assignment statements. As a matter of fact, the programming of the logic of the calculation is generally the most difficult part of the job. The ordinary flow of the program is sequential, but control statements are used to change this ordinary rule and to use the results of previous calculations to control the current stage of calculation.

a. GO TO Statements. There are two forms for the GO TO statement; the first of these is the unconditional GO TO, and has the form GO TO n, where n is a statement number. When the program executes this statement it jumps to the statement indicated and continues execution from that point. Examples:

```
GO TO 25
GO TO 5004
```

The second type of GO TO is the computed GO TO and has the form GO TO ($n_1$, $n_2$, ..., $n_m$), V, where $n_1$, etc., are statement numbers and V is a nonsubscripted integer variable. Examples:

```
GO TO (25, 35, 45), I
GO TO (1225, 504, 350, 25, 10), ITEST
```

The statement to which the jump occurs is determined by the current value of the variable, v: for v = i, the jump is to statement numbered $n_i$. In the first example, if I = 2, then the jump is to statement number 35.

b. IF Statements. There are two types of IF statements in FORTRAN, the arithmetic IF and the logical IF. The form of the arithmetic IF is

IF(e) $n_1$, $n_2$, $n_3$

where e is any arithmetic expression and $n_1$, $n_2$, $n_3$ are statement numbers. The jump which is taken is determined by the rules:

| If the expression e is | the jump is to statement number |
|---|---|
| negative | $n_1$ |
| zero | $n_2$ |
| positive | $n_3$ |

Examples:

```
IF(A-B) 25, 25, 30
IF(CAT *DOG) 1505, 1510, 1515
```

The form of the logical IF is

```
IF(e) s
```

where e is a logical expression and s is a statement. If the value of the logical expression is .TRUE. the statement s is executed; otherwise execution is continued at the next statement after the IF. Examples:

```
IF(A-B .GT. TEST) A=C
IF(SUM .LE. STEST .AND. I.LT. ITEST) GO TO 25
```

c. DO Statement. It regularly occurs in a program that the same operation is to be performed several times. This may be done by setting up a counter and incrementing it each time the series of operations is performed, and at the end of the loop testing the value of the counter to determine whether to jump back to the beginning of the loop or to continue with the rest of the program. An example of a loop set up this way would be:

```
    I=1
25  A(I)=5. *B(I)+C(I)
     .
     .
     .
    B(I)=A(I)-I
    I=I+1
    IF(I .LE. N) GO TO 25
```

In this example the loop will be executed N times, with values of I = 1, 2, 3, etc. A generally more convenient way to set up a loop of this sort is the DO statement. This statement has the form

$$\text{DO n, V}=m_1,\ m_2,\ m_3$$

where n is a statement number, v is an integer variable, and $m_1$, $m_2$, $m_3$ are the integer constants or nonsubscripted integer variables. The DO statement is the beginning of the loop, and the statement number gives the last statement of the loop. v is the counter; $m_1$ is the initial value which the counter takes, $m_2$ is the final value, and $m_3$ is the increment which is added to the counter each time the loop is executed. $m_3$ may be omitted, in which case it is assumed equal to 1. The program segment shown above could be expressed with a DO loop as

```
      DO 35 I=1,N,1      (or DO 35 I=1,N)
      A(I)=5.*B(I)+C(I)
      .
      .
      .
   35 B(I)=A(I)-I
```

There are cases in which it may not be desired to execute the last statement in a DO loop on every pass through the loop; it might be desired to set B(I) equal to A(I) only when certain conditions are fulfilled. In these cases a dummy statement which does nothing but signal the end of the DO loop is used. This statement has the form

```
   n CONTINUE
```

where n is a statement number. Examples:

```
      DO 35 I=1,N
      .
      .
      .
   35 CONTINUE
```

A jump from outside a DO loop to the inside is forbidden; in case of such a jump the value of the counter is undefined. One DO loop may be nested inside another, but it must be completely inside — it may not overlap, half inside, half outside. Figure 5 illustrates correctly and incorrectly nested DO loops.

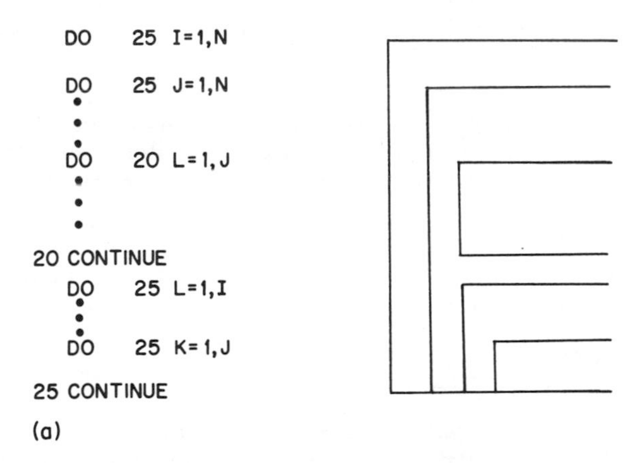


Fig. 5. (a) Correctly nested DO loops.

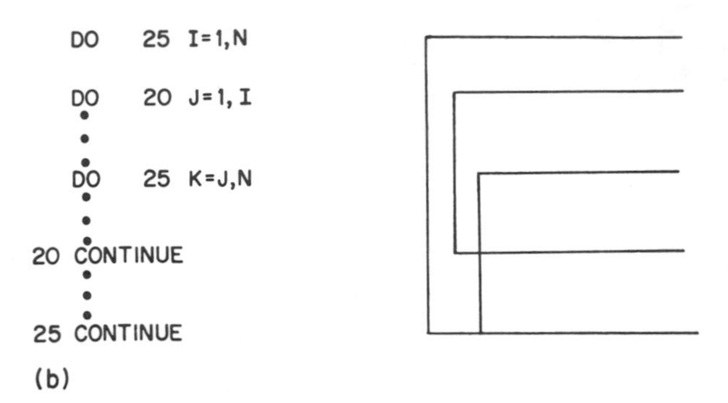
DO 25 I=1,N
DO 20 J=1,I
DO 25 K=J,N
20 CONTINUE
25 CONTINUE
(b)

Fig. 5. (b) Incorrectly nested DO loops. The range of the DO 25 K=J, N loop is partly within and partly without the range of the DO 20 J=1, I loop.

## 8. Subroutines

It is often desired to perform the same set of operations for many different sets of variables; and this set of operations may be needed in several different programs. For example it might be necessary to find the roots of a polynomial for many different polynomials which occur at different places in the main program. In order to avoid programming a complete routine for solution at each occurrence of the problem, FORTRAN provides for writing these routines once and calling them whenever there is a need for them. There are three types of subroutines in FORTRAN; the Arithmetic Statement Function, the FUNCTION subprogram, and the SUBROUTINE subprogram.

a. Arithmetic Statement Functions. The arithmetic statement function is used when the desired quantity can be calculated in a single expression. This type of subroutine has the general form

$$f(x_1, x_2, \ldots, x_n) = e$$

where f is the name of the function, $x_1, x_2, \ldots, x_n$ are the formal parameters of the function, and e is an expression involving $x_1$, etc. Examples:

```
FUNC(X, Y, Z)=X**2+Y**2+Z**2
JOE(A, B)=3. *A-2. *(B-A)
```

The arithmetic statement functions for a program are placed immediately before the first executable statement in a program. (For more details on ordering of statements in a program see Sec. III.A.10). The statement function is called by a statement in the program

$$v = e[f(a_1, a_2, \ldots, a_3)]$$

where v is a variable, e is an expression involving the function, f is the name of the function, and $a_1$, etc., are the actual parameters which are to be used in calculating the function. Examples:

```
FUNC(X, Y, Z)=X**2+Y**2+Z**2
.
.
.
SUM=SUM*FUNC(A, B, C)
.
.
.
TERM=5.*FUNC(ROOT1, ROOT2, ROOT3)
```

b. FUNCTION Subprogram. It may not be possible to calculate the desired function in a single expression. In this case the FUNCTION subprogram is used. The FUNCTION subprogram is entirely outside any other program; it is called in exactly the same way as an arithmetic statement function. The first statement of the subprogram must be

FUNCTION $f(x_1, x_2, \ldots, x_m)$

where f is the name of the subprogram, and $x_1$, $x_2$, etc., are the formal parameters. Somewhere in the subprogram statement

f=e

must occur, where f is the subprogram name and e is any expression. The last executable statement must be

RETURN

which causes the value of the function calculated to be returned to the calling program. Example:

```
   FUNCTION DOTP(X, Y, N)
   DIMENSION X(N), Y(N)
   SUM=0
   DO 15 I=1, N
15 SUM=SUM + X(I)*Y(I)
   DOTP=SUM
   RETURN
   END
```

The above subprogram would be called by a statement such as:

```
SIZE=DOTP(A, B, I)
```

(The END statement must be the last statement of any main program or subprogram.)

FORTRAN provides several built-in FUNCTION subprograms, for trigonometric functions, absolute values, square roots, exponentials, etc. A complete list of those available for a given computer can be found in the manual for the computer.

c. SUBROUTINE Subprograms. If more than one value needs to be returned to the calling program the FUNCTION subprogram is not suitable, and the SUBROUTINE subprogram must be used instead. The first statement of the subprogram has the form

SUBROUTINE s($x_1$, $x_2$, ..., $x_n$)

where s is the name of the subprogram, and $x_1$, etc., are the formal parameters. The SUBROUTINE subprogram is called by a statement

CALL s($x_1$, $x_2$, ..., $x_n$)

The SUBROUTINE subprogram must contain at least one RETURN statement. The following example is a subroutine which will multiply two matrices together, forming each element of the product matrices

$$C_{ij} = \sum_k A_{ik} B_{kj}$$

Example:

```
      SUBROUTINE MATMUL (A, B, C, N)
      DIMENSION  A(N, N), B(N, N), C(N, N)
      DO 25 I=1, N
      DO 25 J=1, N
      SUM=0
      DO 20 K=1, N
   20 SUM=SUM + A(I, K) *B(K, J)
   25 C(I, J)=SUM
      RETURN
      END
```

This subprogram could be called by the statement

CALL MATMUL(SAM, CAR, TOM, IK)

Data may be communicated to subprograms by another method besides the argument list. This second method involves reserving a section of memory for a particular set of data and referring to these same memory locations in different subprograms of the whole programs. These memory locations are reserved by a statement of the form

COMMON list

or

COMMON/Label/List

There may be only one unlabeled COMMON block in a program; there may be as many labeled COMMON blocks as are convenient for programming. Examples:

```
COMMON HI(20, 20), LO(20, 20)
COMMON/SCRATCH/AVD(500), SUMS(500)
```

In addition to using COMMON to communicate data from one program segment to another, it is also useful for conserving memory in large programs by using the same memory area for more than a single variable. Suppose that in one subroutine a 40 by 40 matrix was needed to perform calculations within the subroutine, but was needed nowhere else; while in another subroutine a vector of length 2000 was needed. The same memory area could be used for both of these by having in the one program a COMMON declaration

```
COMMON/SCRATCH/MATRIX(40, 40), DUMMY(400)
```

In the other subroutine the declaration

```
COMMON/SCRATCH/VECTOR(2000)
```

In the first declaration an extra 400 spaces are reserved for the variable DUMMY which is not used in the subroutine; this is done so that the length of the common block /SCRATCH/ will be the same in both subroutines. It is not necessary to use the same name or dimensions for the variables in different subroutines; it is necessary, however, that a given block have the same length in every subroutine in which it occurs. When the common blocks are used to transfer data it is a good idea to give the variables the same names and dimensions in every block; otherwise errors are very likely to creep in, and errors of this sort are very difficult to detect.

## 9. Input/Output

It is in the area of I/O implementation that the greatest divergences occur among the various dialects of FORTRAN. In the short synopsis given here only the I/O forms recognized in American Standard FORTRAN will be discussed, though there are dozens of extra options, some of which have been generally implemented by most computers.

The general form of an I/O statement is

```
READ(n, m) List
WRITE(n, m) List
```

where n is an integer constant or variable which specifies the number of the I/O device which is to be used, and m is an integer constant referring to a FORMAT statement which tells the computer the form in which the data are to be input or output. Examples:

```
READ(5, 1105) A, B, C
WRITE(6, 2500) NAME, DAT(I)
READ(NTAPE, 3040) RATE, TIME
```

When the data are read or written each variable is matched to a format code, which indicates the number of spaces which the variables occupies on the I/O device, the type of variable, and the location of decimal points (for

TABLE 7

Format Code Specifications

| Specification | Meaning | Example | I/O Data [a] |
|---|---|---|---|
| Iw | Integer in field of w spaces | I5 | bb325 |
| Fw.d | Real number in field of w spaces; d digits to right of decimal point | F10.4 | bbb25.3872 |
| Ew.d | Real number in field of w spaces; d digits to right of decimal point; exponential form of number, 3 must be ≥d + 8 to provide spaces for exponent, minus signs and leading zero | E18.8 | bbb-0.42874913E-12 |
| Dw.d | Same as Ew.d, but double precision numbers | D20.10 | bbb-0.6284152183D-02 |
| Gw.d | General purpose for real numbers. If number will fit in field Gw.d is equivalent to Fw.d; otherwise it is equivalent to Ew.d | G20.10 | |
| Lw | Logical variable in field of w spaces. Result is T or F right justified in field | L5 | bbbbT |
| Aw | Hollerith field of w spaces | A4 | SUM= |
| nH | The n characters following H are to be transmitted | 3H A= | bA= |
| nX | The next n spaces are to be skipped | 5X | bbbbb |
| / | Space to new record. This may be be next line of printer, or next data card | | |

[a] b indicates a blank space.

real numbers). In addition the FORMAT statement may be used to control the spacing on a printer and to provide the messages to label data. The format codes which are recognized in American Standard FORTRAN, and their meanings, are given in Table 7. Any of the format specifications may be preceded by an integer, which indicates the number of times that specification is to be repeated. Thus, 3F8.3 is equivalent to F8.3, F8.3, F8.3. A group of specifications may be enclosed in parentheses, and preceded by an integer which will indicate the number of times the whole group is to be repeated. Thus 2(I5, F8.3) is equivalent to I5, F8.3, I5, F8.3.

A convenient feature of the I/O statements is that implied DO loops may be included in the list of data items. For example, to read in the elements of a vector one could write:

```
READ(NREAD, 1105) (VEC(I), I=1, N)
```

Several DO loops may be used in one statement, for example:

```
READ(NREAD, 1105) ((MAT(I, J), J=1, N), I=1, N)
```

For reading and writing on tapes or other devices where it is not necessary that the data be formatted in any special way, unformatted I/O may be used, and is generally preferable, as it is faster. The I/O statements are the same, except that the FORMAT statement number is omitted. Examples:

```
READ(NTAPE) (A(I), I=1, N)
WRITE(NDISK) (A(I), I=1, N)
```

### 10. Ordering of Statements in Program Units

Within a given main program or subprogram a certain ordering of statements must be followed. The order in American Standard FORTRAN is

1. SUBROUTINE or FUNCTION statement, if not main program.
2. Specification statements, such as DIMENSION and COMMON.
3. Arithmetic function definitions, if any.
4. Executable statements and FORMAT statements. The FORMAT statements may be inserted anywhere in the program. They are generally placed either with the first I/O statement for which they give the format or else all together at the beginning or the end.
5. END statement. The END statement signals the compiler that all of the statements of the program have been read and compilation may begin.

## B. ALGOL

ALGOL (ALGOrithmic Language) was developed in Europe in the late 1950's. A conference held in Paris in 1960 produced a Report on the Algorithmic Language ALGOL 60. Minor revisions have been made since that time, but the language remains essentially that of the 1960 Report. One of the main concerns of those developing ALGOL was that it should not be dependent on the computer on which the program was run. Since I/O is very dependent on the machine available, no I/O statements were defined for the language. This has turned out to be one of the weaknesses of the language;

since each computer has a slightly different I/O routine for ALGOL programs, and since all programs involve some I/O, it is not possible to transfer programs from one machine to another without changes in the I/O statements. It was also intended that the language be used in publishing new algorithms developed for computer use. For this purpose I/O is unnecessary; and, in fact, the language is widely used as a publication medium.

ALGOL does not reflect the limitations of any particular machine since it is defined by the report of 1960, and not by the compiler for a particular machine (as FORTRAN originally was). The basic symbols of ALGOL consist of certain groups of letters; to distinguish these from the letters used individually they will be underlined.

## 1. Basic Symbols

The basic symbols which may be used in an ALGOL program are listed in Table 8. Both upper and lower case letters are defined for the language;

TABLE 8

The Basic Symbols of ALGOL [a]

Numeric:
0 1 2 3 4 5 6 7 8 9

Alphabetic:
A B C D E F G H I J K L M N O P Q R S T U V W X Y Z
a b c d e f g h i j k l m n o p q r s t u v w x y z

Arithmetic operators:
- \+ addition
- \- subtraction
- × multiplication
- / division
- ÷ integer division; remainder is dropped
- ↑ exponentiation

Relational operators:
- < less than
- ≤ less than or equal
- = equal
- ≠ not equal
- ≥ greater than or equal
- > greater than

Boolean operators:
- ¬ not
- ∧ and
- ∨ inclusive or
- ⊃ implies
- ≡ equivalent

Special symbols:
- . period or decimal point
- : colon
- , comma
- ; semicolon
- $_{10}$ exponent to base 10
- := assignment
- [ ] brackets
- ( ) parentheses
- ' quotes

[a]Key words (explained in text): go to if then else for step until while do begin end comment true false real integer Boolean array switch procedure own string label value

but for implementation on most computers only the upper case letters may be used. Likewise, if the 48 character set of the IBM 26 keypunch is used there must be some transliterations for the special symbols, relational operators, and Boolean operators. Some of these symbols are available on the IBM 29 keypunch, but not all. Therefore the symbols which are used in the printed descriptions of the language or in printed programs written in the language must be transliterated to be acceptable to most computers.

## 2. Form of Statements, Comments

There are no special conventions with regard to columns or length of statements which must be followed in ALGOL. Statements may begin anywhere, and more than one statement may appear on a line. The end of a statement is signaled by a semicolon. If it is desired to label a statement this may be done by placing the name identifying the statement first, then a colon, the statement, and a semicolon to signal the end of the statement.

Comments may be indicated in two different ways in ALGOL:

(a) The symbol comment may be followed by a sequence of characters up to, but not including, a semicolon, which signals the end of the comment. Any characters other than a semicolon may be used in the comment. The symbol comment may occur after a semicolon or after the symbol begin.

(b) A comment may be inserted after an end. In this case the comment is assumed to extend until an end, else or semicolon is encountered. end, else and semicolon may not be part of the comment.

## 3. Constants

ALGOL constants may be of three types: numbers, logical values, and strings. Some examples of numbers:

325
325.0 (note that an ALGOL constant may not end in a decimal point)
$3.25_{10}2 = 325$
$-35592_{10}-4 = -3.5592$

The two possible logical values are represented by the basic symbols true and false. Strings are any sequence of characters enclosed between quotation marks. Strings may be used in output for labeling results or providing messages to indicate the progress of the program.

## 4. Variable Types and Identifiers

Three variable types are recognized in ALGOL: integer, real, and Boolean. An integer variable may have a value of a positive or negative integer (up to the largest number which the computer memory can hold).

The identifiers for integer variables must be declared at the beginning of a program segment. Examples:

integer i, j, cat, rest

A real variable is represented in the computer memory as a mantissa and the power of a base. The number of digits in the mantissa depends on the computer; for some computers it is possible to specify to the compiler whether real numbers shall be represented in a single-precision or double-precision form in the computer. However, it is not possible to declare some real numbers to be double precision while leaving others single precision. Real variables are declared at the beginning of a program segment by the declaration real. Example:

real income, expenses, totals, net gain

As in FORTRAN, Boolean variables may have only the values true and false, generally represented by the binary digits 1 and 0. The declaration Boolean is used to specify Boolean variables. Example:

Boolean test, acute angle, obtuse angle, right angle

Many of the restrictions placed on choosing the names of variables in FORTRAN are removed in ALGOL. The identifiers may be of any length, and formed of any combination of letters and numbers, so long as the first character is a letter. Every identifier must be declared at the beginning of the program segment in which it is used. There are no possible implicit declarations possible, as in FORTRAN.

## 5. Arrays

Arrays in ALGOL are similar to those used in FORTRAN, but some of the restrictions on FORTRAN arrays are relaxed. The subscripts of arrays are not limited to positive integers, but may be zero or negative. Any valid ALGOL expression may be used as a subscript of an array; real expressions which are not integers are rounded to the nearest integer to determine the subscript. The array may have as many subscripts as desired. The arrays must be declared by an array statement. If the type of array is not specified it is assumed to be real. If several arrays have the same bounds and dimensions these arrays may be declared together. Examples:

integer array i, j, k(-10:10, -5:5);
Boolean array jumps(0:25);
array a, b, c(-5:25, -5:25, 0:50), d(1:15)

In the above examples the integer arrays i, j, k all have the same dimensions and bounds; the first subscript may be in the range -10 to 10, and the second subscript may be in the range -5 to 5. The dimensions of the array will therefore be 21 by 11.

The block structure of ALGOL (which will be discussed at greater length in Sec. 9) permits the bounds to be given as variables, so that the dimensions of an array may change during the course of the running of the program. This feature is not available in FORTRAN and provides another example of the greater flexibility of ALGOL.

## 6. Expressions

There are two types of expressions in ALGOL, arithmetic and Boolean. The rules for forming these expressions are virtually identical to those for FORTRAN though some of the restrictions found in FORTRAN are lifted in ALGOL.

a. Arithmetic Expressions. The arithmetic operators of ALGOL and their meanings have been given in Table 8. The hierarchy of operations is

1. ↑
2. × / ÷
3. + -

In the case of two operations having the same rank the ordering is from left to right. Thus

$$A/B \times C \quad \text{is evaluated as} \quad \frac{AC}{B}.$$

Parentheses may be used to enclose operations to be performed first. Real and integer numbers may be combined in any way. Examples:

A + B/C - 2.0 × (B + C)
B - C + D ↑ Z - 5 × A

b. Boolean Expressions. There are two Boolean operators in ALGOL not encountered in FORTRAN; these are the symbol ⊃ (implies) and ≡ (equivalent). The statement a ⊃ b is false only if a is true and b is false. The statement a ≡ b is true if a and b are both true or both false, otherwise

TABLE 9

Multiplication Tables for ⊃ and ≡

| b \ a | a ⊃ b | | a ≡ b | |
|---|---|---|---|---|
| | T | F | T | F |
| T | T | F | T | F |
| F | T | T | F | T |

it is false. The multiplication tables for these operators are given in Table 9. These operators are useful chiefly for problems in mathematical logic and would rarely be needed to control the logic of a program written for numeric calculations.

The hierarchy of operations in Boolean expressions is the same as that for FORTRAN; after all the arithmetic operations have been performed the order is:

1. $< \leq = \neq \geq >$
2. $\neg$
3. $\wedge$
4. $\vee$
5. $\supset$
6. $\equiv$

Examples:

$A - B > A \vee A - C \leq A$
atest $<$ a $\wedge$ btest $<$ b $\wedge$ ctest $<$ c
$\neg a \quad \neg b \supset \neg c$

## 7. Assignment Statement

The symbol indicating the assignment of an expression to a variable has the form

v: = e;

The same expression may be assigned to several variables in the same assignment statement:

$v_1 := v_2 := v_3 := v_4 := v_5 := v_6 = e$;

The use of the symbol := emphasizes the fact that the assignment is not the same thing as an algebraic equality. Examples:

Jim:=Joe:= 5 × A + C;
Sum:=Sum + A(X + 5, X + Y - Z);
Test 1:=Test 2:= 5 × (Bigsum - Smallsum);

## 8. Control Statements

ALGOL has the same types of control statements as FORTRAN, though several of these allow for options not available in FORTRAN. In the following discussions we will emphasize these extensions.

a. go to Statement. The forms of the unconditions go to statement is

go to l;

where l is a label for a statement. Example:

```
start: a:= a + b;
       c:= a/b + sum;
       .
       .
       .
       go to start;
```

b. if Statement. The general form of the if statement is

```
if B then u;
```

where B is any Boolean statement and u is an unconditional statement, that is, it may not contain another if clause. Examples:

```
if x > y then sum:=sum + y;
```

Another form of the if statement is possible which allows for the execution of a statement whether the condition is found to be true or false. This form is:

```
if B then u else s
```

where B is a Boolean expression, u is an unconditional statement, and s is any statement, conditional or unconditional. Examples:

```
if x > y then max:= x else max = y;
if x max > x then z:=x
              else z:=y;
```

The restriction that the first statement may not be a conditional statement may be overcome by the use of the begin and end symbols. These symbols act as the opening and closing parentheses for a group of statements. The statements between the begin and end are then treated as a single statement. Examples:

```
if x max ≥ x then begin if x = xmax then z:-w
              else z:=x end else z:=y;
if w > wmax then begin x:=y; z:=a + b end
             else begin x:-z; z:-a - b end
```

c. for Statement. The for statement provides for the iteration of a sequence of statements in much the same way as the DO loop in FORTRAN. However, the restrictions on the variables which are acceptable in the DO statement are all lifted in the for statement in ALGOL. The general form of the for statement is:

```
for v:=L do S;
```

where v is a variable (integer or real), L is a list of elements (for-list elements) which are separated by commas if there are more than one, and S is the statement to be iterated. S may be a compound statement, enclosed by begin and end, and may contain other for loops within it. The forms which may be taken by the list elements in L are:

1. e
2. $e_1$ step $e_2$ until $e_3$
3. e while B

where the e's are arithmetic expressions and B is a Boolean expression. The values of $e_2$ and $e_3$ may be altered within the statement S to affect the further repetitions of S or may be such that S is never executed. In the first form of L the list elements are simply arithmetic expressions. Some examples of this form are:

```
for x:= .5, 3.5, 5. do y:=y + x ↑ 2;
for z:= x + y, x, x - y do sum:=sum-z/y;
for i:= 1, 2, 3 do
begin x(i)=x(i) + z(i); y(i)=y(i) + a end;
```

The second form has the same effect as the DO statement of FORTRAN. The initial value of the variable is $e_1$, and this is incremented by the step value, $e_2$, on each iteration, until the value of the variable exceeds $e_3$. Example:

```
for i:= 1 step 1 until 100 do
begin
sum1:=sum2:=0;
for x=a(i)step(b(i)-a(i))/100 until b(i)do
    begin
    sum1:=sum1+ z(x)/x
    sum2:=sum2- y(x) + z(x)
    end;
end;
```

In the third form of the list, the statement continues to be repeated until the Boolean expression is false. Examples:

```
for x=a while test > 10-6 do
begin ... ; test:=(x + y)/z end;
```

d. switch Declaration. The functions in FORTRAN which are handled by the computed GO TO are performed in ALGOL by the switch declaration in conjunction with a go to statement. The switch declaration has the form

switch $S:=l_1, l_2, \ldots, l_n$;

where S is the name of the switch and the l's are labels of statements. Examples:

```
switch SW1;=15, next, start, finish;
switch last:=15, 25, D22
```

The effect of the switch is to set up an array in which the elements are the labels in the switch declaration. In the program body, control is transferred using the switch by the statement

```
go to S(i);
```

where S is the name of the switch and i is an expression. This expression will be evaluated and rounded to the nearest integer; this integer will give the subscript in the switch array determining the label of the statement to which control is to be transferred. Example:

```
        switch first:= L1, L2;
            .
            .
        go to first (a + b↑2);
            .
            .
L1:     term:= x↑2 - 3/y
            .
            .
L2:     term:= x↑2 + 5/d
```

## 9. Block Structure

An ALGOL program may be divided into segments, called blocks, which serve to limit the range of validity of a variable. A given variable is recognized only in the block in which it is declared. The general form of a block is:

```
begin D; ... D; S; ... ; S end
```

where the D's are declarations of variables and the S's are statements. In particular one of the S may be another block, so that blocks may be nested within one another. The variables used within a given block may be either local (declared within that block) or global (declared outside the block). When a block is entered the memory location for the local variables are assigned; when the block is left these memory locations are freed for use as storage for other variables, so that upon reentry to the block the values will have been destroyed. This structure provides for a dynamic allocation of memory space; the dimensions given for variables within a block may be variables which are calculated when the block is entered. Figure 6 shows an example of how the

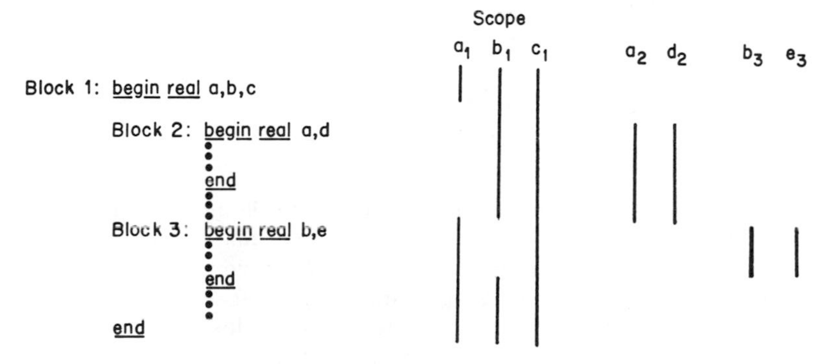


Fig. 6. Block structure of an ALGOL program.

block structure of a program might be arranged. In this figure the variable c of Block 1 is global to the entire program. In Block 2 a new variable with the same identifier, a, is declared; within this block $a_1$ is not accessible; any reference to a will refer to $a_2$. Thus, although it is not generally a good idea, the same identifier can be used for different variables declared in different blocks. The variable d is local to Block 2 and is not recognized elsewhere in the program.

## 10. Procedures

The FUNCTION and SUBROUTINE statements are very important in FORTRAN to provide for routines that are going to be used many times; the procedure statement in ALGOL serves the same functions. There are two types of procedure statements in ALGOL; the first corresponds to the FUNCTION statement in FORTRAN, and the second to the SUBROUTINE statement. The general form of the first type is

T procedure name ($x_1$, $x_2$, ..., $x_n$);

where T is the name of the type of function which will result from the procedure, and may be real, integer, or Boolean, name is the name of the procedure, and the x's are the formal parameters of the procedure. Examples:

```
real procedure term (x, y, z);
term:=x + y - z↑2;
Boolean procedure test (a, b, n);
real a, b, d; integer n;
d:= a↑n + c; test:= d ≥ b;
```

It should be noted in the second example that the variable c does not occur among the formal parameters of the procedure. In this case the variable c will be that declared in the block from which the call to the procedure is made. In fact it is not necessary to include any parameters in the procedure statement; they may all be taken from the variables declared in the block. A reference to the procedure name in the program will serve to call up the procedure. For example, the procedure term might be called by any of the following statements:

```
big:= big + term(a+b, sqrt(y), m);
inst:= term(c, cos(c+b), y-2×Z);
a(n+1):= term(term(a, x, c), a, b);
```

It should be noted that the actual parameters in the procedure call may be constants, variables, or even procedures; as a matter of fact the procedure may call itself. This great generality makes for very difficult translation of a procedure into machine language, since all of these various options must be acceptable. However in many cases it is only desired to transfer a number whose value will not change in the course of the procedure. For example a procedure which was taking a sum of several numbers might have as a parameter the number of items in the sum. This number could be transferred very simply; but in the procedures given above the compiler has no way of knowing that one or more of the parameters does not need to be considered in full generality. Considerable simplification in the translated program can be accomplished by specifying which parameters are to be transferred as numerical values. This is done by putting the symbol value after the procedure declaration, followed by the names of the variables which are to be transferred as numbers. Example:

```
real procedure fact (n); value n; integer n, i; real term; term = 1;
for i ; 1step1 until n do
term:= term × n; fact:= term;
```

The translation of a program into machine language can also be considerably simplified if certain of the variables are known to always be of a certain type. This information may be given to the compiler by including specification statements after the value declaration (if it is used). The types which may be specified are: label, switch, string, real, integer, Boolean, procedure, real procedure, Boolean procedure, array, real array, integer array, Boolean array. Omission of the specification statements can lead to a very much longer program than necessary, which has a great deal of generality which is never used. Some ALGOL compilers require that all parameters be specified. Example:

```
real procedure sum (a, b, x, n); value n; real array a, b; real x;
```

The second type of ALGOL procedure, corresponding to the FORTRAN SUBROUTINE, is declared by the statement

procedure Name ($x_1$, $x_2$, ..., $x_n$)

where Name is the name of the procedure, and the x's are the formal parameters. All of the rules regarding function-type procedures hold equally for subroutine-type procedures. These procedures are called by a statement consisting of the name of the procedure and its actual parameters. Example:

```
procedure matmul (a, b, c, n)
value n; real array a, b; integer i, j, k,
real term;
for i:= 1 step 1 until n do
begin for j:= 1 step 1 until n do
           begin term:= 0
           for k:= 1 step 1 until n do
           term:= term + a(i, k) × b(k, j);
           c(i, j) = term
           end;
end;
```

11. Input/Output

As previously mentioned, the input/output forms for ALGOL were not specified in the report of 1960, so that each manufacturer was free to choose a different I/O system. However, the I/O functions are generally specified by procedures, and data transfer is accomplished by calling the proper procedure in the usual way, and using the names of the variables to be transferred as the actual parameters for the procedure. The form of the call for these procedures are generally similar to:

```
inreal(n, x);
outreal(n, x);
```

where n is the number designating the I/O device and x is the variable being transferred. For the procedures inreal and outreal x would be of type real; other procedures would be supplied for variables of other types.

## C. PL/I

PL/I (Programming Language I) was developed by IBM for use on its 360 series of computers. This language contains many of the features of the languages FORTRAN, ALGOL, and COBOL (Common Business Oriented Language). The language is intended to be used for any of the purposes for which the three older languages are used, as well as including extensions of these languages. The language has been designed so that a programmer need learn only as much of it as necessary for his application. The instruction set is much larger than those of FORTRAN or ALGOL. We will consider those features of PL/I which are most useful in programming scientific problems. The basis for this discussion is the PL/I (F) Language Reference Manual, IBM File No. GC28-8201-3.

1. Character Set

PL/I makes use of the 60 characters of the IBM 29 keypunch. The characters of this set and their use in PL/I are given in Table 10. It should be noted that the dollar sign ($), the number sign (#), and the commercial "at" sign (@) are considered to be alphabetic characters and to precede the 26 letters of the English alphabet. The total set is therefore composed of 29 alphabetic characters, 10 numeric characters, and 21 special characters.

2. Program Format and Comments

PL/I programs are written in free format, just as ALGOL programs. Semicolons are used to separate statements from one another, and colons to separate labels from their statements. The program is arranged with respect to indenting, number of statements per line, etc., to make it as readable as

TABLE 10

Character Set of PL/I

1. Alphabetic

   $ # @ A B C D E F G H I J K L M N O P Q R S T U V W X Y Z

2. Numeric

   0 1 2 3 4 5 6 7 8 9

3. Special

| | |
|---|---|
| | Blank — separates elements of statement |
| = | Equal sign or assignment symbol |
| + | Plus sign |
| - | Minus sign |
| * | Asterisk or multiply symbol |
| / | Slash or divide symbol |
| ( | Left parenthesis |
| ) | Right parenthesis |
| , | Comma — separates elements of a list |
| . | Decimal or binary point — period |
| ' | Single quote or apostrophe — encloses string constants |
| % | Percent symbol |
| ; | Semicolon — terminates statements |
| : | Colon — separates label from statement to which it refers |
| ¬ | Not symbol |
| & | Ampersand — and symbol |
| l | Or symbol |
| > | Greater than symbol |
| < | Less than symbol |
| _ | Break character — underline on typewriter; can be used to improve readability |
| ? | Question mark |

possible. Comments may occur anywhere in a statement that blanks are permitted, that is, around identifiers or operators, but not inside identifiers or compound operators. The general form of a comment is:

```
/*COMMENT */
```

The comment may contain any characters except */ which would be interpreted as the end of the comment. The composite symbols /* and */ must not contain blanks.

## 3. Constant and Variable Types

There are three types of data which we will consider: arithmetic, string, and label data. Variables used to represent data may have identifiers of 1 to 31 characters; the first must be a letter, the others letters or numbers. Variables may be declared to be one of these types, or may assume a type by the default option, as in FORTRAN. The general form of the statement used to declare a variable type is

DECLARE name attribute

where name is the name of the variable, and the attributes which the variable is to have are placed after the name. If several variables are to have the same attributes they may be declared together:

DECLARE ($name_1$, $name_2$, ...) attributes;

a. Arithmetic Data. Data which are integers or in which the decimal point will always be at the same digit (for example dollars and cents for business calculations) are declared as fixed point variables. The declaration for these have the form

DECLARE name FIXED DECIMAL (m, n)
and
DECLARE name FIXED BINARY (m, n)

where m is the maximum number of digits in the number and n is the number of digits to the right of the point. The fixed point numbers may be either decimal or binary. Examples:

DECLARE COST FIXED DECIMAL (6, 2)
This declaration would handle variables ranging in size from 0000.00 to 9999.99 in steps of .01.
DECLARE TEST FIXED BINARY (6, 6)
This declaration would handle variables ranging in size from .000000 to .111111 in steps of .000001.

For the case of integers the declaration may be shortened to

DECLARE name FIXED BINARY
or
DECLARE name FIXED DECIMAL

The number of digits in the integer is determined by the particular computer on which the program is run. Examples:

DECLARE (ITEST, J, RST) FIXED BINARY;
DECLARE ABS FIXED DECIMAL;

Variables which are not declared and begin with I, J, K, L, M, N are declared by default to be of type FIXED BINARY (n, 0) where n is the maximum number of digits which the particular computer can hold in a memory location.

If a larger range of values for the variable is needed than can be provided by a fixed point number, floating point numbers are used. These are declared by the statements:

```
DECLARE name FLOAT DECIMAL (m)
DECLARE name FLOAT BINARY (m)
```

where m is the number of digits (decimal or binary) which should be carried by the variable. Variables beginning with $#@ A-H, O-Z are assumed to be of type FLOAT DECIMAL (m), where m is typically 6 to 8, depending on the computer. It is not necessary to specify both the type of variable (FIXED or FLOAT) and the base (DECIMAL or BINARY). If only the type is specified, the base is assumed to be DEDIMAL; if only the base is specified, the type is assumed to be FLOAT.

If it is desired to have a variable which is complex this may be declared along with the other attributes of the variable. Examples:

```
DECLARE CAB FLOAT DECIMAL COMPLEX;
DECLARE ITEM FIXED BINARY COMPLEX;
```

Both the real and imaginary part of the complex number will have the attributes specified or assigned by default in the DECLARE statement.

b. String Data. Character string constants are sequences enclosed between apostrophes. Examples:

```
'THE ROOTS OF THE EQUATION ARE:'
'LOGARITHM'
```

Variables may be declared to be character strings by the declaration:

```
DECLARE name CHARACTER (m);
```

where name is the name of the variable and m is the maximum number of characters in the variable. Examples:

```
DECLARE HEADING CHARACTER (25);
DECLARE ADDRESS CHARACTER (10);
```

Strings of binary digits may also be declared; the form of the declaration here is

```
DECLARE name BIT (m);
```

where m is the number of binary digits in the string. Bit strings of length 1 are used to represent logical values; these have only the two values 1 and 0 corresponding to the logical true and false. Examples:

```
DECLARE TEST (1);
DECLARE BITSTRING (25);
```

Bit constants are represented by the bit string enclosed in apostrophes and followed immediately by B (no intervening blank). Examples:

'11001011'B
'1'B equivalent to logical true
'0'B equivalent to logical false

c. Label Data. This type of variable is similar to that used in the switch declaration in ALGOL. When a variable is declared to be a label then the values which it may take are labels which identify statements in the program. Example:

DECLARE NEXT LABEL;

After this declaration the variable NEXT could be assigned the values of labels in the program by a statement such as

NEXT=START or NEXT=LAB2

The statement for a transfer could then be written as GO TO NEXT;.

4. Arrays

Arrays are declared in PL/I in much the same way as in ALGOL. Arrays may have any number of dimensions, and the subscripts may be in any range. Arrays are indicated by giving the dimensions in parentheses after the variable name in a DECLARE statement. Examples:

DECLARE A(-5:5, -10:10);
DECLARE B(5, 5, 5, 5);

If the lower bound for a subscript is omitted the value is assumed to be 1. The rest of the DECLARE statement is the same as for an unsubscripted variable.

5. Expressions

The operational expressions of PL/I are virtually identical to those of ALGOL. Likewise the hierarchy of operations is the same. There are compound relational operators formed from pairs of basic characters which are not encountered in the same form in the other languages. The arithmetic operators are (in order of priority):

1. ** (exponentiation)
2. */ (multiplication, division)
3. +- (addition, subtraction)

The next priority is given to the relational operators:

4. < (less than)
   ¬< (not less than)
   <= (less than or equal to)
   = (equal to)
   ¬= (not equal to)
   >= (greater than or equal to)
   > (greater than)
   ¬> (not greater than)
5. & (and)
6. | (or)

## 6. The Assignment Statement

In PL/I the equality sign is used in two different meanings; the first is that discussed above in the relational operator, in which it has the ordinary algebraic meaning. The second meaning, which it has in the assignment statement, is the same as the meaning of the equality sign in FORTRAN. Unlike ALGOL, PL/I does not have a special symbol for assignment. Examples:

```
X = A + B;
B = A(I) - N/2;
TEST - A>B & AMAX>A;
LAST = 2*N + 1;
```

The assignment statement may be used with arrays; in the first example above, if A, B, and X were arrays with the same specifications and dimensions the assignment statement would take each element of A, add it to the corresponding element of B, and store the result in the corresponding element of X.

## 7. Control Statements

a. GO TO Statement. The point to which control is to be transferred may be specified by a label or a label variable. Label variables may be subscripted, so that quite a variety of switching procedures may be accomplished by the GO TO statement. Examples:

```
GO TO L3;
GO TO SWITCH (I);
```

b. IF Statement. The form of the IF statement in PL/I is the same as in ALGOL. The general form is:

IF B THEN $s_1$ ; ELSE $s_2$;

where B is a logical expression and $s_1$ and $s_2$ are statements. The ELSE clause may be omitted; both $s_1$ and $s_2$ may be compound statements and may contain further conditional clauses. Examples:

```
IF A >= B THEN GO TO NEXT;
IF NT > T
THEN A = B + 2*C;
ELSE A = B - 2*C;
```

c. DO Statement. The DO statement in PL/I has many features of the DO statement in FORTRAN and the for statement in ALGOL but is somewhat more flexible. A DO group in PL/I starts with a DO statement and ends with an END statement. The general form for a group of statements is then:

```
DO;
.
.
.
END;
```

In this form the statements contained between the DO and END are not to be executed several times; the DO and END serve as opening and closing parentheses for the group, in the same way as the begin and end in ALGOL. The group of statements following the THEN or ELSE of a conditional are enclosed by a DO group in this way. Example:

```
IF IT ≤ N
THEN DO;
IT = IT + 1;
.
.
.
END;
ELSE DO;
IT = IT - 1;
.
.
.
END;
```

The DO and END may also be used to enclose a group of statements that are to be executed repeatedly. In this context the form of the DO is

DO WHILE (B);
DO v = $e_1$ TO $e_2$ BY $e_3$;
DO v = $e_1$ TO $e_2$ BY $e_3$ WHILE (B);

where v is a variable, the e's are arithmetic expressions, and B is a logical expression. The values of $e_1$ and $e_2$ may be such that the group of statements is never executed. If $e_3$ is not specified it is assumed to be +1. Examples:

```
DO WHILE (A < ATEST);
.
.
.
END;
```

```
DO A = B-C TO B+C BY C/100;
.
.
.
END;
DO I = 1 to 10, 20 to 30;
.
.
.
END;
DO I = 1 to 10 WHILE (A > B);
.
.
.
DO I = 1 to 10, 11 WHILE (A > B)
.
.
.
END
```

In the first example the group of statements will be executed repetitively until the value of A is no longer less than that of ATEST. In the second example the value of A will initially be set to B-C and then incremented on each iteration by C/100 until A is greater than B+C. In the third example I will initially be equal to 1 and will be incremented by 1 until it is equal to 10; then it will be increased to 20 and incremented again by one on each pass through the group until it is equal to 30. In the fourth example I will initially be set to 1 and incremented by 1 until it is equal to 10; however if A is greater than B at the beginning of the loop on any pass, control will pass out of the loop. In the last example I will again be set to 1 initially and will then be incremented unconditionally until it is 10; after this the loop will be executed repetitively with i equal to 11 until A is found to be less than or equal to B.

## 8. Block Structure

A block of statements begins with the BEGIN statement and ends with the END statement. A label may precede the BEGIN and follow the END to make it more apparent to anyone reading the program when blocks begin and end. However, the labels are not mandatory. Example:

```
BLOCK 1: BEGIN;
         .
         .
         .
         END BLOCK 1;
```

Storage is allocated for variables which are declared in blocks in the same way as in ALGOL; the rules governing the allocation and freeing of storage locations are essentially the same. However, PL/I has the possibility of saving the allocation for a variable after the block in which it was allocated has been left. This is done by specifying that, in a DECLARE statement, the variable is to have the CONTROLLED attribute, meaning that allocation for this variable is controlled. When storage for this variable must be allocated this is done with the statement ALLOCATE name, where name is the name of the variable. Once storage has been allocated it will not be freed for other use until the statement FREE name is encountered. In this way more control over the dynamic allocation of storage space is possible in PL/I than in ALGOL.

## 9. Procedures

As in ALGOL, there are two types of procedures available in PL/I: function procedures and subroutine procedures. Function procedures return the value of a function and are called by referring to the name of the function procedure in an expression. Subroutine procedures may calculate many different quantities, and they must be specifically called by the program in a separate statement.

a. Function Procedures. Function subroutines might best be illustrated by a sample procedure. The following procedure will calculate the length of a three-dimensional vector with components x, y, z. Example:

```
LENGTH: PROCEDURE (X, Y, Z) FLOAT;
/* THIS PROCEDURE RETURNS THE LENGTH OF A VECTOR WITH
    COMPONENTS X, Y, Z */
DECLARE (X, Y, Z) FLOAT;
RETURN (SQRT (X**2 + Y**2 + Z**2));
END LENGTH;
```

The name of the procedure is LENGTH, and FLOAT specifies that the value of the quantity to be returned by this procedure will be a floating-point number. The expression following the RETURN is the value which will be returned to the calling program. There could have been many statements preceding the RETURN so that the calculation could have been quite complicated, but only one value, which is placed in the RETURN statement, can be returned to the calling program. It is necessary that the PROCEDURE statement be labeled, since the label is used to call the procedure. The above procedure might be called by a statement

```
NORM = LENGTH (A(I), B(I), C(I));
```

b. Subroutine Procedures. Subroutine procedures have the same forms as function procedures; however, no value to be associated with the name of the procedure is placed with the RETURN statement. A subroutine procedure is called by the statement

CALL name ($x_1$, $x_2$, ..., $x_n$);

where name is the name of the procedure, and the x's are the actual parameters of the procedure.

Procedures in PL/I may call themselves, just as in ALGOL; multiple entries to a procedure are also possible. Variables in procedures may be either local or may be referenced in other procedures in the entire program. The main program itself is a procedure, and each main program must be started with the statement

```
name: PROCEDURE OPTIONS(MAIN)
```

where main is the name of the main program.

## 10. Input/Output

PL/I is intended to deal with a great many different types of computing problems, from those encountered in research laboratories to those of inventory checks for a business. The I/O functions needed for these different types of applications vary greatly, and we shall discuss here only those options which will be of most interest to those involved in scientific programming problems.

The statements used for input and output in PL/I are GET and PUT. A collection of data on an external device is called a file. If only the standard I/O devices are used, card reader for input and printer for output, it is not necessary to specify the file number, so that very simple I/O statements are possible, for example,

```
GET LIST (A, B, C);
PUT LIST (A, B, C);
```

would suffice to read the values of A, B, C from cards or print them on a printer. Three types of representation of the data on the I/O files may be used. Table 11 outlines these and shows examples of each.

TABLE 11

Data Representation in a File

| | Form of data constants | Items delimited by |
|---|---|---|
| I. LIST directed | 3, 5, 1, 5, 4298.71<br>2.9, 1.3, .9528 | Comma or blank(s) |
| II. DATA directed | Variable = constant<br>A = 3.5, B = 1.5, C = 4298.71;<br>L = 2.9, G = 1.3, M = .9528; | Comma or blank(s) and semicolon after last item in list |
| III. EDIT directed | String of characters<br>0.325E 02 ¦ -0.29E-05 ¦ 2.5 ¦ 05 | Format specifications<br>E 9.2, E 10.3, E 9.2, F 3.1, I 2 |

In LIST-directed I/O the simple form of GET and PUT statement previously given may be used. The variables may be input on cards without any formatting; the values may be separated by commas or by one or more blanks. On output the value will be printed in a format which is suitable for the type of variable.

DATA directed I/O use the statements

```
GET DATA;
GET DATA (list)
PUT DATA (list)
```

On the data card for input each data item is associated with the proper variable; for example a data card for a program requiring three variables called X, Y, Z might be

```
X=1.2935, Z=-8.99274, Y=1.5E*09;
```

Note that the ordering of the data on the card is not important; although the variables might be requested in the order X, Y, Z the values will still be correctly transmitted by the above card. For input it is not necessary to specify the list; new data items will continue to be read until a semicolon is encountered signaling the end of that data set. The next GET DATA; will activate the transmission again, and transmission of the next set of data will commence, continuing until another semicolon is encountered. In output the name of the variable will be printed, followed by an equals sign and the value of the variable in suitable format.

EDIT-directed I/O statements have the general form

```
GET EDIT (list) (formats)
PUT EDIT (list) (formats)
```

The format specifiers are most useful on output when it is desired to place the results on the printed page in a certain fashion to improve readability, etc. On input it is much easier, and much less likely to lead to errors, to use either LIST or DATA directed input.

The format specifiers are the same as those used in FORTRAN (see Table 7). The specifiers are placed with the parentheses after the list of data and do not have to be written in a separate FORMAT statement, as is required in FORTRAN. However, if the same specifiers are to be used many times in a program it is more convenient to write them only once and then to refer to them each time they are needed. This may be done by writing a FORMAT statement, just as in FORTRAN, and then referencing it by the specification R(label). Example:

```
F2:FORMAT (6E20.6);
PUT EDIT (A, B, C, X, Y, Z) (R(F2))
```

## D. Some Comparisons of FORTRAN, ALGOL, PL/I

Probably the greatest advantage of FORTRAN is its general availability on almost all computers of any size. Programs written in FORTRAN, if the programmer is careful to use only the options of American Standard FORTRAN, can be run at present with almost no changes on a wide variety of machines. Although ALGOL has many advantages over FORTRAN it has never been as popular as FORTRAN; difficulties with I/O routines prevent general exchange of ALGOL programs without considerable changes. It has been widely accepted as a medium for communicating algorithms between humans, but as a medium for communicating with machines it has not been as successful as its developers had hoped. At present PL/I is not as generally available as FORTRAN; but it is being pushed by IBM which will insure its availability at a large number of computer installations in the United States, if not at a majority. The format-free I/O is a great improvement over the formatted I/O required for FORTRAN IV. In addition it has the general structure of ALGOL, rather than FORTRAN, so that it is more suitable for many scientific calculations than FORTRAN. Shorter programs which require no sophisticated features, such as dynamic allocation of storage, can be equally well programmed in FORTRAN or PL/I; the programmer familiar with FORTRAN will probably not want to bother with another language. The features of PL/I serve to make the programming of complicated problems somewhat similar. In addition PL/I has other features, not discussed here, which are possessed by neither FORTRAN nor ALGOL. For example, its multitasking capability makes it possible to execute two parts of a program simultaneously, rather than sequentially. These features make PL/I very attractive, and in a limited number of cases a necessary choice, for a programming language.

## BIBLIOGRAPHY

The references listed below are all of a very general nature, and not specific for any single computer. They have been chosen on the basis of being good general introductions to the specific area of programming which they discuss. All of them should be read in conjunction with the manuals for the specific computer which is being used.

Bates, F. and Douglas, M. L. Programming Language One, Prentice-Hall, Englewood Cliffs, N.J., 1967.

Blatt, J. M. Introduction to FORTRAN IV Programming. Goodyear Publishing Co., Pacific Palisades, Cal., 1971.

McCracken, D. D. A Guide to ALGOL Programming, Wiley, New York, 1962.

Scheid, F., Introduction to Computer Science, Schaum Outline Series, McGraw-Hill, New York, 1970.

Wirth, N. "An Introduction to FORTRAN and ALGOL Programming" in Mathematical Methods for Digital Computers (A. Ralston and H. Wilf, eds.), Vol. II, Wiley, New York, 1967.

Chapter 5

# SIMULATION TECHNIQUES

Vincent A. LoDato

The RAND Corporation
Santa Monica, California 90496



## I. INTRODUCTION

In the last twenty years computers have been used extensively in the simulation of various processes. The process might be physical, such as eddy diffusion of gases from a smoke stack [1], or it might be a process which occurs in the business environment, as in the control of bank statements. Myriads of other examples can be cited that are found in physical, biological, and social sciences. Another novel example is when a computer is used to simulate the behavior of another machine.

In all these processes one tries to duplicate the process in question. The tool which the originator uses is the computer and the method of communication is the language; the tool can be mathematical analysis when the process is deterministic, probability and statistical analysis when the process is stochastic, or a series of statements that call for a logical conclusion.

Most processes are often very complicated. In many typical situations one can describe individual segments of the process but the entire process does not have an analytic solution. Nevertheless, with the aid of a computer, one can in some cases simulate the process.

The simulation techniques can be compared to props that the analyst and/or programmer use. If one has good "props" and ingenuity, the phenomena can be simulated or acted out with a fair degree of accuracy.

An important characteristic of most simulation models is that they are partly determined by chance. When noxious gases come from smoke stacks, the concentrations are only probabilities of what will occur. Hence one of the most important features in simulation is the generation of random numbers. The use of random numbers coupled with initial conditions and constraints along with logic commands and numerical analysis are found in most simulators. In many cases the programing of the process for a digital computer has been aided considerably by the development of special purpose programs, assemblers, interpreters, and high level languages. The basic purpose of all these language and software packages is to provide a means of communicating a problem to the machine without doing so in machine language. These packages take care of many of the mechanical details of programing, such as memory location assignment and basic operation (e.g., addition input-output commands and communication among program segments). The languages help the analyst and the programmer communicate with each other.

It is important at the outset that we define some terms. The term "language" is used here to mean the specification of the language without regard to implementation; the term "software package" or "package" means the set of material available to the user. The usefulness of languages and software packages has been great, and their number has increased greatly. For a given process the user must choose from among several competing packages, each of which offers several features that may be suitable to his particular problem. The same is true for a given language. For example we shall consider the operation of a supermarket. Before a customer begins shopping at the supermarket he is obliged to take a shopping cart. There are a limited number of carts. If the customer arrives and is not able to find a cart, he leaves without shopping. If he gets a cart he does his shopping and checks out at one of the four counters, returns the cart, and leaves the supermarket. Since most scientists are familiar with the FORTRAN language one would attempt to develop an algorithm that simulates this process and use the FORTRAN language to communicate with the computer. In the algorithm one may use a random number generator to generate the probabilities of a certain occurrence, say finding a cart. All in all this may develop into what may be considered time consuming and sizable work, since some commands and software packages (random number generator) are not built into the language. However instead of using FORTRAN one could use the language GPSS (General Purpose Simulation System). One could simulate the process using in the neighborhood of twenty machine instructions in this language. Hence the choice of a language is an extremely important part in the technique of simulation. However, one may argue that most physical scientists do not worry about supermarket simulation and that FORTRAN is the panacea for a scientific simulation. However this is not true. In many instances, using a combination of languages is the most optimal in large scale simulation. In the development of a transient circuit analysis program may be coupled with an input language and FORTRAN to produce an optimal code.

In most approaches to physical simulation, one usually encounters six steps: (1) develop a mathematical model based on data and experience; (2) represent the model in terms of suitable analog, digital, or hybrid computers; (3) run a computer simulation for a number of situations where the real system behavior is known; (4) adjust the model parameters and structures until it acts like the real system; (5) execute the simulation to predict the system behavior in new systems; (6) from the output results develop a closed feedback loop and let the new output be input until the desired results are obtained. These six steps constitute what is termed "simulation techniques."

In this chapter we shall discuss the concepts of discrete and continuous simulation. Then we shall compare a number of simulation languages and elucidate on their merit and shortcomings. This is followed by three examples: (1) a circuit analysis simulator, (2) the BIOMOD simulator, which is an interactive computer graphic system for modeling, and (3) an example of a discrete simulator.

## II. DISCRETE TIME EVENTS

Changes that are discontinuous in character are characterized by discrete systems. The concept of the term event in such systems describes the occurrence of a change at a given time. The event causes a change in the value of some attribute of the system. It may stop or start an activity or create or destroy an entity of the system. Thus the task of discrete system simulation is the temporal evolution of a sequence of events.

The evolution of time is recorded by a number referred to as clock time. At the beginning of the simulation the clock is set to zero and allowed to run. Thus we have the term simulation time, which indicates the clock time and not the time that a computer has to carry out the simulation. In discrete simulation the clock is advanced and then the next event occurs, and so on. Thus a discrete sequence of events occur and the clock is forwarded. This method is called event oriented and is usually characterized by discrete simulation.

Another aspect of this type of simulation is the generation of exogenous arrivals, i.e., the event and the arrival of time of the next entity is recorded as one of the event times. When the clock time approaches this event time, the event of entering the entity into the system is executed and the arrival time of the following entity is immediately calculated from the interarrival time distribution. Thus we have a process in which one entity creates another, and the process is projected in time via a boot-strapping method.

There exists a number of programing languages which are suitable to discrete simulation. All twenty-two discrete simulation languages can be divided into two categories, particle oriented or event oriented. In a particle-oriented language the emphasis is focused on the entities in the system, and the simulation is regarded as the task of following the changes that occur as the entities move from activity. In an event-oriented language attention is focused on the activities and the times at which system changes occur are treated as attributes of the entities, and the simulation follows the history of the activities as they are applied to different entities. In an event-oriented language the times at which system changes occur are treated as characteristics of the activities. An example of a particle-oriented language is GPSS (General Purpose Simulation Systems) and an event-oriented language is SIMSCRIPT.

Thus, a discrete change model is characterized by three points: (1) the system contains components each of which performs definite and prescribed functions; (2) the items flow through the system from one component to another, requiring the performance of a function at the component before the item can be moved to the next component; and (3) the components have finite capacity to process the items and, consequently, items may have to wait in queues or waiting lines before reaching a particular component.

The computations in this type of simulation are basically temporal inventory and moving items from a queue to components and timing the necessary processes or functional transformation, and the result of a simulation run is a set of statistics describing the behavior of the simulated system during the run.

## III. CONTINUOUS TIME EVENTS

If the information or material flow is continuous, these models can be represented by differential or difference equations, which describe the rates of change of the given processes. Models of this sort are found in abundance in chemistry, physics, biology, and engineering. Such models have also been applied to economics and the social sciences.

In some cases it is not possible to have analytical or numerical techniques to solve continuous changes in systems. At this point the analyst may use mechanical or electronic analog computations. Unfortunately analog machines cannot be used because the system has too many variables, and some of the variables may be random while others may require difficult operations which are not suitable to analog machines; also some of the functions may be discontinuous.

However with a digital machine it is possible to use finite difference equations to represent differential equations. These finite difference equations approach the differential equation in a limit which is dictated by stability and error conditions.

If X (t) is a state vector which describes the system at some time, then the representation at some later time t is represented by

$$x(t + \Delta t) = g(\bar{x}(t), \; z(t), \; w)$$

where $\bar{x}(t)$ represents the state vector for all previous values of t and z (t) represents the vector of values of exogenous variables for all relevant values of t and w is a vector representing the parameter values and g specifies the behavior of the system.

Some of the continuous model simulation languages are DYNAMO and JANIS, which are differential equation solvers, DAS which is a digital analog simulator, and FORBLOC which is a FORTRAN complied block-oriented simulation language.

However languages such as FORTRAN, MAD, ALGOL, PL/I, and assemblers are general purpose packages and can be used to describe a continuous simulation. Although many simulation languages are available, many simulation studies are carried out in FORTRAN. The reason is that it is a general purpose language for scientific and engineering problems and is the most familiar language known in these communities.

It is very difficult to draw conclusions from time studies in comparing simulation programs written in different languages, since the time studies depend on the programmer's efficiency and the nature of the model. However, there is little doubt that general purpose simulation languages have great capability in the execution time and storage allocation for "certain" simulation processes.

## IV. COMPARISON OF SIMULATION LANGUAGES

In this section we would like to describe and compare four simulation languages; SIMSCRIPT, GPSS, DYNAMO, and FORTRAN. SIMSCRIPT and GPSS are discrete time simulation languages, while DYNAMO is used for continuous systems and FORTRAN is a multiple purpose language.

### A. SIMSCRIPT

SIMSCRIPT [2] has its foundations in the notion that the state of the system can be described in terms of entities (i.e., things or objects of which the system is composed, attributes which are associated with entities) and sets which are groups of entities. The programmer is asked first to explicitly specify all entities with a complete list of attributes and possible set membership as the initial steps in the simulator's development. When the programmer defines the entity, he is actually defining a class of things in which each member has the same attributes. For simplicity, we shall refer to the class as the entity and the member of the class as an individual entity. For example, the entity might be man, the attributes his weight, height, and age. These entities may be temporary or permanent depending on whether they exist through the whole simulation process.

A SIMSCRIPT simulation program is divided into well-defined events. Each separate occurrence of an event occurs in zero simulation time, at a prespecified clock time, under the control of SIMSCRIPT. It is feasible for a given event to cause itself, and it is also feasible for more than one occurrence of an event to be on the schedule of events concurrently. The output is obtained primarily through a special class of report subroutines which are coupled by a report generator. The reports read out descriptions of the state of the model at any time with specific details with flexible and arbitrary formating.

## B. GPSS

In GPSS [3] the basic elements are blocks, transactions, and equipment. The simulation model in this language is described by block diagrams, and the blocks are linked together by physical and logical flows of transactions in the system. The transactions are only temporary elements which move from one block to another. And each transaction has eight parameters associated with it which list its priority in the system. As it moves through the system and enters a given block, the priority of the transaction can change, and thereby its execution is altered. The only output of such a system is usually a standard set of statistics which gives averages, maxima, and minima, distribution functions, etc.

## C. DYNAMO

The DYNAMO [4] language deals with continuous, closed loop information-feedback systems. And the system may be nonlinear and not necessarily stable. The mechanism employed by DYNAMO is to approximate the continuous process by a set of first-order difference equations, and the level at $x_i$ depends on $x_k$ only by the increment $\Delta x$, i.e.,

$$x_i = x_k + \Delta x$$

DYNAMO operates simply by performing a sequential solution of all the equations describing the system. Given an initial state $t_o$ of the system it advances the system $\Delta t$ and solves the system for $t_o + \Delta t$. Then, using this as an input, it computes the system for a $\Delta t$ increment and continues on in time. Hence the output of DYNAMO forms a time series for any desired variable.

## D. FORTRAN

FORTRAN is by far the most widely used language for scientific programing.

This language is a general purpose language, i.e., it can be used in the simulation of discrete and continuous systems. Some of the simulation languages are written in FORTRAN. However general purpose simulation languages like CSMO, GPSS, and SIMSCRIPT collectively have the potential of programing almost all systems.

The FORTRAN language is quite versatile since it can be used for continuous systems whose expressions are difference equations and for discrete systems which deal with queues and stochastic processes. Random number generators are found in well-known subroutines in FORTRAN (known as RANDU and GAUSS). Thus this language is self-contained.

If a model is to be run many times and remains virtually unchanged in a system, time can be gained by writing the program in FORTRAN. It is for this reason that FORTRAN should be used in simulation studies. However for complicated processes the time of programing in another language would be considerable compared to a FORTRAN model.

The structure of FORTRAN is demonstrated in many books. The power of this language should not be underestimated since the following discrete simulation packages are FORTRAN based: CSL, FORSIM IV, GPSS II, SIMSCRIPT, SIMTRAN.

Other languages which are also in the general purpose package category are MAD, ALGOL, PL/I, and Assemblers. By far the most versatile is the Assemblers or machine language. However the programing aspects are extremely difficult.

In the early 1960's, FORTRAN and the various ALGOL-type languages were used for most simulation programs, and to this date most simulation studies can be achieved using the FORTRAN language.

# V. SOME EXAMPLES OF COMPUTER SIMULATORS

## A. Circuit Analysis Simulation

The simulation program for circuit analysis is written in MAP and FORTRAN. The general program consists of the following main segments. First, the program reads the numerical values of circuit elements, function parameters, and functions supplied in tabular form; once a complete set of such data is entered, analysis can proceed. Second, the loop and node equations are machine generated from the network topology. The set of nonlinear differential equations are reduced to a set of nonlinear algebraic equations, and these are repeatedly solved by a system of linear algebraic equations. A symbolic generator [5] is machine produced to solve the algebraic equations, and a permutation is performed to separate the nonlinear from the linear portion [5-7]. Since the most time-consuming part of this simulation is the solution of linear equations, an optimal generator has been designed and can be developed as a simple special purpose high speed compiler.

The numerical integrator used in this simulation is an implicit scheme since there are advantages for stiff systems of differential equations.

The output of such a simulator is the current and voltage of each component. Hence, by varying some parameters, one can observe the current and voltage characteristics in the circuit.

This simulator is more than a regular program in that most of the decisions are machine made. The program inputs what components are connected to each other and their appropriate characteristics, and the computer does the remainder of the work:

(1) It analyzes the input data to see that they are correct. If there are any errors in the input it returns a diagnostic.
(2) It symbolically generates the loop and node equations.
(3) It chooses the numerical algorithm to solve the set of nonlinear differential equations.
(4) It generates a symbolic solver for a set of linear equations.
(5) It outputs the results of state vectors of current and voltage c/v characteristics.
(6) A feedback loop is developed to see the effect certain circuit parameters have on the c/v characteristics.

As an example network, consider Figs. 1 and 2. The value of $R_2$ is tabulated as a function of $I_D$, and the diode is defined by

$$I_D = F_1(V_D) = P_1(e^{P_2 V_D} - 1) \tag{1}$$

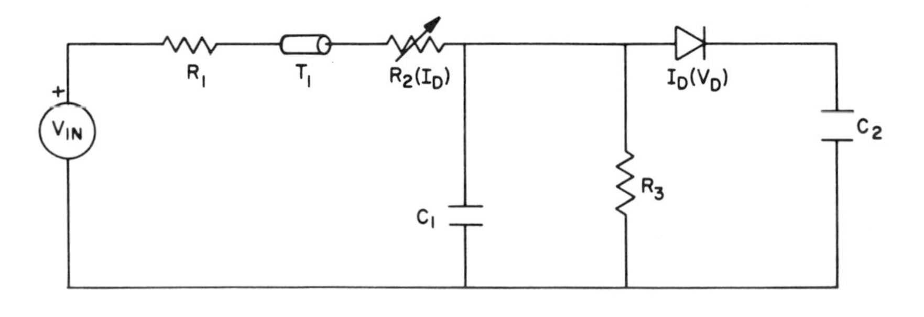


Fig. 1

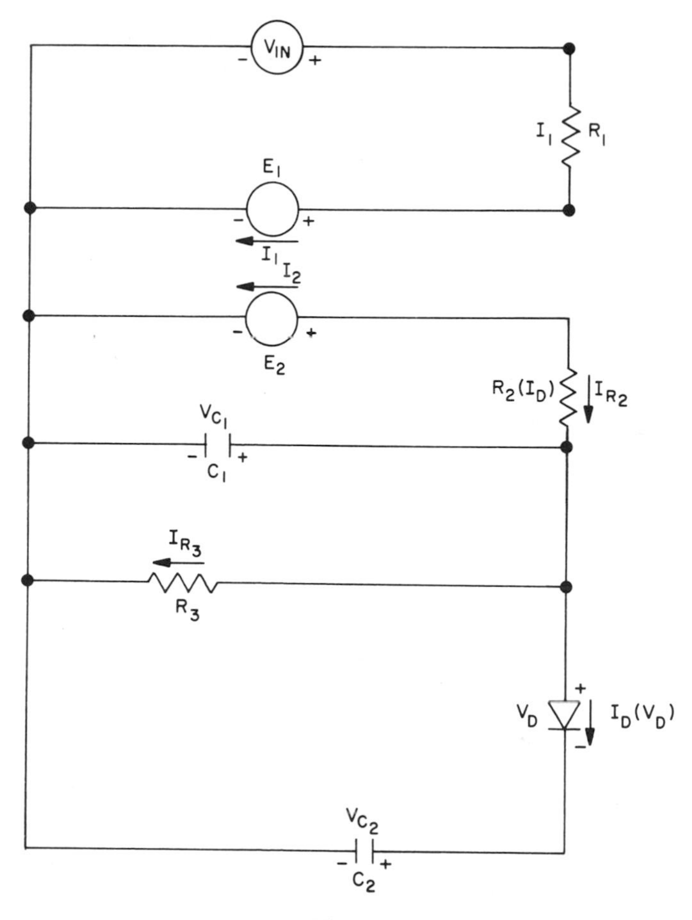


Fig. 2

The thirteen simultaneous equations to be solved at each time step have been ordered as follows:

$$V_{C_1} - \tfrac{1}{2}\Delta t \dot{V}_{C_1} - V_{C_1}(t - \Delta t) - \tfrac{1}{2}\Delta t \dot{V}_{C_1}(t - \Delta t) = 0 \tag{2}$$

$$V_{C_2} - \tfrac{1}{2}\Delta t \dot{V}_{C_2} - V_{C_2}(t - \Delta t) - \tfrac{1}{2}\Delta t \dot{V}_{C_2}(t - \Delta t) = 0 \tag{3}$$

$$E_1 - z_o I_1 - E_2(t - \tau) - z_o I_2(t - \tau) = 0 \tag{4}$$

$$E_2 - z_o I_2 - E_1(t - \tau) - z_o I_1(t - \tau) = 0 \tag{5}$$

$$I_D = F_1(V_D) \tag{6}$$

$$I_{R_1} - I_1 = 0 \tag{7}$$

$$I_{R_2} + I_2 = 0 \tag{8}$$

$$I_{R_2} - C_1\dot{V}_{C_1} - I_{R_3} - I_D = 0 \tag{9}$$

$$I_D - C_2\dot{V}_{C_2} = 0 \tag{10}$$

$$R_1 I_{R_1} + E_1 - V_{1N}(t) = 0 \tag{11}$$

$$R_2(I_D)I_{R_2} + V_{C_1} - E_2 = 0 \tag{12}$$

$$V_{C_1} - R_3 I_{R_3} = 0 \tag{13}$$

$$V_D + V_{C_2} - V_{C_1} = 0 \tag{14}$$

Equations (2) and (3) may be noted as the integration formula application; Eqs. (4) and (5) define the transmission line; Eq. (6) defines the dependent function arising because both the voltage and current of the diode are state variables; Eqs. (7)-(10) are simple node equations, and the last four are loop equations. Both the diode voltage and current are state variables, because each has a function dependent on it; $I_D = I_D(V)$ and $R_2(I_D)$. At interpretation time, such a device is probably most easily specified as a dependent current source, although other definitions are possible.

Now the integration formula is switched to a backward Euler for the determination of initial conditions, because an asymptotically vanishing amplification factor is highly desirable there.

To build in all these features, as one can see, is an extremely complicated task since all this is machine generated.

These equations are then machine transcribed to a set of linear equations which are solved iteratively for each time set until a prescribed error is met in the Jacobian matrix.

## B. BIOMOD

The BIOMOD simulator [8] developed at RAND is designed for the unsophisticated computer user to study models of continuous time systems. The basic features of the simulation system are (1) a highly graphic interactive system and (2) flexibility in model structuring.

The BIOMOD user may represent his model by drawing diagrams and typing or hand-printing text. Each component of a model block diagram may be defined by another block diagram. This divides the model into meaningful substructures. The user then defines each component block by analogous computer-like elements, algebraic, differential, or chemical equations, and/or FORTRAN statements. Since this simulator is highly interactive the user may stop the simulation, plot different variables, change scales, alter parameter values, and then either continue simulating or revise the description of his model.

The system operates on an IBM 360 Model 40 machine; it utilizes approximately 228, 000 bytes, and the operating system may be either MFT II or MVT version of OS/360. This is supplemented by the video graphic packages of IBM and RAND.

The user of this system has applicability in medical fields or the solution of a set of chemical rate equations. In the latter case the user writes down the symbolic set of chemical equations, say,

$$\begin{aligned} A + B &\rightarrow C + D \\ C + F &\rightarrow A + G \\ &\vdots \\ G + Z &\rightarrow P \end{aligned}$$

This is accomplished by using a light pen; he also inputs the appropriate rate constants along with initial conditions. The simulator then sets up the differential equations and chooses an appropriate numerical method of solution (the user also has the option if he has a particular scheme to either use it if it is available or add it). The output is then displayed on the CRT display and an option is there to obtain a hard copy (or picture) by a button. There is also a closed feedback loop to vary the effect of certain parameters on the concentration distribution.

## C. A Discrete System Simulation

In discrete simulation one is concerned with a system and a particular response variable $\mathcal{R}$, which is influenced by other variables, say x, y, ..., which are collectively represented by the vector X. For example, $\mathcal{R}$ might be the waiting time of an aircraft at an airport; then X represents the interarrival times of the planes appearing previously on that day. The modeling process involves relating $\mathcal{R}$ to X. As another example, consider the formation of a chemical species which is normally distributed with a mean value of n and a standard deviation of m. Also, assume the destruction of the chemical species is normally distributed with a mean of $n'$ and a standard deviation of $m'$. One is now asked to predict the formation or destruction of the chemical species and calculate the rate of change for some increment of time, given distribution data as a function of time.

All these modeling processes involve relating $\mathcal{R}$ to X; the latter is often taken to be a random variable. In discrete simulation, one is interested in investigating the distribution of $\mathcal{R}$ in terms of X. Thus one seeks to find the characteristics of

$$\mathcal{R} = f(x)$$

where f is presumed to be known and is usually a complicated function. Thus a discrete simulation output is a statistic and its variations due to input parameter.

In most scientific simulation processes, a discrete simulator can provide variations of initial and boundary conditions. Thus it can be coupled with a continuous system output. For example, a set of chemical reactions can have associated with it a set of rate constants which can be varied via a discrete simulator, and the solution of the chemical kinetics can occur via continuous simulation.

## V. CONCLUSION

In conclusion, computer simulation techniques provide a means of studying systems. Some of the positive aspects of a simulation study are: (1) it is ideal for the collection and processing of quantitative data, (2) it is free from physical limitations on the system being studied, and (3) it is completely repeatable. Simulation used as a research technique can help the investigator establish estimates and guidelines and, at the same time, can aid experimental system design and perform an evaluation of performance.

The techniques of discrete simulation can be used by the analytical chemist in the analysis of a large sample gathering and in the coupling of discrete and continuous system simulation in an area which has not been fully explored in chemistry.

## REFERENCES

[1]. V. A. LoDato, Simulated Environmental Physics, Wiley-Interscience, New York, in press.

[2]. H. Markowitz, B. Hausner, and H. Kan, SIMSCRIPT: A Simulation Programming Language, Prentice-Hall, Englewood Cliffs, New Jersey, 1963.

[3]. R. Efron and G. Gordon, A general purpose digital simulator and examples of its application, IBM Sys. J., 3, 1, 22-34 (1964).

[4]. A. L. Pugh, DYNAMO User's Manual, MIT Press, Cambridge, Massachusetts, 1961.

[5]. V. A. LoDato, The permutation of a certain class of matrices, The Comp. J., 13, 405-410 (1970).

[6]. V. A. LoDato, The numerical solution of a nonlinear master equation, J. of Comp. Phys., 6, 105-112 (1970).

[7]. V. A. LoDato, D. McElwain, and H. O. Pritchard, The master equation for the dissociation of a dilute diatomic gas. I. A method of solution, J. Am. Chem. Soc., 91, 7688 (1969).

[8]. R. L. Clark, G. F. Groner, and R. A. Berman, The BIOMOD User's Reference Manual, The RAND Corporation R-746-NIH, July, 1971.

Chapter 6

# ANALOG RESPONSE BY DIGITAL COMPUTERS

J. G. Sellers

Imperial Chemical Industries Limited
Corporate Laboratory
The Heath
Runcorn, Cheshire, England



## I. INTRODUCTION

Workers in analytical chemistry have for many years been using small analog computers to study the ordinary differential equations representing reaction kinetic schemes. In the late 1950's and early 1960's, analog computers were also used extensively for studies in process control [1] and aerospace, but interest in these fields has now largely switched to the use of digital computers.

Analog computers are very suitable for developing small models, where the complexity of the model is not very high and a large degree of involvement by the experimenter is required. These small computers are inexpensive, readily portable, and fairly easy to program.

When problems increase in size or in demands on specialized equipment, the analog computer is not nearly so attractive because medium or large machines are expensive and cannot be remotely operated (or transported).

Thus the users with large simulation problems started to investigate the use of digital computers, as they became more powerful.

The digital computer is inherently far less suited to the solution of sets of simultaneous differential equations than is the analog computer, as the digital method can only make stepwise approximations to the integration process. However, it has many attractions, not the least being the availability of digital computing facilities almost universally (and not purchased by the simulator). The digital method usually requires that the simulation be performed on a batch computer system, where the experimenter has no interaction with the solution during a run, or on a time-sharing system, where the input/output facilities offered by a typewriter terminal are much inferior to the control and graphical output features of analog computers. The more recent introduction of small on-line computers into research laboratories offers the research chemists, such as reaction kineticists, an opportunity to utilize digital methods to tackle problems beyond the capability of analog computers available to them, but providing input/output facilities more comparable with the analog machine. However, the program for a small digital computer must be written in a different way from that used for the programs on large digital computers, otherwise the problem size allowed is very restricted and solution is extremely slow.

The use of analog computers to simulate kinetic models is thoroughly covered in Volume 3 of this series [2]. The aim of this chapter is to indicate the alternative, digital, methods of carrying out "analog simulation" when an analog machine either is not available or is inadequate to solve the experimenter's problem. A brief comparison is made between analog and digital computers, but no mention is made of hybrid computers which would not normally be required or justified for the problems of the analytical chemist.

## II. DIGITAL SIMULATION PRINCIPLES

A general purpose digital computer can be programed to model processes which would often be simulated on analog computers. As the analog computer consists of a number of discrete components — adders, integrators, multipliers, etc. — these are connected together to form the electrical equivalent of the system under study and, when set to run, can be expected to behave more or less like that system, depending on how close the equivalence is. The digital computer, however, contains only one calculating item, the <u>central processing unit</u> (CPU); and although this is extremely fast it can only carry out one operation at a time. Consider the solution of the single second-order sine wave equation

$$\ddot{x}(t) = -x(t)$$

Using the symbols in Fig. 1, the analog computer represents this (Fig. 2) by the connection of two integrators and a sign inverter. The two integrators

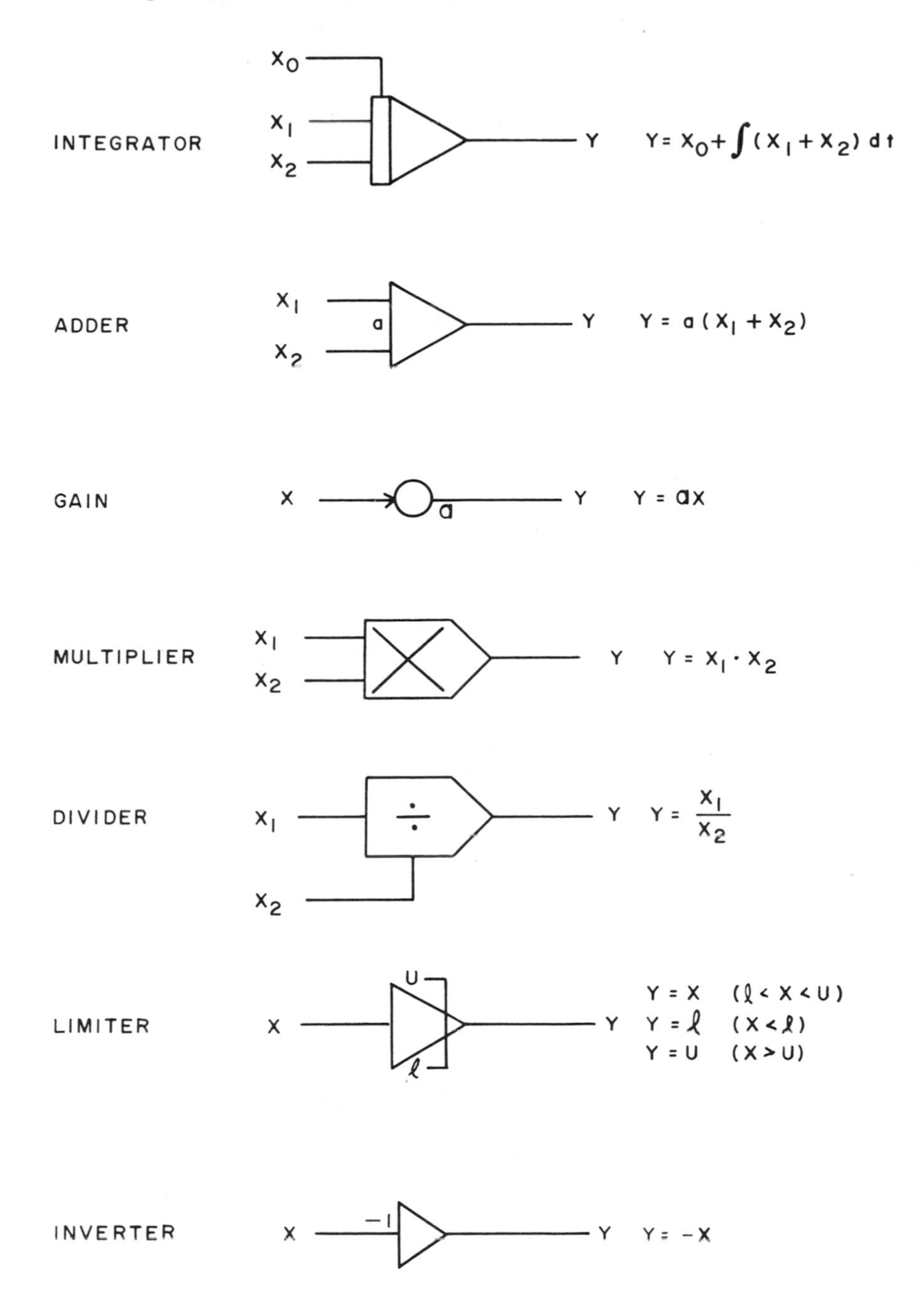


Fig. 1. Commonly used digital/analog symbols.

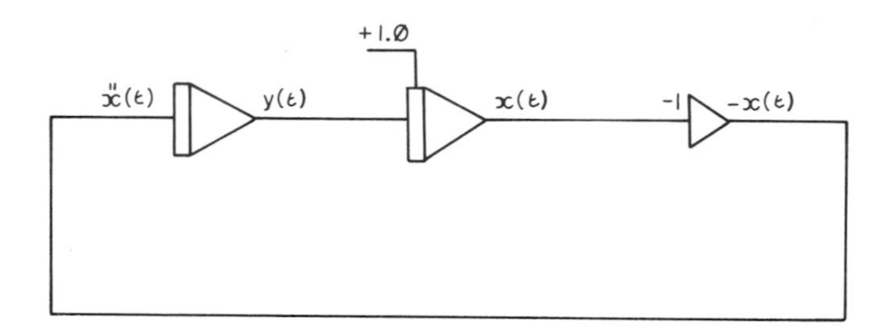


Fig. 2. Patching diagram for simple sine:
$y(t) = -\dot{x}(t)$ for analog computer (sign reversal in integrators)
$y(t) = +\dot{x}(t)$ for digital/analog (no sign reversal)

are switched from HOLD to COMPUTE simultaneously at time $t = 0$ and thereafter compute the values of $\dot{x}(t)$ and $x(t)$. The digital computer cannot exactly represent the equation

$$x(t) = \int_0^t \dot{x}(t)\,dt$$

as it cannot operate continuously. Instead it can make the approximation

$$x(t) \simeq x(n\Delta) = \sum_{i=0}^{n} \dot{x}(i\Delta) \cdot \Delta \qquad \text{for} \qquad t = n\Delta$$

which is true as $\Delta \to 0$ $(n \to \infty)$.

In any closed loop system such as this, it is not possible to calculate the value of one variable only from $t = 0$ to $t = T$ because the new value at the output of integrator 1 is fed round to alter the value at its input. Thus the digital program must step round the loop to calculate $\dot{x}(\Delta)$ and $x(\Delta)$, then repeat this for simulated time $t = 2\Delta$ to calculate $\dot{x}(2\Delta)$, $x(2\Delta)$, etc. A very important factor in this calculation is the time step size $\Delta$. If it is too small, then, for a given time t, n will be large, and the number of steps taken to calculate the value x(t) correspondingly large, thus using substantial time in the CPU. If on the other hand $\Delta$ is too large, then the approximation for integration is no longer valid and the solution for x(t) incorrect. This stepwise approximation to integration is the major limiting factor in digital simulation, and much effort has been directed to finding integration algorithms, more sophisticated than the simple rectangular approximation, which permit the use of larger integration steps without degrading accuracy. References [3-8] give a broad view of the techniques which have been developed, but there is little consensus of opinion on the best "all round" method, mainly because particular examples can always be found where a specific algorithm performs well, even if it does not do well in other cases. Briefly, the more complex methods seek to use information about the way in which $\dot{x}(t)$ is changing over the interval $\Delta$ to calculate the increment in x more exactly.

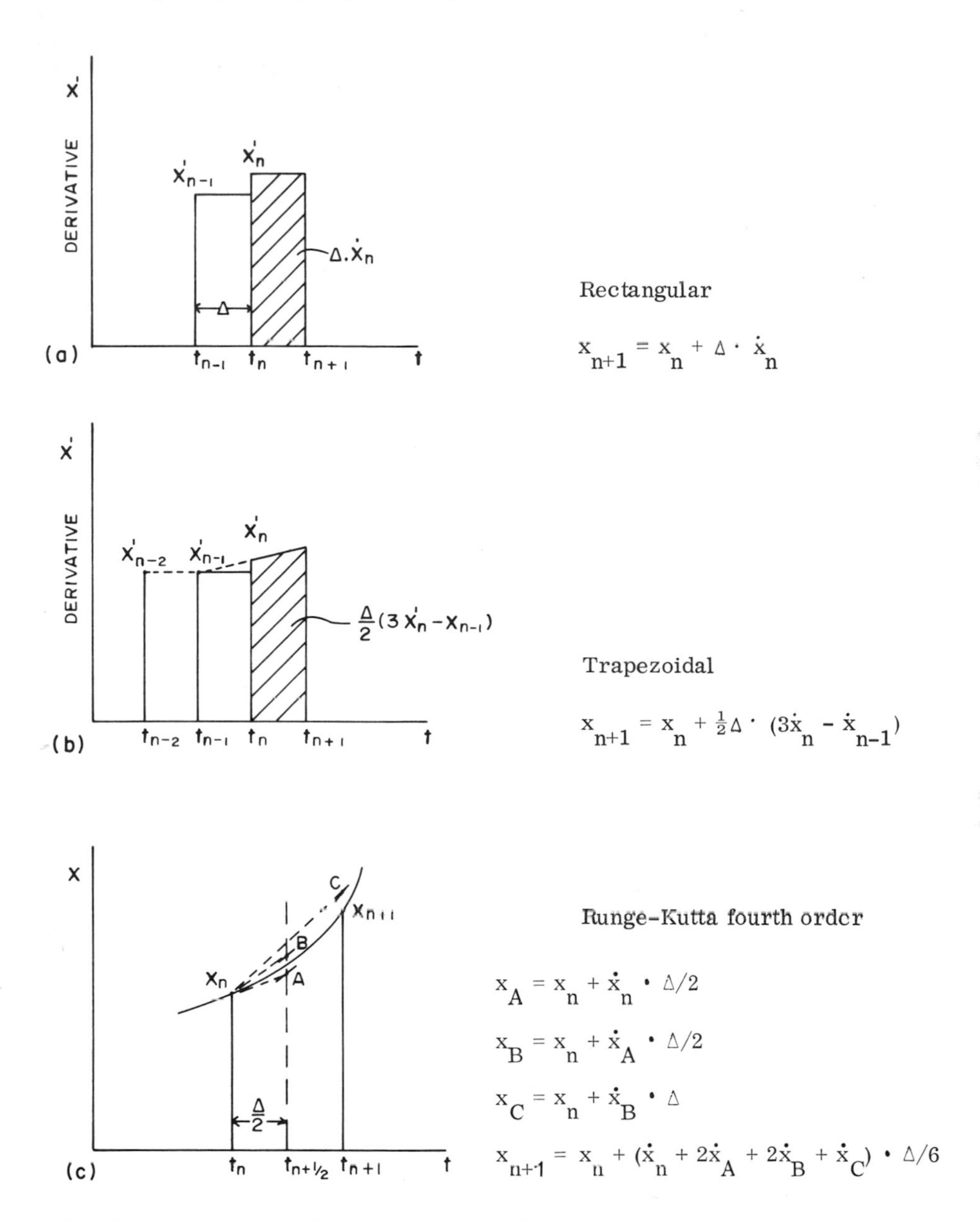


Fig. 3. Integration methods: (a) Rectangular, (b) Trapezoidal, (c) Runge-Kutta fourth order.

Figure 3 illustrates some of the methods. The rectangular method uses the equation

$$x_{n+1} = x_n + \Delta \cdot \dot{x}_n$$

thus making the approximation that the slope of x is constant throughout the integration step, as in Fig. 3(a).

The value of $\dot{x}_{n+1}$, for use on the next step, is then calculated from the equations of the model which define $\dot{x}_{n+1}$ in terms of $x_{n+1}$ and the other variables at time n + 1.

A more refined method is to use past knowledge of the way in which the slope has been changing to predict how it will behave during the next step. Figure 3(b) illustrates the trapezoidal (or 2nd-order Adams-Bashforth) method which uses the slopes at $t_{n-1}$ and $t_n$ to predict the slope at $t_{n+1}$, which is then used to update $x_n$ to $x_{n+1}$, i.e.,

$$x_{n+1} = x_n + \tfrac{1}{2}\Delta(3\dot{x}_n - \dot{x}_{n-1}).$$

The predicted value of $\dot{x}_{n+1}$ is used only to calculate $x_{n+1}$, the value of $\dot{x}_{n+1}$ actually used to calculate the next step being found, as before, from the model equations. Formulas are available to include many values of past slopes, a popular one being the fourth-order Adams-Bashforth formula:

$$x_{n+1} = x_n + \frac{\Delta}{24}(55\dot{x}_n - 59\dot{x}_{n-1} + 37\dot{x}_{n-2} - 9\dot{x}_{n-3}) \qquad (1)$$

This class is known as multistep predictor methods.

An extension is to use predictor-corrector methods. Here the value $x_{n+1}$, as calculated in (1), is regarded only as a predicted value $x^P_{n+1}$, which is inserted into the model calculations to give the predicted value of $\dot{x}^P_{n+1}$ This is then used in a formula which includes the predicted value of $\dot{x}_{n+1}$ as well as the actual values of $\dot{x}_n$, $\dot{x}_{n-1}$, etc. Equations of this type are known as Adams-Moulton formulas, the equations for the fourth-order Adams-Moulton predictor-corrector being

$$x^P_{n+1} = x_n + \frac{\Delta}{24}(55\dot{x}_n - 59\dot{x}_{n-1} + 37\dot{x}_{n-2} - 9\dot{x}_{n-3})$$

$$\dot{x}^P_{n+1} = f(x^P_{n+1}, t_{n+1})$$

$$x_{n+1} = x_n + \frac{\Delta}{24}(9\dot{x}^P_{n+1} + 19\dot{x}_n - 5\dot{x}_{n-1} + \dot{x}_{n-2})$$

$$\dot{x}_{n+1} = f(x_{n+1}, t_{n+1})$$

The multistep methods have the disadvantage of being difficult to start, as they require past values of slopes. This difficulty also applies if the step length $\Delta$ is changed. A class of methods known as "single-step" are self-contained and, although sometimes used to start off a multistep method, are often used to generate the whole solution. One of the most commonly used methods is the fourth-order Runge-Kutta, which operates as in Fig. 3(c). The basis of the method is to carry out preliminary explorations of the area, evaluating the derivative from trial values of x and t, before carrying out the final accurate step.

The slope $\dot{x}_n$ is first used to go forward $\Delta/2$, i.e., half a step, to the point A, where the slope is evaluated:

$$x_A = x_n + \tfrac{1}{2}\Delta \cdot \dot{x}_n$$

$$\dot{x}_A = f(x_A, t_{n+\frac{1}{2}})$$

The slope $\dot{x}_A$ is now used to repeat the half-step to point B, where the slope is again evaluated:

$$x_B = x_n + \tfrac{1}{2}\Delta \cdot \dot{x}_A$$

$$\dot{x}_B = f(x_B, t_{n+\frac{1}{2}})$$

The slope $\dot{x}_B$ is used to make a full step forward to point C, then the slope is evaluated:

$$x_C = x_n + \Delta \cdot \dot{x}_B$$

$$\dot{x}_C = f(x_C, t_{n+1})$$

Weighted means of these slopes are then used to carry out the final, accurate, step forward:

$$x_{n+1} = x_n + \frac{\Delta}{6}(\dot{x}_n + 2\dot{x}_A + 2\dot{x}_B + \dot{x}_C)$$

$$\dot{x}_{n+1} = f(x_{n+1}, t_{n+1})$$

It will be realized that all the predictor methods must make dummy steps, at each of which the derivative term is evaluated (i.e., the value inserted in the model equations and the values of all integrator inputs recalculated). Thus the fourth-order Runge-Kutta method makes four derivative evaluations per time step $\Delta$. However, the increased accuracy of this method over simple methods (for some types of problem) allows the step size $\Delta$ to be greatly increased for the same accuracy of solution.

Variable step lengths may be introduced into the integration to allow the calculations to proceed as fast as the problem will allow. This can be extremely valuable in cases where the system proceeds at different rates at different times in the simulation. The classic example is the bouncing of an elastic ball where the dynamics of the problem fall into two distinct regions — free fall, and elastic bounce. During the free fall, motion is steady and large integration steps are quite adequate. However, as soon as the ball contacts the ground, it begins to deform and the equations of motion are extremely rapid, requiring small integration steps for accuracy. Thus it is possible to run with a large step in the first mode and then "change down" to a short step for the bounce, changing up again once the ball has left contact with the ground. This gives a much more rapid solution than if the whole problem is run with the small step demanded by the elastic bounce. Automatic step size selection is possible in multistep predictor-corrector and in single step exploration methods, using techniques to assess the error at each step and adjust the step size to keep this at an acceptable level. However, this can result in the algorithm spending a long time choosing the optimal sized step and very little time using it. Tramposch and Jones [9] report a simple method of changing the step length when it is known, from the characteristics of the model, where this will be necessary, e.g., in the case of the bouncing ball, the smaller interval is selected when the height of the ball is nearly zero. It would appear that sophisticated automatic step length routines, although useful in some cases, are misused in the majority of problems where the dynamics do not change significantly, and serve as an excuse to forget about step length. However laudable this aim, step length control is inescapable in digital simulation, and in the variable step cases the user is still required to choose the "error per step" on specific variables. Unless the most sensitive variable is chosen, this can lead to either instability or long computing times. In the majority of practical problems, an adequate method is to choose a "sensible" step size (say 1/25 of the fastest time constant) and run the problem, then experiment with halved and doubled step sizes until the largest value is found which gives results agreeing within the desired accuracy with the results from smaller steps. This should then be used for the experiments with the model.

To use these techniques to simulate a problem on a digital computer, one could write a program (probably in FORTRAN) which incorporates the chosen integration algorithm and use this to solve the simultaneous differential equations representing the problem. However, many standard programs are in existence which will accept the definition of the problem as data and perform the simulation, giving good error reports and good presentation of results. Their use simplifies the task of the experimenter to providing a correct model, rather than writing a complex program, and so this approach will generally be taken.

## III. DIGITAL SIMULATION PROGRAMS

The first digital simulation program was reported by Selfridge [10] in 1955. Since then many programs have been developed — even in 1964, Linebarger and Brennan [11] reported 28 different programs. Lately there has been some consolidation and the number of distinct programs in wide-spread use is much smaller, but these fall into two types: block structured and procedural.

### A. Block Structured Programs

These programs are the direct digital equivalent of analog computers. In a widely used program [12] the user sets up his problem using a block diagram to identify the interconnections between integrators, adders, etc. These functional blocks almost exactly parallel those available in the analog computer. Instead of patching up this block diagram using patch wires, the user provides a set of data cards to the simulation program specifying the interconnections required, for example, in the case of the simple sine model:

| Block | Type | Input 1 | Input 2 | Input 3 |
|---|---|---|---|---|
| 1 | I | 3 | 0 | 0 |
| 2 | I | 1 | 0 | 0 |
| 3 | + | 2 | 0 | 0 |

| Block | Parameter 1 (Initial condition) | Parameter 2 | Parameter 3 |
|---|---|---|---|
| 1 | 0.0 | — | — |
| 2 | 1.0 | — | — |
| 3 | -1.0 | — | — |

The user then controls the program, by typing on a keyboard or by operating monitor switches, in a similar way to running an analog computer.

The big advantage of such a program over the analog computer is that, within the limits set by the store size of the computer, a wide variety of types and numbers of computing blocks can be used depending on the requirements of the problem. Output of results is normally tabular onto a typewriter or lineprinter, the latter sometimes being used also for "lineprinter graphs." The printed output is very useful for detailed analysis of results, but not nearly so helpful as a multipen recorder output from an analog computer if the experimenter is interested in trends. This type of program is typically found on small in-house computers and on some time-sharing services. Several block structured programs have been available on batch computing systems, but in recent years these have been overtaken by the procedural programs, which are more convenient for remote batch use.

## B. Procedural Simulation Programs

This type of program is moving further away from the direct replacement of the analog computer in that input to the problem is now a set of equations, very similar in most cases to FORTRAN (rather than the specification of a patching diagram). Generally speaking, these programs are used in batch mode on a powerful computer, the user submitting a series of experiments and then receiving the results back some time later, normally as lineprinter output, but in some cases also as graph plotted records. The analog block concept is not entirely lost in that analog facilities are available as standard functions in the program language, e.g., typical statements for the simple sine would be

```
XDOT  = INTGRL (0.0, XDDOT)
X     = INTGRL (1.0, XDOT)
XDDOT = -X
```

In this case, INTGRL represents the integration function, the first term in the brackets being the initial value of the integral and the second term being the name of the derivative parameter.

Brennan and Silberberg [13] compare a procedural language [14] with a block structured language [12] and emphasize the different ways in which the two types are used.

An important step in the development of procedural languages came in 1967 when Simulation Councils Inc. published the specification for a Continuous System Simulation Language (CSSL) [15]. This provides for a basic language which should be used by all developers of continuous simulation programs, thus allowing problems developed on one program to be readily run on another program (in theory at least; in practice no doubt, local differences will arise, as currently occurs in FORTRAN, making the transfer from one computer to another not a simple task). At least one program has been developed to the CSSL specification and is widely available [16]. Hopefully, most other manufacturers will follow suit. Recent developments allow a model to be developed ("debugged") at a time-sharing terminal and then production runs made under batch operation, offering lower processor charges and better output facilities such as line printout and graphical plots.

## C. Discussion

The user of analog computers will, for a variety of reasons, often get to the stage where he has to use digital computers. These reasons include:

(1) Problems too big for an available analog computer.
(2) Poor access to an analog computer because it is being used for a large complex simulation.
(3) No analog computer within easy reach.

He is likely to turn to one of the block-oriented digital simulation languages, on a small computer or time-sharing system, to which he transfers his analog patch diagram and which he operates himself. The important advantages he will find are:

(1) No need to scale problem.
(2) Readily repeatable results (no trouble with faulty amplifiers, etc.).
(3) Record of operations and results for future reference.
(4) Many special purpose functions readily available.
(5) Able to program special functions, not build them electronically.
(6) Rapid mounting of problem, and changeover to other users.
(7) No need to justify computing equipment solely on use for simulation.

On the other hand there are a number of disadvantages:

(1) Much poorer graphical output facilities.
(2) Simulation speed decreases as the problem size and complexity grows.
(3) The cost of CPU time can become significant.

While the great majority of programs written or used by scientists and engineers use only trivial amounts of CPU time compared with the printing time, simulation remains an exception because of the time spent integrating simultaneous differential equations. This means that for nontrivial problems the user soon finds it tedious to sit while the computer laboriously types out the responses of his model, and he often moves on to the next stage of simulation, a remote batch-service procedural program. Here he can plan a series of experiments with the simulated system and send them off to the computer, receiving back the results some hours later. There are two major disadvantages:

(1) The system is remote so programing errors cannot be corrected on the spot.
(2) The user is no longer part of the experiment and cannot modify each run of the simulation in the light of the previous response. Thus tuning or curve-fitting problems become very difficult.

However this approach works very well for complex models, the results from which need careful and concentrated study before carrying out a further series of tests. In general, for a given problem, this procedure will raise lower computer charges per run than a block-oriented program operated by the user, as the computer can be used more efficiently. When however the lack of access causes the user to do unnecessary runs, this cost advantage may be reversed.

It is often found that scientists who have not used analog computers tend to regard the block structured programs as requiring an irrelevant knowledge of analog computing, but by doing so deny themselves the hands-on operation so fruitful for small and medium sized problems.

## IV. SIMULATION USING A SMALL COMPUTER

The availability of a small on-line computer in a research laboratory has led to the development of a simple but effective simulation program for that machine [17]. Although the particular computer (a Ferranti Argus 300/500) will not be available to most other workers, some details of the program may be of interest to those who find themselves without analog or digital computing services, but have a small on-line computer which is available for off-line use at some periods of the day or evening. A more complete description may be found elsewhere [18]. The computer in our case was originally an ARGUS 300 with 8K of 24 bit store, coupled to an AEI MS-9 high resolution mass spectrometer [19], although it has since been upgraded to a more powerful Argus 500 time-sharing system to service a number of experiments.

### A. Techniques

The requirements formulated for the program were that it should give rapid, accurate simulations for small to medium models of the type encountered in the chemical industry, while occupying a minimum core store area in the computer. The simulation should be controlled by the user, with facilities to change parameters between runs and to modify the simulation model quickly.

From the techniques available for producing simulation programs, the appropriate ones must be chosen in the following areas:

(1) Type of simulation program: block structured or procedural.
(2) Arithmetic: fixed point or floating point.
(3) Source language: assembly code or high level.
(4) Integration method.
(5) Problem order sorting: manual or automatic.

First, we consider the type of simulation program in the light of resources required rather than the intrinsic appeal of blocks or equations. The procedural program normally consists of a "precompiler" which converts the simulation language statements into a standard high level language (generally FORTRAN). The resulting "FORTRAN program" is then compiled into an executable program by the FORTRAN compiler. Such a FORTRAN compiler will normally require at least 8K core store and two passes of paper tape or cards through the computer, if the resulting code is to be even reasonably efficient (this discussion precludes the use of backing storage — magnetic tape, drum or disk — which would not normally be a feature of the size of computer we consider here). Because of the tape handling, this process is unlikely to take less than 30 minutes per assembly, and would have to be repeated to remove any errors or to make any changes other than trivial ones. However, a block structured program will have been compiled or assembled previously and simply requires a new data set to specify a new problem. This data set will be loaded by a single pass of tape or cards into the computer, requiring at most a 500-word loader for this purpose. The choice of a block structured program then becomes obvious in this application.

The Argus, like the majority of on-line computers, does not incorporate hardware for manipulating floating point numbers (i.e., numbers stored in the computer as argument A and exponent E, such that $y = A \times 2^E$). The very great advantage of floating point working is that scaling can practically be forgotten as the system will represent numbers in the range (typically) $10^{-76}$ to $10^{+76}$, to a precision of 10 decimal digits. However, the lack of floating point hardware means that all operations on these numbers must be carried out using special floating point subroutines, which greatly increases the computer time compared with that required to carry out equivalent calculations in fixed point arithmetic (in typical cases by a factor of 20:1). As has already been stated, central processor time is usually significant in simulation programs, so this means a straight increase in running time of the program. Thus there is a conflict between operating speed and convenience of use, because scaling is a definite headache on the analog computer. However, an examination of precisions shows that, even with fixed point arithmetic, the scaling task on a digital computer can be much simpler than on the analog. Due to the low precision of many operations in the analog (typically 0.1% f.s.d.), it is essential that variables be scaled to keep their computer representation with a value between say 10% and 90% of full scale. As well as straight scaling, offsets are also often employed, thus a temperature of 70-90°C might well be represented by 1°C equivalent to 4% computer unit, i.e.,

10% equivalent to 70°C
90% equivalent to 90°C

This obviously makes scaling a tedious and error-prone task. However the 24-bit word used to hold each data value in this computer can hold fractions in the range -1.0000000 to +0.9999999 to a precision greater than 1 in $10^6$ [0.0001% f.s.d. (full scale deflection, i.e., accuracy is a fixed percentage of the full range, not of the current value)]. Thus one is able to scale much more generously without introducing significant errors, and the normal practice is to use only factors of 10 in the scaling and to avoid offsets completely, representing the above example by

0.7 equivalent to 70°C
0.9 equivalent to 90°C

As the difficulty of scaling is so much reduced, it was decided, regretfully, to accept limited scaling in exchange for a very desirable increase in problem solution speed. On the subsidiary question of where to fix the decimal point, there can be several proposals, e.g.,

range ±0.9999999 (Fraction)
±127.9999 (Per cent)
±8388607 (Integer)

For our type of work, fractions seemed the best choice, particularly for calculations where multiplication is involved as two fractions will always multiply to produce another fraction without overflowing.

The source language in which the simulation program should be written is relatively unimportant and will not affect the user, who simply supplies data to a working program. Obviously a high-level language such as FORTRAN increases intelligibility of the program and eases any attempts to translate the program for another computer. On the other hand, simple FORTRAN compilers rarely generate code which is anywhere near as compact or as efficient as that produced by a good assembly language programer. Also techniques of packing several items of information (such as status bits) into one computer word are readily available in assembly language but not generally in FORTRAN. The choice will depend heavily on the high-level language facilities in the computer and the programing expertise of the user; in our case we decided to use assembly language, particularly so that we might minimize the core size required by the program.

It is difficult to choose an integration algorithm because of the wide variety available. However, having decided to use fixed-point arithmetic and assembly language, it becomes a tedious task to code up several complicated algorithms, so it was initially decided to use a simple method which could be replaced by a more sophisticated one at a later date. In fact, this simple one has proved itself satisfactory for the problems so far encountered, and there has been no incentive to replace it as yet.

The method used is trapezoidal, with one interesting modification in that all computing blocks are sorted into order whereas other programs known to the author sort the elements between integrators. If the order of updating the computing blocks for a given cycle is chosen such that, before any given block is updated, all inputs to it have already been updated, then before calculating $x_{n+1}$ the value of $\dot{x}_{n+1}$ is known. If the value of $x_n$ was remembered, then the simple equation

$$x_{n+1} = x_n + \tfrac{1}{2}\Delta(\dot{x}_n + \dot{x}_{n+1}) \qquad (2)$$

represents $x_{n+1}$ to a good degree of accuracy, completely avoiding the multiple searches of a Runge-Kutta method or the long memory and self-starting problem of an Adams method. Obviously, in a model containing closed loops, it is not possible to order the blocks such that no block is updated until all its inputs have been updated. However, when the block "patching diagram" is used, it is a trivial task to order the blocks such that the number where this occurs is small. In many cases, the delay of one sampling interval introduced by this technique will be smaller than actual delays omitted from the model for simplicity. On the few occasions where the delay has a significant effect, it may be greatly reduced by inserting a predictor block at the inputs of the affected block, which simply predicts the value of $\dot{x}_{n+1}$ from the values of $\dot{x}_n$ and $\dot{x}_{n-1}$. Initially, the integration was carried out using single length (24-bit) arithmetic, but it was soon found that, as the step length was reduced in an attempt to improve accuracy, slow moving variables stopped moving at all because the increment in x from (2) is so small that its 24-bit representation is zero. This is readily avoided by calculating the integral to double length, although only the most significant 24-bits are carried on to the next block in the calculation.

The problem of sorting the blocks into the correct order for computation, so as to minimize the introduction of sampling delays, is thought to be so trivial by eye, when the problem is laid out in diagram form, that no program has been produced to do it automatically. However, this would not be true for a procedural simulation language, where the interconnections between different parts of the calculation are not normally displayed graphically.

In summary, the simulation program implemented in a laboratory computer was of block structured type, carrying out fixed point arithmetic calculations on scaled equations, written in assembly code, using a simple trapezoidal integration method, and requiring that the user should manually sort the computing blocks into an acceptable order for calculation. Within this structure, the program has proved fairly simple to write and in basic form it occupies only 1400 words of 24-bit core store. Each data block — the equivalent of an analog computing element — requires a further 8 words of store. Print routines and a simple loader for a stand alone system occupy a further 1000 words, thus in 4K core store it is possible to represent an analog computer with about 200 computing elements, i.e., adders, multipliers, gains, integrators, logic functions, etc. Many of the blocks provided in the basic system offer very useful facilities, and within the total limit of blocks for which there is room, there are no restrictions on the numbers of blocks of any one type in use. Examples of these blocks are: limiter, time delay, 3-term controller, maximum hold, square root, exponential, comparator, AND gate, function generator, and backlash.

The program is operated by the user via handswitches on the computer panel, which are used for STOP/GO, special operations (e.g., print all block outputs for debugging), initiating the loading of new data, etc. Output is normally in tabular form to a teletype, with graphical output to an incremental plotter, where available. If the computer is provided with analog or digital outputs, graphical displays can readily be produced on X-Y plotters or oscilloscopes.

The program speed is such that, on the Argus 500 (core cycle time = 2 $\mu$sec), it will normally update 5000 blocks per second. Thus a 200-block problem will be updated by 25 time steps per second. If the sampling interval $\Delta$ is 0.04 sec then the problem runs in real time. As the value of $\Delta$ will typically be 1/25 of the fastest time constant in the system, this means that a 200-block problem with fastest time constant 1 sec will be simulated in real time.

## B. Solution of a Kinetic Model

A wide variety of problems has been simulated by this program, ranging from process control loops (for which it was designed) to kinetic reaction schemes. An example of the latter is described here.

The scheme involves five major components, $X_1$ to $X_5$, reacting in solution. $X_5$ is a dissolved gas. Components $Y_1$ and $Y_2$ are present in excess, so do not figure in the rate equations:

$$X_1 + Y_1 \xrightarrow{k_1} X_2 \tag{3a}$$

$$X_2 + Y_2 \xrightarrow{k_2} X_3 \tag{3b}$$

$$X_3 + X_5 \xrightarrow{k_3} X_4 \tag{3c}$$

$$X_5 \text{ (gaseous)} \xrightarrow{k_4} X_5 \tag{3d}$$

The corresponding differential equations are:

$$\frac{dX_1}{dt} = -k_1(X_1) \tag{4a}$$

$$\frac{dX_2}{dt} = k_1(X_1) - k_2(X_2) \tag{4b}$$

$$\frac{dX_3}{dt} = k_2(X_2) - k_3(X_3)(X_5) \tag{4c}$$

$$\frac{dX_4}{dt} = k_3(X_3)(X_5) \tag{4d}$$

$$\frac{dX_5}{dt} = k_4[(CS - X_5) + 0.5(X_3)] - k_3(X_3)(X_5) \tag{4e}$$

In Eq. (4e), CS is the saturation value of $X_5$ in solution. The term $0.5(X_3)$ is empirical and represents the observation that the rate of solution of $X_5$ apparently depends on the rate at which $X_5$ is reacting, as well as on its dissolved concentration.

The computer equations which result from converting the differential equations into integral form and scaling variables to lie in the range $\pm 0.9999999$ are as follows (scaling factors omitted for clarity):

$$X_1 = CO - 0.01 \int k_1 X_1 \ dt \tag{5a}$$

$$X_2 = 0.1 \int (k_1 X_1 - k_2 X_2) \ dt \tag{5b}$$

$$X_3 = \int (k_2 X_2 - 10 k_3 X_3 X_5) \ dt \tag{5c}$$

$$X_4 = 0.1 \int k_3 X_3 X_5 \ dt \tag{5d}$$

$$X_5 = CS + 500 \int (0.001 k_4 (CS - X_5) + 0.05 k_4 X_3 - k_3 X_3 X_5) \ dt \tag{5e}$$

where CO is the initial concentration of $X_1$.

After scaling, one expects any constants to be of one or two orders of magnitude at most, as in (5a)-(5d). However (5e) contains some of three orders, indicating either poor scaling or a "stiff" model — in this case, the latter. A stiff model is one where the time constants involved cover a wide range, here this occurs because the unscaled rate constants $k_3$ and $k_4$ are very different ($k_3 = 10^5 k_4$). The patching diagram is drawn (Fig. 4) and the blocks numbered to give a correctly sorted order for computation, from which a data set is prepared to specify block types, parameters, and interconnections.

A typical set of results from this model is shown in Fig. 5. It is found necessary to use a small sampling interval in this problem or the high gain loop 9-16-17-18-19-21-23-9 goes into oscillation. The loop gain is linearly dependent on the value of $X_3$, which reaches a peak around time 0.06 and then decreases rapidly. Using this fact, it proved easy to patch together further computing blocks which adjusted the sampling interval inversely as the value of $X_3$. This allowed a total run time less than half that obtained by using a fixed step length throughout the simulation. This particular model, due to its unusual stiffness and resulting small step length, was slow to simulate (6 minutes per run) but would also be difficult to implement on an analog computer due to the wide range of values covered by several of the significant variables (e.g., $X_5$ is around 1% of its maximum value during much of the simulation).

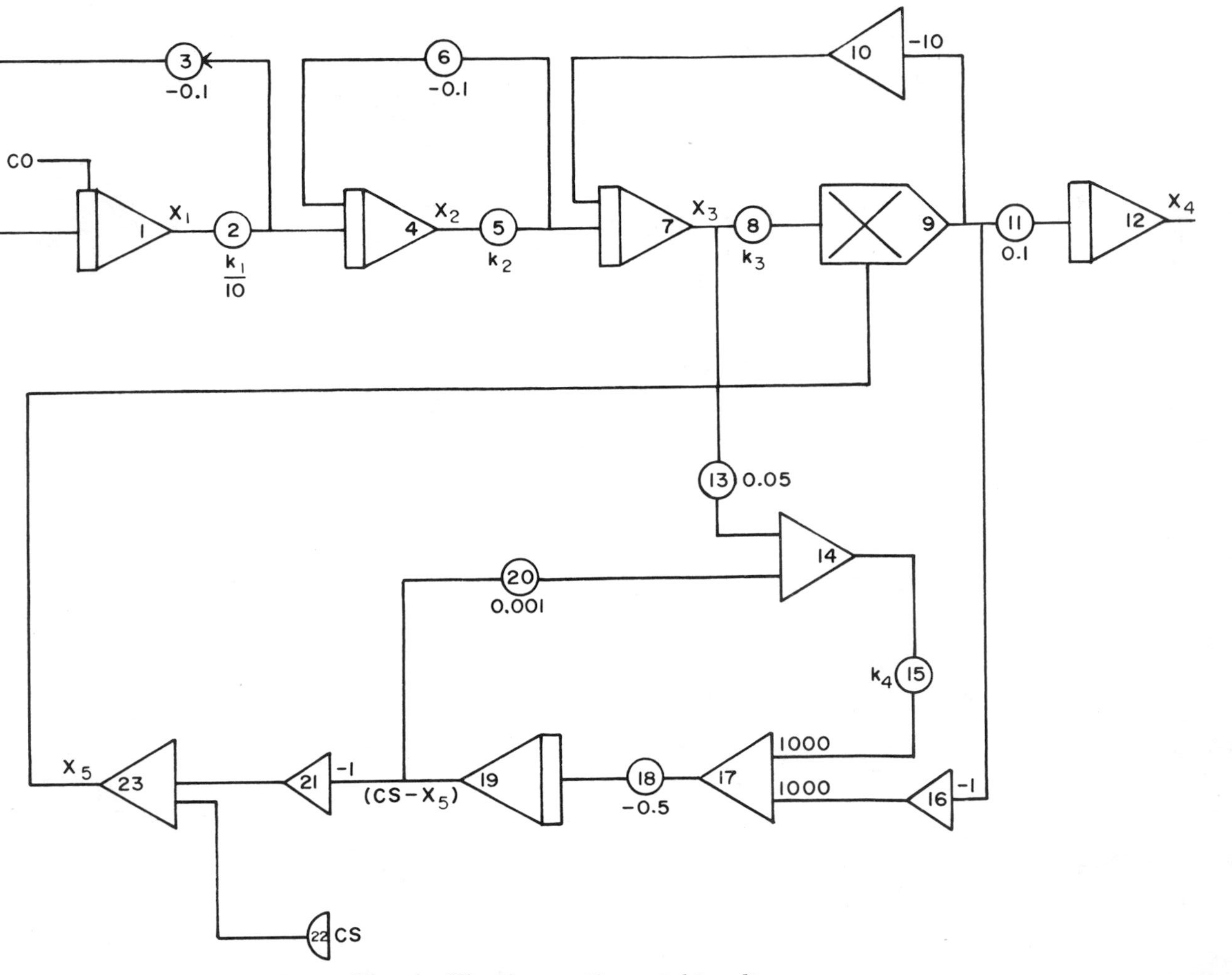


Fig. 4. Kinetic reaction patching diagram.

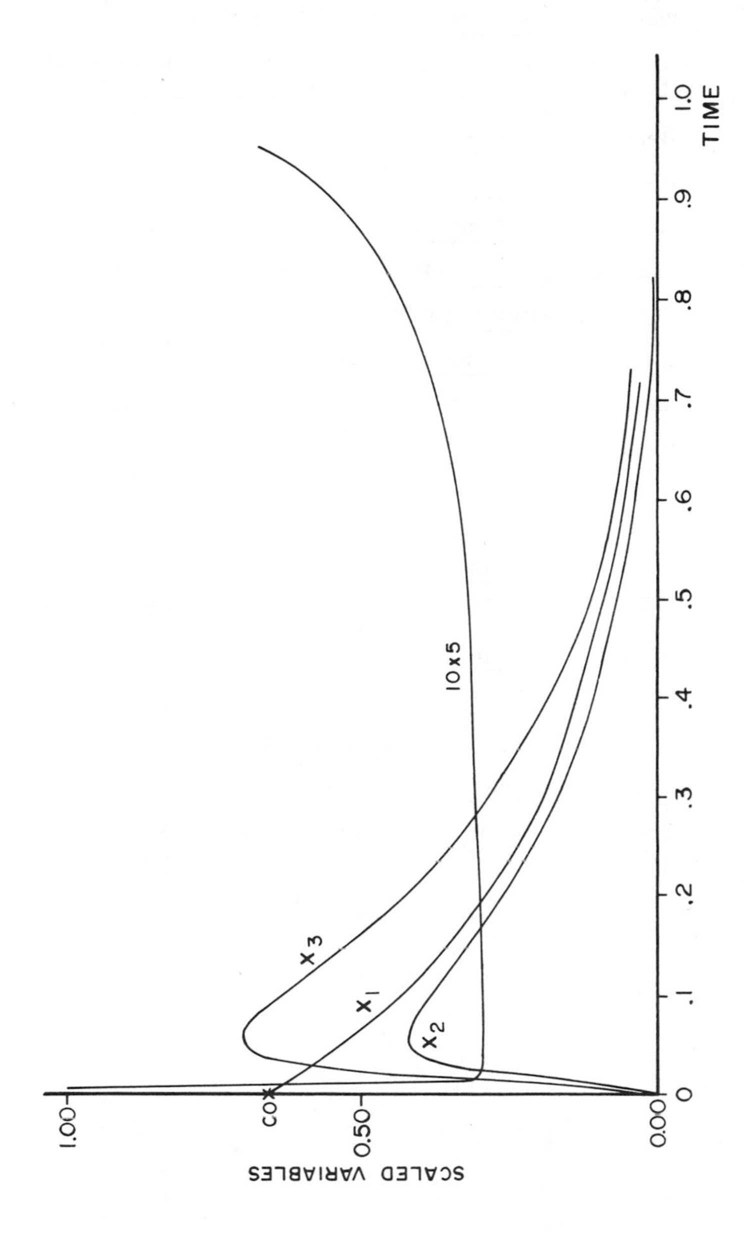


Fig. 5. Kinetic reaction responses.

## REFERENCES

[1]. J. G. Thomason, Computer J., 3, 4, 211 (1960).

[2]. J. Janata, in Computers in Chemistry and Instrumentation (J. Mattson, H. Mark, and H. MacDonald, eds.); Vol. 3, Dekker, New York, 1971.

[3]. P. R. Benyon, Simulation, 219 (Nov. 1968).

[4]. C. F. Haines, Computer J., 12, 183 (1969).

[5]. D. H. Brandin, Fall Joint Computer Conference, 345 (1968).

[6]. R. England, Computer J., 12, 166 (1969).

[7]. A. S. Chai, Simulation, 221 (May 1969).

[8]. H. R. Martens, Simulation, 87 (Feb. 1969).

[9]. H. Tramposch and H. A. Jones, Jr., Simulation, 73 (Feb. 1970).

[10]. R. G. Selfridge, Western Joint Computer Conf., 82 (1955).

[11]. R. N. Linebarger and R. D. Brennan, Simulation, 3, 6, 22 (Dec. 1964).

[12]. IBM Corporation, Form H20-0209-1 (1966).

[13]. R. D. Brennan and M. Y. Silberberg, IBM Systems J., 6, 242 (1967).

[14]. IBM Corporation, Form GH20-0367-3 (1969).

[15]. J. C. Strauss, et al., Simulation, 281 (Dec. 1967).

[16]. Xerox Data Systems, Manual 90 16 17A (1970).

[17]. J. G. Sellers, Electronic Letters, 4, 371 (1968).

[18]. J. G. Sellers, Ph.D. Thesis, University of Surrey, 1972.

[19]. H. C. Bowen, T. Chenevix-Trench, S. D. Drackley, R. C. Faust, and R. A. Saunders, J. Sci. Instr., 44, 343 (1967).

Chapter 7

# INTERFACING EXPERIMENTS TO COMPUTERS

D. K. Means

Research and Development Center for Electronics
Reliance Electric Company
Ann Arbor, Michigan 48106



## I. INTRODUCTION

Minicomputers are finding increased usage within individual laboratories, although the specific applications are as varied as the number of laboratories in which these computers are used. The needs for such a device fall into two broad categories. Conceptually, the simplest usage, and historically the earliest application of minicomputers, was for data acquisition and data reduction from experiments which typically produced more data than could be conveniently handled by a human being. When minicomputers took over this essentially repetitive task, they relieved a great burdern of data collection from the experimenter. This freed him for more productive work in designing new experiments and developing theories to unify the results produced by the combination of a sophisticated experiment and minicomputer logging system.

However, minicomputers soon outgrew this task and became capable of performing much more complex and time-consuming tasks. When this occurred some experimenters began to turn over, not only the data acquisition aspects of their experiments to tbe minicomputers, but also many of the control functions required during the experiment. Two important things resulted from this new close relation between the experiment and the minicomputer. First, more precise control of experimental parameters was achieved, since the minicomputer was not subject to fatigue as the human being is or to mechanical failure such as often happens when a home built control system is constructed for a specific experiment. Second, experimental parameters can be varied at much higher speeds than is possible with manual control or with speed control loops for each of the experimental variables. This high speed was a direct result of the assumption by the minicomputer of nearly all the control over the experiment.

For the purposes of experimental data logging and experimental control two types of computing systems need to be distinguished. Time-shared computing systems are generally available either through a captive large computer or through a commercial time sharing service. These systems offer extremely powerful programming capabilities but suffer from two serious drawbacks. First, the rate at which data (or control information) can be transferred to and from the computing system is relatively slow. Second, and potentially most troublesome, an unpredictable time delay can occur between the request for service and data output from the computing system. This time delay is a function of the total demand on the system resources, and this information is generally not available to an individual user. The combination of these two drawbacks severely limits the usefulness of the time-shared system for control purposes. However, for many experiments where the computer is to be used only for data collection this alternative is extremely useful.

In situations where high speed response and predictable time delays must be maintained, the dedicated computing system can be used to good advantage. This system is generally a minicomputer whose only job is to supervise one experiment and collect data from it. Such close association between the computer and the experiment leads to a very efficient experimental situation. The primary drawback to this happy state is that the minicomputer generally does not offer the very flexible and powerful programming capability that is offered through the time-sharing service. This problem significantly increases the programming tasks.

## II. DATA LOGGING SYSTEMS

When the task assigned to the computer is strictly one of collecting data and reducing it to a form that is useful to the experimenter, the requirements of the interface between the experiment and the computer are fairly simple. The interface must translate experimental measurements to the format which the computer understands. This often means that the interface must convert information from either an analog domain or a time domain to a digital form.

These conversions are performed by devices called analog-to-digital converters and frequency-to-digital converters, respectively. In addition to this translation function the interface must accept data upon a signal from the experiment and store those data until they can be accepted by the computer. This storage function can often be performed by a single memory device. Such devices are available from several manufacturers of Metal Oxide Semiconductors (MOS) memory chips. When a time-shared computer is used for data collection, the transmission of data to and from the computer is performed serially, generally over telephone lines. The use of these lines imposes several constraints on the interface at the experimenter's end of the telephone line. First, the bandwidth of these lines is limited, so that data rates must be relatively slow and special signaling systems must be employed in order to insure that the signal transmitted over the telephone line is within the proper frequency range. Since data from the experiment are almost always available in parallel form, one major function of the interface must be the conversion of this parallel data to a serial form acceptable for transmission over telephone lines. For those who enjoy circuit design this conversion can be built up using standard TTL elements. However, several manufacturers have recently introduced components which perform the parallel to serial and serial to parallel conversion within a single device. This has been made possible through large scale integration of MOS devices upon a single silicon chip.

To complete this type of interface, a control and timing section is almost always required. This section must generate timing signals to control the data conversion from the experiment, to control the transfer of data to and from the temporary storage, and to control data input and output from the parallel to serial and serial to parallel converter. A block diagram of the interface to a time-sharing system which performs a data collection function is given in Fig. 1.

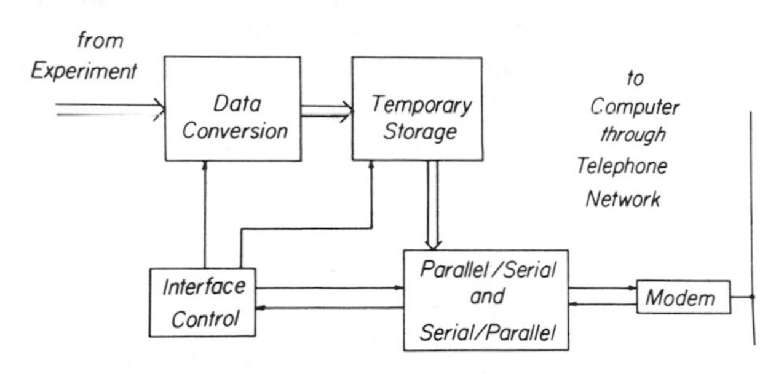


Fig. 1

When the volume of data to be collected from the experiment is extremely large or when the rate at which data must be collected from the experiment is extremely high the transmission characteristics of a time-shared computer link are totally unsatisfactory. In these cases a dedicated computer is required to handle data collection. A block diagram of the interface to such a computer is shown in Fig. 2. Since data are transferred to and from the computer in parallel fashion and since, in general, the computer can accept data as fast as they can be generated, the interface to this system is greatly simplified from that required for a time-share system.

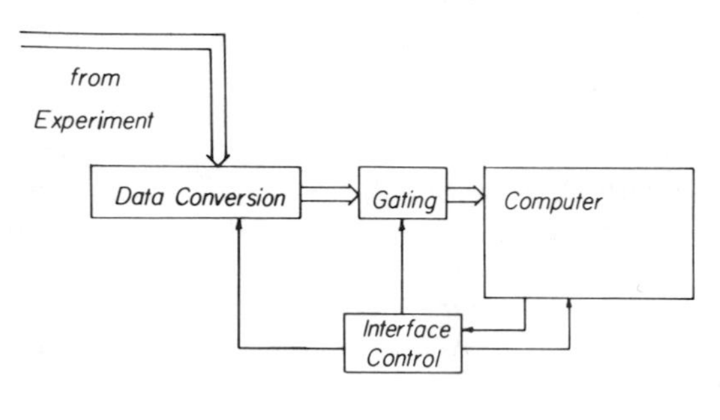


Fig. 2

Data conversion from the analog domain and time-frequency domain is still required. But, the output from these conversions can be gated directly into the computer memory either on a direct access basis or under program control.

## III. COMPUTER CONTROL SYSTEMS

Although the functions performed by the computer in a data logging and data reduction application are extremely valuable and can save a great deal of an experimenter's time, the full capability of a modern computer is not exploited in these applications. Much more of the capability available from these computers is exploited when responsibility for control of the experiment is assigned to the computer rather than to external hardware. Figure 3 shows in block diagram form how computer control of the experiment may be implemented. Data (including measurement of controlled variables) are

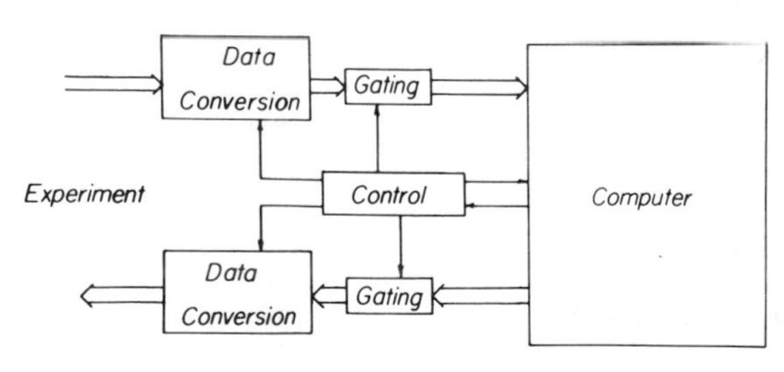


Fig. 3

converted by the input data conversion electronics to digital format. This information is then transferred into the computer and used in the control algorithms or data reduction algorithms. The results of the control algorithms are then gated back out to the computer interface and converted back to the proper form for direct control of experimental parameters. This last conversion often takes the form of either digital-to-analog conversion or digital-to-time conversion.

## IV. INTERFACE DESIGN

Conceptually, the most difficult problems associated with computer-controlled experiments are the problem of deriving a control algorithm and the problem of control loop stability. A complete body of theory has been developed covering these problems. * Therefore, we will pass these problems by and go on to the simpler problem of circuit design in the interface. The first problem is that of data conversion from either the analog domain or the time/frequency domain. Several techniques are available for converting analog signals to digital form, but in terms of circuit complexity and in terms of cost the successive approximation technique is clearly the leader for most applications. The circuits in this type of converter compare the incoming analog signal with an approximation to that signal which is successively refined. If the digital output contains n bits, then n comparisons are made between the incoming analog signal and the synthesized analog quantity, as indicated in Fig. 4. After each comparison a decision is made as to

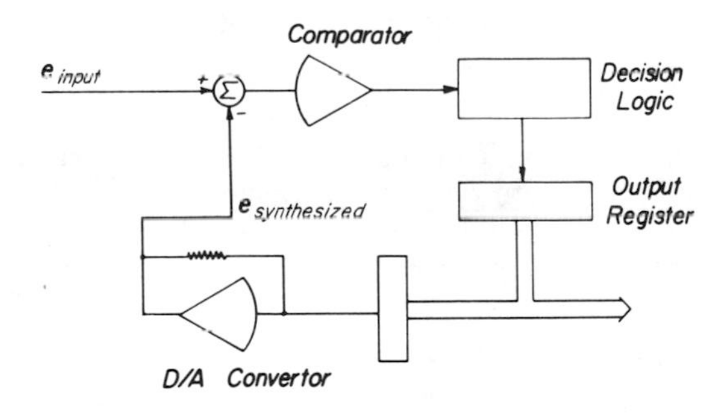


Fig. 4

whether the next digital bit is to be 1 or 0. After this decision is made, another comparison is carried out between the analog input and the partial result. This process is repeated until n comparisons have been made. The

*See for example Saucedo and Schiring, Introduction to Continuous and Digital Control Systems, MacMillan, New York, 1968.

result is then available in the output register and may be transferred to the computer. The individual components for this subsystem are readily available and quite cheap. Alternatively, the complete subsystem can be bought for a reasonable amount of money.

Conversion from the time or frequency domain is most readily implemented through counting techniques. When a time interval is to be measured a counter is set to 0 at the beginning of the interval and clock pulses (available from the computer with a well-defined frequency) are counted until the end of the time period. The result of this counting operation is a digital representation of the time interval to be measured. Conversion from a frequency domain requires two counters. One of these counters counts a predetermined number of clock pulses (thereby establishing a fixed time period). The second counter counts cycles of the input signal for the time period determined by the first counter. The result contained in the second counter at the end of the counting interval is then a direct digital representation of the input frequency.

Conversions in the other direction are handled in much the same way: digital-to-analog conversion is simply one part of the succession approximation analog-to-digital system, and digital-to-time conversion is part of the frequency-to-digital system. Direct conversion from the digital domain to the frequency domain can be accomplished by a circuit called a rate multiplier. This device accepts a digital word and a clock signal and produces from them a wave form whose average frequency is the product of a clock frequency and the digital word taken as a binary fraction less than 1. If a single frequency output is required, the simplest technique is to convert the digital data to an analog signal and then use that analog signal to control the frequency of a voltage-controlled oscillator.

The control section of any computer interface is potentially the most troublesome. Unfortunately, a design is subject to wide variations depending on the particular computer to which the experiment must be interfaced. Most computer manufacturers provide some clue as to how this control should be accomplished and, in fact, several of them provide general purpose interfaces for their own computers.

Broadly speaking, parallel transfer to the input-output systems fall into two categories: synchronous or clocked systems and asynchronous or handshaking systems. The first type provides several pulse lines to control data transfer in and out of the computer. The computer expects that the interface will either provide or accept data at certain well-defined times relative to these control pulses. This sometimes puts a severe constraint on the design of the interface and often requires temporary storage in order to avoid loss of data between the computer and a data source which operate at different speeds. The second type of input-output system is inherently more flexible since the computer presents a command to the interface and then waits while the interface either fetches data for input to the computer or accepts data from the computer. The interface is then responsible for signaling the computer when the data transfer is complete. This "hand-shaking system" provided for more secure data transfer and at the same time reduces the complexity of the interface between a computer and the outside world.

From another point of view minicomputer parallel input-output data transfers can be classified according to the type of programming required to support the transfer. The most basic form of data transfer, which all minicomputers are capable of, is program-controlled data transfer. In this case data are transferred to or from the outside world at times determined by the normal execution of program steps in the minicomputer. This type of data transfer, while logically the most straightforward, requires that the programmer pay strict attention to the sequence of events which must occur as a result of data transfers both into and out of the computer.

When data transfers are required at long or random intervals, minicomputers that use this type of data transfer exclusively spend a great deal of effort keeping track of the proper times to initiate the required transfers. A more efficient scheme for handling these transfers requires that the interface provide a signal to the computer when a data transfer is required. This signal causes a processor interrupt. When such an interrupt occurs the minicomputer suspends its normal activity (i.e., data reduction or control computation) and handles the required data transfer. When this chore is completed the computer returns to the original task it was performing. This technique for initiating data transfers is quite efficient for relatively low frequency requirements when extremely large amounts of data must be transferred in or out of the computer or when data must be transferred extremely rapidly.

A third technique may be employed at the expense of additional hardware. This technique is called direct memory access transfer. In most minicomputers the computer memory is idle for some fraction of each instruction cycle. During this time the computer processor electronics are calculating intermediate results to complete the instruction. During this idle time direct memory access electronics can take control of the memory and cause data transfers to occur directly between memory and an external data source. (In some newer minicomputers the memory speed is the limiting step in instruction times. In these machines direct memory access causes the processor simply to wait for one memory cycle time while the data transfer occurs. This delay causes a small increase in program execution time.)

Let us examine in more detail the mechanics of data transfer in the two major types of minicomputers. When the transfer occurs synchronously the computer raises a set of address lines which identify the input-output device (in this case the experiment interface) and enables a set of data lines. A short time later one or more control pulses is generated to define the direction of data transfer or other interface control function which is to take place. The interface must respond by either accepting or providing data to the computer at the proper time (which is a predefined function of the specific computer). Generally, data transfer in this type of computer interface is to and from one or more storage registers which are part of the experiment interface. Data and control signals then proceed from these registers to the experiment, perhaps by way of one or more conversion circuits.

When data transfer occurs asynchronously, the mechanics of transfer is simpler. The computer raises a set of device address lines and a set of control lines which indicate the type of transfer that is to occur and, a short time later, raises a line to indicate the beginning of the data transfer. At this point the interface takes over the timing of the data transfer. It either fetches data or accepts data from the computer, and when the data are either accepted or ready for transmission to the computer the interface raises a second line which indicates to the computer that data transfer is ready for completion. The computer electronics then regains control and completes the data transfer leaving the interface in its rest condition. Since time is allowed in this scheme for the interface to fetch data from a remote source or to transmit data to a remote destination a storage register for data is often not required. Data simply flow through the interface directly between the data conversion circuits and the computer electronics.

Interrupt service requests from the interface to the computer can vary from very simple single operations to extremely complex sequences of interchanges between the interface and the computer. In general, the more complex the hardware that is assigned to this task the simpler the software will be and conversely. At one end of the spectrum a single line going into the computer is available for requesting interrupt service. If this line is raised the computer will respond at some later time with a sequence of input and output commands generated through a special sequence of the storage program, called an interrupt handler. This software goes through a polling sequence in which the computer examines in turn each potential interrupting source to determine which source requires service. This scheme requires only minimal hardware but requires considerable computer time for each interrupt service request. Most very small minicomputers use this scheme for handling interrupt requests.

The software burden can be reduced in a number of ways. First, rather than requiring the processor to conduct a polling sequence, each interface can transmit a unique identification code along with its interrupt request. This code indicates to the computer where the corresponding interrupt service routine begins in the computer memory. Therefore, this type of interrupt is called a vector interrupt. As implied above, this scheme completely eliminates the polling sequence and directs the computer automatically to the specific interrupt service routine for the device requiring service. This procedure speeds up considerably the handling of interrupts from several sources.

In many situations where several interrupt devices are connected to the computer some devices must be serviced more rapidly than others. In order to speed up interrupt service to these high priority devices it is desirable that the computer be able to ignore certain interrupts selectively. Two schemes are fairly common in this area today in medium-sized minicomputers. One provides a separate connection for each interrupt device and a corresponding bit in an interrupt mask word. Normally all interrupts are enabled, but when a high priority interrupt request is received the program substitutes an interrupt mask word which disables all lower priority interrupts. This technique is extremely flexible, since it allows any device to be assigned in priority relative to the rest of the system. However, this flexibility requires some software support since a new mask word must be fetched from memory and deposited in the interrupt mask register each time an interrupt occurs.

A second scheme for providing several different priorities for interrupting devices utilizes a fixed number of predefined priority levels. Each interrupt source is assigned to one of these levels, and when an interrupt request occurs on any one of these levels the computer automatically raises its interrupt priority to match that of the interrupt to be serviced. While this interrupt service is going on all interrupt requests on low priority levels are delayed until the high priority interrupt service is completed. While this technique does not permit reordering device priorities during the program execution, it does relieve almost totally the software support required for servicing interrupts in a highly effective way.

## V. SUMMARY

Many problems associated with minicomputer interfaces can be understood on the basis of the foregoing discussion. The areas of difficulty are generally either a result of vague system planning (which results in an interface that "just grew"), or a result of timing conflicts between the requirements of the computer and the requirements of the experiment. By paying attention to the details supplied by the computer manufacturer, many of these pitfalls can be avoided.

The result of this care will be a computer/experiment interface that simply and efficiently performs, rather than a gadget that requires constant maintenance.

Chapter 8

# LEARNING MACHINES

T. L. Isenhour

Department of Chemistry
University of North Carolina
Chapel Hill, North Carolina

and

P. C. Jurs

Department of Chemistry
The Pennsylvania State University
University Park, Pennsylvania



## I. ANALYTICAL DATA TREATMENT

The interpretation of experimental data, and the corresponding establishment of cause and effect relations, is an essential aspect of experimental chemistry. Perhaps the interpretation of experimental data is the most pressing current problem in analytical chemistry. The advent of the high speed digital computer along with instrumentation that produces massive volumes of data are making information theory and related topics more and more essential to modern chemistry.

In general the investigator has data that he wishes to place into categories. For example, infrared spectra can be used to place compounds into categories defined by functional groups, or $pK_a$ values can be used to define the degradation products of certain protein reactions. Emission spectra can be placed in categories indicating the presence of certain trace elements, and a variety of techniques have been used to classify geological and cosmological aspects of the lunar samples collected in the Apollo program. The results of various electrochemical measurements can be categorized to indicate whether certain mechanisms are acceptable for given reactions. Placing data into specific categories, then, is very often the basis of interpretation of experimental results.

Two approaches can be used to relate data to categories, theoretical and empirical. Theoretical methods of data interpretation are usually preferred because they are based upon explicit causal relations derived from earlier observations or from logically constructed models. That is, scientists normally prefer interpretations based on theory, because they feel they understand the measurement process and underlying chemical phenomena in some or even all aspects. However, not even the most ardent theoretician would be likely to attempt the interpretation of the dc arc emission spectra of an iron alloy starting from first principles. Empirical methods are, however, readily applied in many common analytical situations; and, most frequently, some combination of the theoretical and empirical approaches is used. For example, while most scientists are satisfied with current theories of light absorption by molecules, it is standard procedure to measure the spectrum of a new compound and select a desirable absorption wavelength empirically in order to develop a colorimetric method.

The learning machine method, which is one approach to the overall subject of pattern recognition, is a totally empirical method of data interpretation. The sole assumption is that a relation between the data and the defined categories exists; that is, that the experiment measured something related to the property of interest. This assumption is made, of course, in any interpretive process. Even this assumption will be investigated by the empirical method itself. Hence, the learning method does not depend upon established theory and, while this is disadvantageous in that accepted hypotheses may not be used, it is simultaneously advantageous in that interpretation will not be restricted to current accepted schools of thought. The ultimate application of this approach may be in coupling it with current chemical theories and experience.

The term "learning" is used in this context to refer to a decision process which improves its performance of a task as its experience at performing the task increases. The application of negative feedback causes the decision process to be modified so as to discriminate against wrong answers, and therefore it improves its performance with time. In the general case, empirical relations are established between available inputs and desired outputs. In this chapter the inputs will be chemical measurements and the outputs will be the previously mentioned data categories.

## II. PATTERN RECOGNITION

Starting in the late 1940's a great many books, papers, and conference reports have dealt with the various phases of the theory, design, development, and use of learning machines (e.g., [1-15]). Such studies have been the province of applied mathematicians, statisticians, computer-oriented engineers, and others in several disciplines investigating biological behavior on the neural level. A recent review by Nagy [16] demonstrates the amorphous nature of the subject. Applications have appeared in such divergent scientific areas as character recognition (alphabetic and numeric), particle tracking (cloud, bubble, spark), fingerprint identification, speech analysis, weather prediction, medical diagnosis, and photographic processing (cell images and aerial photography). Recently, chemical applications have started to appear in a number of areas of spectroscopy [17-33].

The pattern recognition process will be described as four stages:

(1) Measurement
(2) Feature Selection and Preprocessing (Transformation)
(3) Decision Development (Discriminant Training)
(4) Generalization

### A. Measurement

The measurement process is generally not a problem in chemical applications. Indeed it has been said that the modern problems of data interpretation have been generated by the incredible rate at which modern instruments can produce data. The quality of data is generally excellent in the physical sciences and, in most cases, meaningful limits can be placed on accuracy and precision, and experiments can be repeated to check reproducibility.

Data to be used in pattern recognition studies are represented as vectors ($\underline{X} \equiv (x_1, x_2, \ldots, x_d)$). The utility of this representation is demonstrated by the following example. Figure 1 shows a two-dimensional plot of the melting points and boiling points of several organic compounds. A's represent organic acids and K's represent ketones. Note that each point in the two-dimensional space completely defines the two pieces of information and, furthermore, the points could be represented as two-dimensional vectors from the origin. It is clear that acids are high boiling and high melting, while the ketones are low boiling and low melting. Hence, from this figure an investigator who knew nothing of chemistry would immediately recognize that the acids and ketones cluster on this plot, and furthermore, that the experimental data available suggested a good method of distinguishing between

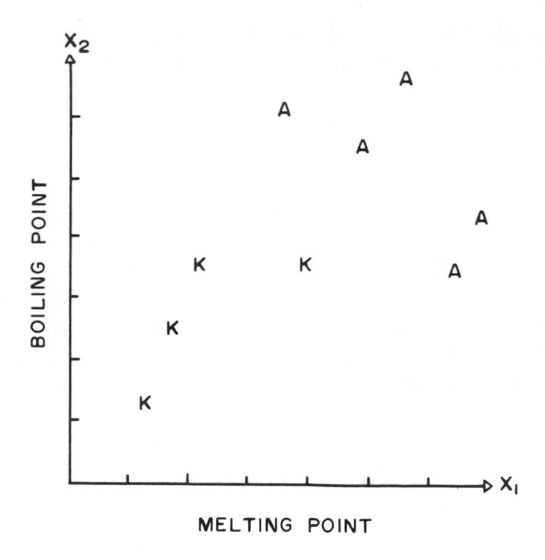


Fig. 1. Melting and boiling points of organic acids and ketones plotted in 2-dimensional space.

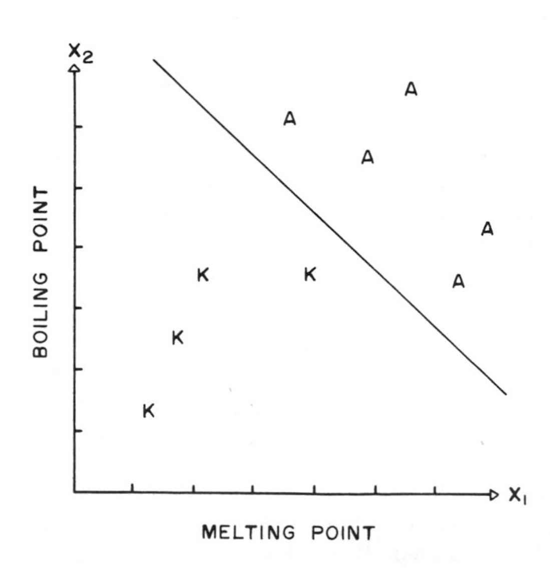


Fig. 2. A linear decision surface for organic acids and ketones.

the two categories, acids and ketones. Figure 2 makes clear the notion of linear separability, which just means that the pattern points can be placed into their two classes by a linear decision surface — a line for this two-dimensional case.

Most types of data can be represented in vector form by providing a sufficient number of dimensions. Chemical spectroscopy data, for example, are usually a spectrum of intensities versus frequency or wavelength. (Mass spectrometry is a notable exception where the abscissa is mass-to-charge ratio.) If the abscissa is quantized, the number of dimensions can simply be the range of the abscissa divided by the resolution. For example, an infrared spectrum recorded from 2.0 to 14.9 μm can be represented by a 130-component vector or a single point in 130-dimensional space without information loss. Approximately 150 dimensions are sufficient to represent low resolution mass spectra of simple organic compounds with up to ten carbons. The learning machine method is one approach to finding decision surfaces in such a multidimensional space where the experimenter can no longer plot the data and simply look for clusters or other trends.

## B. Feature Selection and Preprocessing (Transformation)

The basic overall objective of the pattern recognition method is to classify the patterns into the desired categories. Preprocessing of the data includes algebraic transformations such as the extraction of roots and taking of logarithms, feature selection, or changes in variables through transforms such as the Fourier transform. Such preprocessing can be useful for the following two reasons. First, some transformations can spread the clusters of the patterns in the two categories further apart in the pattern space, thus making discrimination easier. Second, preprocessing can reduce the dimensionality of the pattern space, either by discarding dimensions deemed expendable or by combining dimensions (possibly in very complex ways). The advantages gained by reducing the dimensionality will be discussed later. In general these two goals will not be served simultaneously by a given preprocessing method, and a compromise must be made.

Unfortunately, it is usually not possible to separate the preprocessing stage from the decision stage in pattern recognition systems. This adds to the complexity of studying the overall process. In addition any preprocessing necessarily imposes some bias on the method of interpretation and some of the advantages of the empirical method are thereby diminished.

Preprocessing of data before classification has received considerable attention (e.g., [34, 35]). In specific studies of chemical data several preprocessing operations have been investigated, including converting spectroscopic peak intensities to their square roots [18], or their logarithms [25], generating cross terms [28], and using Fourier transforms [27, 33]. A very promising transform method using Walsh functions for generating complex-valued nonlinear discriminate functions has recently been reported [36].

### C. Decision Development (Discriminant Training)

The widely accepted optimum method for making pattern classification decisions, known as Bayes strategies, depends on having the probability density functions for the classes. Suppose that it is desired to classify patterns represented by d-dimensional vectors $X \equiv (x_1, x_2, \ldots, x_d)$ into one of two possible categories. Let $F_1(X)$ and $F_2(X)$ be the probability density functions for the two categories, let $L_1$ and $L_2$ be the losses associated with misclassifying a member of category one or category two, and let $P_1$ and $P_2 = 1 - P_1$ be the *a priori* probabilities of occurrence of patterns in categories one and two. Then it can be shown [2] that the Bayes strategy says to make the decisions as follows:

$$\begin{aligned} &\text{If } P_1 L_1 F_1(X) > P_2 L_2 F_2(X) \text{ then classify } X \text{ in category } 1 \\ &\text{If } P_2 L_2 F_2(X) > P_1 L_1 F_1(X) \text{ then classify } X \text{ in category } 2 \end{aligned} \tag{1}$$

This procedure can be generalized to allow decisions among more than two classes. In order to use the optimum Bayes strategy, the probability density functions, loss functions, and *a priori* probabilities of each class must be either known or estimated.

If the distribution is not known, or cannot be approximated accurately, then one must either estimate the distribution and proceed accordingly or apply some nonparametric method. For the data produced by most chemical experimentation, systems are so complex that rarely is the distribution function known or easily estimable.

In most chemical experimentation, particularly that in the domain of chemical analysis, it is rare that any appreciable fraction of the universal set of data has been collected under controlled conditions. In mass spectrometry, one of the areas where considerable attention has been paid to the collection of data, large files typically contain several thousand entries, far short of the more than a million known chemical compounds, and much smaller in comparison to the imaginable number of compounds. Hence, at most times we are working with a very small subset of the universal set. Any direct assumption of the universal set from such a small subset could be misleading. For these reasons we resort to an empirical method for developing decision makers.

The principal decision process to be described in this chapter is the threshold logic unit (TLU) [5]. We will be concerned with TLU's which are binary pattern classifiers capable of placing a pattern in one of two categories. (This can, however, be made into a complete solution because a series of binary pattern classifiers may be used to subdivide data to any desired degree.)

The original data pattern is denoted by the vector $\underline{X}$. The TLU implements a plane of the same dimensionality as the patterns which will separate the data into the desired two classes. The 2-dimensional data shown in Fig. 1 may be separated into the desired categories by any of a family of straight lines (planes in two dimensions), one of which is shown in Fig. 2. In order to cause the decision plane to pass through the origin, an extra degree of freedom is added by augmenting the original d-dimensional pattern vector $\underline{X}$ by a (d + 1)-st dimension (which has the same value for every pattern) to give a new vector $\underline{Y}$. Usually an arbitrary value of 1 is given for the d + 1 component of each pattern. (This value can, however, have some effect on the development of decision makers [20], although it does not affect the separability of pattern sets.) Hence

$$\underline{X} \equiv (x_1, x_2, \ldots, x_d) \quad \text{and} \quad \underline{Y} \equiv (y_1, y_2, \ldots, y_d, y_{d+1}) \tag{2}$$

Figure 3 shows the effect for the 2-dimensional case given in Fig. 2. Now a three-dimensional plane which passes through the origin may be used to separate the pattern sets.

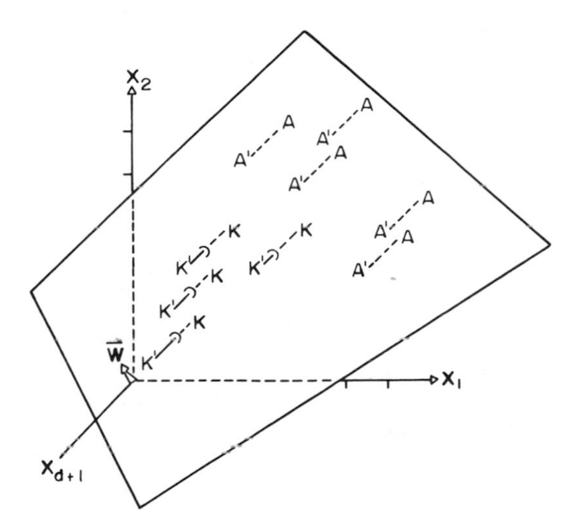


Fig. 3. Addition of a (d + 1)st dimension to allow the decision surface to pass through the origin.

A convenient way to determine whether a point lies on one side of the plane or the other is to use a vector normal to the plane at the origin. This vector (called a weight vector $\underline{W}$) may be thought of as defining the locus of points which constitute the plane separating the data classes (see Fig. 3). Because $\underline{W}$ is perpendicular to the plane, the product of $\underline{W}$ with any pattern vector ($\underline{Y}$) will determine whether the pattern lies on one side or the other of the plane.

$$s = \underline{W} \cdot \underline{Y} = |\underline{W}| \; |\underline{Y}| \cos\theta \tag{3}$$

where $\theta$ is the angle between the two vectors. $|\underline{W}|$ and $|\underline{Y}|$ are always positive, and thus

$$\begin{aligned} &\text{for } -90^\circ < \theta < 90^\circ, \quad \cos\theta > 0 \text{ and } s > 0 \\ &\text{for } 90^\circ < \theta < 270^\circ, \quad \cos\theta < 0 \text{ and } s < 0 \end{aligned} \tag{4}$$

Hence for patterns less than $90^\circ$ from $\underline{W}$ (and thereby on one side of the plane) the dot product is always positive, while for patterns on the other side of the plane the dot product is always negative. A further computational convenience is realized in another form of the dot product

$$s = \underline{W} \cdot \underline{Y} = w_1 y_1 + w_2 y_2 + \ldots w_d y_d + w_{d+1} y_{d+1} \tag{5}$$

This sum of the products of components of two arrays is a very easy operation to carry out in a digital computer.

The above derivation generalizes to any number of dimensions and provides a method for determining on which side of a hyperplane a given point lies in hyperspace. Arbitrarily, category 1 can be defined as the positive side of the hyperplane and category 2 as the negative side.

In order to develop a decision maker for a given classification, a training set of patterns, for which the correct categories are known, must be used. The members of the training set are presented to the classifier one at a time, and whenever a misclassification occurs, a correction process (negative feedback) is applied to the weight vector. This process continues until all patterns of the training set are correctly classified. Training is arbitrarily terminated after some preset number of feedbacks if convergence is not obtained in order to conserve computer time.

Several feedback methods have been used. One of the simplest, and most effective to date, is to move the decision hyperplane along the perpendicular axis between the misclassified point and the plane, so that after the correction it is the same distance on the correct side of the point as it was previously on the incorrect side. This movement is accomplished by adding an appropriate multiple of the pattern vector $\underline{Y}$ to the weight vector.

Thus

$$\underline{W} \cdot \underline{Y}_i = s \tag{6}$$

where s has the incorrect sign for classifying $\underline{Y}_i$. It is desired to find a new weight vector $\underline{W}'$ such that

$$\underline{W}' \cdot \underline{Y}_i = -s \tag{7}$$

by combining a fraction c of $\underline{Y}_i$ with $\underline{W}$

$$\underline{W}' = \underline{W} + c\underline{Y}_i \tag{8}$$

Combining Eqs. (7) and (8) gives

$$s' = \underline{W}' \cdot \underline{Y}_i = (\underline{W} + c\underline{Y}_i)\underline{Y}_i \tag{9}$$

which can be solved for c to give

$$c = \frac{s' - s}{\underline{Y}_i \cdot \underline{Y}_i} \tag{10}$$

If, as described, it is desired that $s' = -s$, then

$$c = \frac{-2s}{\underline{Y}_i \cdot \underline{Y}_i} \tag{11}$$

and the new weight vector $\underline{W}'$ is calculated from the equation

$$\underline{W}' = \underline{W} - \left(\frac{2s}{\underline{Y}_i \cdot \underline{Y}_i}\right)\underline{Y}_i \tag{12}$$

Of course, other methods can easily be derived which could also be properly termed negative error-correction feedback.

A slight generalization of this procedure results from the inclusion of a nonzero threshold, z, in the TLU. The classification rules become:

If $s > z$ classify pattern in category 1.
If $s < -z$ classify pattern in category 2.
If $-z < s < z$ do not classify the pattern.

The training procedure is as before with Eq. (10) becoming

$$c = \frac{2}{\underline{Y}_i \cdot \underline{Y}_i}(\pm z - s)$$

in which the sign is determined by the sense of the error being corrected.

## D. Generalization

The application of an error-correction feedback process to a set of data which is not known *a priori* to be linearly separable has some interesting aspects. First, successful convergence demonstrates that such a linear classification is possible, even though it might not have been predicted by theory. On the other hand, lack of convergence in less than an infinite number of feedbacks proves nothing. Programs have been written, however, that can accomplish a large number of investigations for a fairly low expenditure in computer time. It has been the experience of the authors, while using an error forcing computational algorithm, that convergence often occurs within a number of feedbacks that is about twice the number of patterns in the training set. Hence, while inseparability is never proven, it may be reasonably suspected without an inordinate expenditure of computer time.

The use of such empirically developed decision makers can be applied to the routine classification of data. Furthermore, it may be possible to learn something about the chemistry involved in the experimentation process. Once a relation is proved to exist, there is encouragement to try to determine the basis of that relation; and, the constitution of the successful decision maker may give hints as to the nature of the relation. Finally, there is the possibility of learning something about learning itself.

## III. FEATURES OF LEARNING MACHINES

Four parameters which help evaluate the performance of learning machines, and in the particular example used here, threshold logic units, are called recognition, reliability, convergence rate, and prediction.

a. Recognition is the ability of the trained pattern classifier to correctly classify members of the training set. Needless to say, recognition is always 100% for a decision surface which converged to the decision region of a separable set. Hence, once complete training has occurred the decision surface will always be able to correctly classify any member of the training set. This has potential application as a library. In reference [17] 26 weight

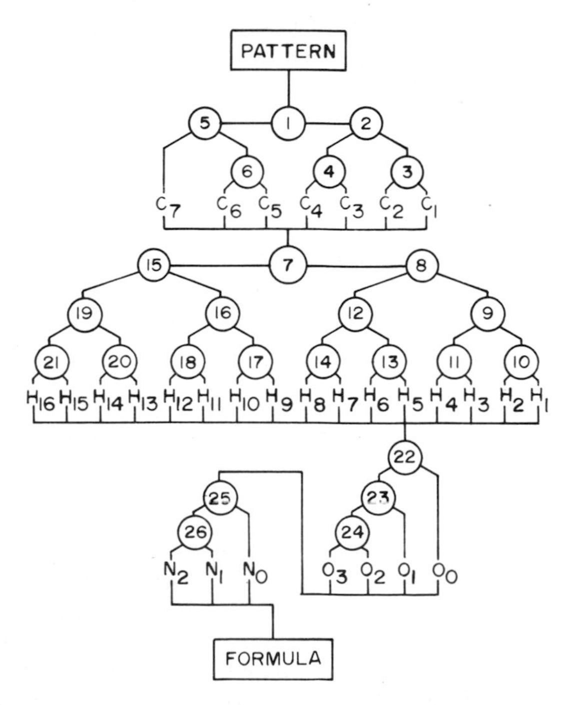


Fig. 4. Branching tree of binary decision makers for determining molecular formulas.

vectors were developed to correctly dichotomize formula subscripts for the mass spectra of 346 compounds from $C_{1-7}H_{1-16}O_{0-3}N_{0-2}$. By storing each of these weight vectors (each requiring the same dimensionality as the original patterns) the molecular formula could be computed using the binary decision tree shown in Fig. 4. This constitutes a storage savings of better than a factor of ten over retaining all the 346 spectra. It is not meant to imply, however, that the entire information content of the mass spectra has been compressed into the weight vectors, but rather the necessary set of information to determine molecular formulas. That is, by specifying the question to be answered, it was possible to reduce the amount of data to be retained. Furthermore, the molecular formula may be "computed" by a series of dot product calculations rather than a library search. (Such computation takes about 50 msec in a second generation computer or 30 min at a desk calculator.)

Other chemical data for which 100% recognition has been obtained for certain questions include: infrared spectra [19], powder diffraction spectra, simulated NMR spectra [31], water resources data, combined infrared-mass spectra [21], gamma-ray spectra [22], and electrochemical data [26].

It is possible that the future may see calculations of weight vectors for large data banks with only the resultant weight vectors supplied to individual laboratories, thereby eliminating the need for repeated development of the same decision makers.

b. Reliability refers to the ability of the decision makers to classify members of the training set which have been distorted. Any data collection process has some noise level. In the case of chemical experimentation it is normal to collect slightly different data each time an experiment is run, even if the source of the data is the same compound or system. If a decision maker is radically affected by small changes in the patterns it classifies, then it is of limited use. In order to test the procedure, its reliability must be known. To determine the reliability of the weight vectors of Fig. 4, the machine was tested on spectra which had been randomly varied to simulate data as they might emerge from a low resolution mass spectrometer. The error in the peak intensities of the spectra was assumed to be Gaussian. After the intensity of each peak was independently varied using a Gaussian distributed random number generator, the molecular formula was determined by the master program. A tabulation was kept of the number of times the master program erred in 1000 such trials for various standard deviations and is presented in Table 1. The stringency of the random variations imposed upon the spectra should be emphasized, because most instrumental variations in a mass spectrometer cause peak intensities to shift in a related fashion rather than randomly. For example, the University of Washington Chemistry Department CEC 21-103 had a relative standard deviation in the ratio of the relative intensities of the masses 45/58 of n-butane of 6.9% over a period of six years during which no particular attention was paid to maintaining long

TABLE 1

Errors in Statistically Varied Data

| σ (%) | Number incorrect out of 1000 | Per cent correct | Error occurrence | | | |
|---|---|---|---|---|---|---|
| | | | C | H | O | N |
| 1 | 0 | 100.0 | 0 | 0 | 0 | 0 |
| 4 | 7 | 99.3 | 2 | 4 | 0 | 1 |
| 10 | 24 | 97.6 | 12 | 9 | 2 | 1 |
| 15 | 30 | 97.0 | 13 | 10 | 2 | 6 |
| 20 | 42 | 95.8 | 19 | 16 | 4 | 5 |

range stability [37]. For a standard deviation of this magnitude, the learning machine can determine molecular formulas with a reliability greater than 98%. It should be noted that the original data are essentially randomly varied because they come from various spectrometers in different laboratories; however, for this learning machine the spectra are standards and thus the reliability testing was in order.

c. Convergence rate is an important component in determining the expense required to develop decision makers. Schemes have been developed for minimizing convergence time, thereby maximizing effective convergence rates, the most promising of which is reducing the dimensionality of the data without losing the important information. This will also be important in the prediction. Examples of convergence rate will be given along with those of prediction. Some training procedures use error forcing procedures to increase convergence rate.

d. Prediction refers to the ability of the pattern classifier to correctly classify patterns which were not members of the training set. Prediction is unquestionably the most interesting and exciting aspect of pattern recognition. If the decision maker can correctly classify patterns that are unknowns, then it is established, or at least strongly implied, that the decision process has learned something of the chemistry of the experiment. Prediction is typically tested by dividing the available known patterns into two sets using a random selection process. One set is used to train the decision vector and the other to test it for prediction. During the prediction test no further feedback is allowed so that the test is fully one of unknown patterns. For a binary decision maker, 50% prediction would be expected from random guessing.

## A. Effect of Training Set Size on Predictive Ability

Table 2 shows the results of training weight vectors to determine oxygen presence while using different sized training sets chosen randomly from an overall set of 630 low resolution mass spectra. Normally the convergence

TABLE 2

Convergence Rate and Prediction of Oxygen Presence as a Function of Training Set Size

| Training set size | Spectra tested[a] | Per cent predicted | Average per cent predicted |
|---|---|---|---|
| 50 | 226 | 88.3 | |
| | 180 | 84.3 | 86.2 |
| | 255 | 85.9 | |
| 100 | 654 | 90.8 | |
| | 561 | 87.7 | 88.2 |
| | 574 | 86.0 | |
| 200 | 1528 | 86.6 | |
| | 1525 | 88.6 | 88.4 |
| | 2189 | 88.1 | |
| 300 | 2590 | 92.7 | |
| | 1959 | 90.6 | 90.6 |
| | 2986 | 88.5 | |

[a] Number of spectra required to produce complete training.

rate decreases as the training set becomes larger. This trend is seen in Table 2, where all weight vectors were trained to complete recognition. It is interesting to note that predictive ability is high even for training set sizes of 50 and 100 where there are more adjustable parameters (weight vector components corresponding to mass positions) than there are patterns in the training set. Predictive ability is expected to increase as the training set size increases, and Table 2 demonstrates this, although the noise level is considerable.

## B. Training for Structural Features of Hydrocarbons

Other studies used a set of 387 hydrocarbon compounds, from which a subset of 200 were chosen randomly to serve as a training set for developing weight vectors and the other 187 were used to test the predictive ability of the weight vectors. Table 3 shows the results of training and testing the 43 weight vectors developed for the determination of structural parameters of hydrocarbons. Each vector was trained to give a binary decision. In all cases, except the carbon:hydrogen ratio, a positive answer indicated the value was greater than the cutoff number, and a negative value indicated that the value was less than or equal to the cutoff. For example, the first vector (carbon

TABLE 3

CH Class

| | | Training set | | | Prediction set | | |
|---|---|---|---|---|---|---|---|
| | Cutoff | Negative category | Positive category | Spectra feedback | Negative category | Positive category | Per cent feedback |
| Carbon number | 9 | 163 | 37 | 227 | 154 | 33 | 89.3 |
| | 8 | 121 | 79 | 167 | 113 | 74 | 92.5 |
| | 7 | 80 | 120 | 185 | 77 | 110 | 93.6 |
| | 6 | 53 | 147 | 99 | 44 | 143 | 94.1 |
| | 5 | 30 | 170 | 71 | 21 | 166 | 97.9 |
| | 4 | 15 | 185 | 42 | 7 | 180 | 97.3 |
| Hydrogen number | 20 | 196 | 4 | 53 | 182 | 5 | 97.3 |
| | 18 | 168 | 32 | 170 | 154 | 33 | 95.7 |
| | 16 | 143 | 57 | 202 | 132 | 55 | 97.3 |
| | 14 | 110 | 90 | 58 | 100 | 87 | 94.1 |
| | 12 | 72 | 138 | 51 | 55 | 132 | 95.2 |
| | 10 | 47 | 153 | 59 | 34 | 153 | 96.8 |
| | 8 | 28 | 172 | 31 | 19 | 168 | 96.8 |
| | 6 | 9 | 191 | 34 | 10 | 177 | 97.3 |
| Carbon:hydrogen ratio | 2n + 2 | 156 | 44 | 25 | 143 | 44 | 96.8 |
| | 2n | 125 | 75 | 28 | 107 | 80 | 96.8 |
| | 2n - 2 | 153 | 47 | 36 | 154 | 33 | 96.8 |
| | 2n - 4 | 191 | 9 | 39 | 180 | 7 | 98.9 |
| | 2n - 6 | 185 | 15 | 13 | 170 | 17 | 98.9 |

| | | | | | | | |
|---|---|---|---|---|---|---|---|
| Methyl | 4 | 191 | 9 | 177 | 165 | 22 | 90.9 |
| | 3 | 160 | 40 | 800 | 136 | 51 | 86.1 |
| | 2 | 114 | 86 | 859 | 97 | 90 | 86.6 |
| | 1 | 62 | 138 | 648 | 47 | 140 | 86.6 |
| | 0 | 29 | 171 | 328 | 24 | 163 | 89.3 |
| Ethyl | 1 | 166 | 34 | 1518 | 153 | 34 | 80.7 |
| | 0 | 104 | 96 | >2000 | 97 | 90 | 73.3 |
| n-Propyl | 1 | 191 | 9 | 211 | 177 | 10 | 90.4 |
| | 0 | 145 | 55 | >2000 | 148 | 39 | 71.7 |
| Largest ring | 6 | 199 | 1 | 11 | 184 | 3 | 97.9 |
| | 5 | 142 | 58 | 141 | 137 | 50 | 89.8 |
| | 4 | 121 | 79 | 149 | 114 | 73 | 90.9 |
| | 3 | 120 | 80 | 194 | 112 | 75 | 91.4 |
| Branch point carbons | 2 | 174 | 26 | 356 | 163 | 24 | 90.9 |
| | 1 | 108 | 92 | >2000 | 90 | 97 | 61.5 |
| | 0 | 42 | 158 | 1486 | 36 | 151 | 87.2 |
| Number of -C=C- | 2 | 171 | 29 | 11 | 165 | 22 | 98.4 |
| | 1 | 159 | 41 | 153 | 158 | 29 | 92.5 |
| | 0 | 102 | 98 | >2000 | 98 | 89 | 77.5 |
| Carbons w/o hydrogens | 1 | 167 | 33 | 432 | 156 | 31 | 83.4 |
| | 0 | 111 | 89 | 1327 | 97 | 90 | 70.6 |
| Benzene ring | 0 | 179 | 21 | 37 | 168 | 19 | 96.8 |
| -C≡C- | 0 | 184 | 16 | 163 | 174 | 13 | 93.6 |
| Vinyl | 0 | 166 | 34 | >2000 | 158 | 29 | 80.2 |

number 9) was trained to give a positive dot product with a mass spectrum of a compound containing ten carbons and a negative dot product with one containing nine or less carbons. The carbon:hydrogen ratio was changed to give a positive result for that particular ratio and a negative result for any other. For example, n-hexane gives a positive dot product for carbon:hydrogen ratio 2n + 2 and a negative dot product for all other ratios. The categories appearing in Table 3 are defined as follows: methyl, ethyl, and n-propyl numbers are the number of each group which can be produced by a single bond rupture, i.e., 3-methylhexane has 3 methyls, 2 ethyls, and 1 n-propyl by this definition; the largest ring classification includes saturated, unsaturated, or aromatic rings; and, branch point carbon number is the number of carbon atoms in the compound which are bonded directly to at least three other carbon atoms. For number of carbon-carbon double bonds, benzene has been classed as three, and the "carbon w/o hydrogen" category refers to carbons which are not bonded to any hydrogens. The final three weight vectors detect presence or absence of benzene rings, acetylenic bonds, and vinyl structural features.

The third and fourth columns of Table 3 indicate the number of compounds of the training set which fell into each class. The fifth column gives the number of feedbacks which were necessary for convergence with >2000 indicated for those that had not been completely trained after 2000 feedbacks. (As indicated earlier, failure to train in some arbitrary number of feedbacks does not prove the given training set to be linearly inseparable, and, indeed, incompletely trained vectors may still have considerable recognition and prediction ability.) The sixth and seventh columns indicate the number of compounds in the prediction set which fell into each class. The final column gives the prediction success for the 187 compounds which were not part of the training set. It should be stressed that these compounds were treated by the machine in every way as complete unknowns. The only difference from real unknowns is that the results of these computations may be evaluated. Predictive ability ranged from 61.5% to 98.9% with an average of 90.3%. The prediction percentage can be used as a gauge of the credibility of an answer when produced for an unknown spectrum. Random guessing would give 50% success. It is seen that in the CH class considerable structural information may be derived with a high confidence level from a completely empirical calculation method.

### C. Interpretation of Infrared Spectra

A study was made of the application of these pattern classification techniques to the interpretation of infrared spectra.

The data used in this study were made available on loan by the Sadtler Research Laboratories, Inc., Philadelphia, Pa., and are part of the same data used in the Sadtler IR Prism Retrieval System. From the 24,142 spectra of standard compounds, the first 4500 were selected that satisfied the requirements of no more than ten carbon atoms, four oxygen atoms, three nitrogen atoms, and no additional elements except hydrogen. These Sadtler spectra are recorded in 0.1-μ bands from 2.0 to 14.9 μ. This gives a total of 130 pattern dimensions, including the d + 1 term. The amplitude of each pattern component is assigned one of four values based on the intensity of the strongest absorption in that 0.1-μ band. For the largest peak in the spectrum, the amplitude is set at 3.0; for the largest peak in a 1.0-μ band, the amplitude is 2.0; for other peaks, the amplitude is 1.0; and for no peak in a given 0.1-μ band, the amplitude is set at 0.0. Because the amplitude data were limited to only three nonzero values, and the majority of the values were zero, it was useful to compress the patterns to save computer storage and decrease computation time. Each pattern was, therefore, rewritten as a series of integers as follows:

$$n_1, p_1, p_2, p_3, \ldots p_{n_1}, n_2, q_1, q_2, q_3, \ldots q_{n_2}, n_3, r_1, r_2, r_3, \ldots r_{n_3}$$

where $n_1$ is the number of peaks with amplitudes of 1.0 followed by the dimensions (or positions) of those peaks, $n_2$ is the number of peaks with amplitudes of 2.0 followed by the dimensions of those peaks, and $n_3$ is the number of peaks with amplitudes of 3.0 followed by the dimensions of those peaks. The resultant patterns required less than one-third the storage of the uncompressed spectra. Furthermore, the dot product process used to form the scalars necessary for classification could be performed as follows. If $\underline{W}$ is the weight vector and $\underline{Y}$ is the pattern for which the dot product is to be formed, then

$$\underline{W} \cdot \underline{Y} = w_1 \cdot y_1 + w_2 \cdot y_2 + w_3 \cdot y_3 \ldots w_{d+1} \cdot y_{d+1} \tag{13}$$

However, if Y contains only values of 0.0, 1.0, 2.0, and 3.0, the dot product may be computed by

$$\underline{W} \cdot \underline{Y} = 1.0 \times \sum_{j-1}^{n_1} w_{p_j} + 2.0 \times \sum_{j-1}^{n_2} W_{q_j} + 3.0 \times \sum_{j-1}^{n_3} w_{r_j} + w_{d+1} \tag{14}$$

For the patterns used, this method of computing dot products decreased computation time by roughly a factor of twenty.

In all parts of this study, the patterns were classified on the basis of the chemical classes supplied with the Sadtler spectrum for each compound.

Each of the nineteen available chemical classes was treated by a linear learning machine to develop weight vectors. Four thousand spectra were used in each case, with five hundred randomly selected as the training set and the other three thousand five hundred used as a prediction set. Table 4 shows the results of this approach. In every case, the learning machine completely converged for each training set of five hundred and thereby had 100% recognition for the patterns with which it was developed. However, it is seen from Table 4 that, in almost every case, the percentage in the negative category is greater than the percentage predictive success. This means that a greater

TABLE 4

Training with Uneven Populations

| | Positive numbers of training set | Number fed back | Per cent prediction | Positive members overall | Total wrong |
|---|---|---|---|---|---|
| Carboxylic acid | 64 | 926 | 78.2 | 583 | 764 |
| Esters of carboxylic acids | 83 | 437 | 75.3 | 480 | 514 |
| Linear amides | 27 | 123 | 86.4 | 272 | 477 |
| Ketones | 32 | 255 | 83.2 | 374 | 587 |
| 1° Alcohols | 33 | 251 | 87.9 | 307 | 423 |
| Phenols | 60 | 286 | 81.8 | 365 | 636 |
| 1° Amines | 50 | 977 | 82.2 | 436 | 624 |
| Ethers acetals | 78 | 3429 | 78.5 | 630 | 752 |
| Nitro, nitroso compounds | 16 | 64 | 91.3 | 268 | 306 |
| Cyclic amides | 7 | 49 | 90.7 | 222 | 325 |
| Nitrites, isonitrites | 25 | 166 | 93.8 | 162 | 217 |
| Urea and derivations | 3 | 45 | 94.3 | 103 | 199 |
| Aldehydes | 21 | 155 | 90.8 | 124 | 323 |
| 2° Alcohols | 31 | 310 | 89.5 | 228 | 369 |
| 3° Alcohols | 20 | 179 | 94.6 | 97 | 189 |
| 2° Amines | 37 | 407 | 82.8 | 334 | 604 |
| 3° Amines | 50 | 661 | 77.1 | 613 | 803 |
| 1° Amine salts | 6 | 61 | 92.9 | 125 | 249 |
| Unsaturated hydrocarbons | 28 | 112 | 93.0 | 115 | 246 |

number of right answers could be obtained by always guessing that the compound was not in the class in question. This again amounts to not answering the question at all. Hence, with such uneven categories there is question as to whether the learning machine approach is useful.

For this reason, subsets of the entire set were chosen randomly to give training sets with equal numbers in each category. For such a set of three hundred compounds, one hundred and fifty of which contained nitro or nitroso groups, complete training occurred and a prediction of 80% was recorded for a different prediction set also randomly chosen and also containing one hundred fifty compounds in each category. Now a random guessing method would only give 50% success and it can be concluded that the classifier is indeed learning about the relation between a compound's infrared spectrum and its chemical structure.

TABLE 5

**Training for Carboxylic Acids with Different Training Set Size**

| Training set size | Number of feedbacks | Prediction percentage |
|---|---|---|
| 10 | 5 | 62 |
| 20 | 10 | 63 |
| 30 | 10 | 61 |
| 40 | 15 | 56 |
| 50 | 22 | 61 |
| 100 | 36 | 67 |
| 150 | 103 | 71 |
| 200 | 167 | 71 |
| 250 | 199 | 72 |
| 300 | 341 | 73 |
| 350 | 398 | 71 |
| 400 | 597 | 71 |
| 450 | 1816 | 68 |
| 500 | a | 71 |

[a] Complete training not obtained after 20, 000 feedback (Recognition = 83.7%).

To determine the influence training set size has on the prediction success, the carboxylic acid group was treated with even training sets varying from 10 to 500 patterns. The results are shown in Table 5. While there are some statistical fluctuations in the data, little improvement is noted for training sets of above one or two hundred. For this reason training sets of three hundred were chosen for the remainder of the investigations.

The nine chemical classes for which there were enough compounds to form balanced training sets of three hundred and balanced prediction sets of three hundred were used to test the learning machine method on infrared date. The results are given in Table 6. For each chemical class two weight vectors were developed, one with all initial components set equal to +1.0 and the other with all initial components at -1.0. Complete training was accomplished in every case, and predictive abilities were all above random with the highest around 80%. Furthermore, it should be realized that complete training means the weight vector is able to perfectly classify spectra of compounds which were in the training set. Hence in these particular cases, a dot product calculation can answer any given question without resorting to the lengthy task of comparing a spectrum to each of the three hundred members of the library.

TABLE 6

Training Using Even Categories[a]

| Chemical class | Weight vector initiated positive | | Weight vector initiated negative | |
|---|---|---|---|---|
| | Number of feedbacks | Per cent prediction | Number of feedbacks | Per cent prediction |
| Carboxylic acids | 341 | 74 | 291 | 74 |
| Esters of carboxylic acids | 304 | 76 | 258 | 76 |
| Linear amides | 153 | 78 | 140 | 80 |
| Ketones | 411 | 73 | 372 | 73 |
| $1^{\circ}$ Alcohols | 1015 | 69 | 708 | 69 |
| Phenols | 205 | 76 | 127 | 82 |
| $1^{\circ}$ Amines | 841 | 71 | 586 | 69 |
| Ethers and acetals | 2291 | 63 | 1595 | 62 |
| Nitro, nitroso compounds | 157 | 81 | 127 | 82 |

[a]Training Set 300 (150/150). Prediction Set 300 (150/150).

As discussed in earlier work, weight vectors from different starting points may have different predictive abilities even though both were produced from the same training set which had been perfectly classified. The hyperplanes represented by the weight vectors so developed act, in effect, to partially described a no-decision region of the pattern space. That is, any pattern which falls inside this portion of the hyperspace was not represented in the training set and will not have as great a probability of correct classification as one falling outside the no-decision region. To verify this concept, the two different weight vectors developed for each chemical class in Table 6 were used to reclassify each member of the prediction set in the following manner. If both weight vectors gave the same classification, then the compound was placed in that category. However, when the weight vectors disagreed, the compound was not classified at all, but rather indicated as a compound for which the classification was outside the capability of the learning machine as presently structured. The results are shown in Table 7. It may be seen by comparing Table 7 to Table 6 that, in every case when prediction was made by the second method, it was better than or equal to the prediction by a single weight vector. However, with the no-decision method fewer patterns are classified. For example, in the case of the nitro or nitroso class, 81% are correctly predicted by the positively initiated weight vector and 82% by the negatively initiated weight vector, while 87% of the predictions using the no-decision process are correct. However, 15.3% of the compounds fell into the no-decision region. It must be realized that

TABLE 7

Results of Prediction Using Two Weight Vectors Simultaneously

Prediction Set 300 (150/150)

| | Per cent performance | | | Per cent correct | Per cent wrong | Per cent decided |
|---|---|---|---|---|---|---|
| | Predicted positive | Predicted negative | Average | | | |
| Carboxylic acids | 75.7 | 81.7 | 78.7 | 66.0 | 18.3 | 15.7 |
| Esters of carboxylic acids | 81.8 | 82.0 | 81.9 | 67.7 | 15.0 | 17.3 |
| Linear amides | 82.8 | 87.1 | 85.0 | 70.7 | 12.7 | 16.7 |
| Ketones | 73.9 | 82.4 | 78.2 | 64.7 | 18.7 | 16.7 |
| $1^{\circ}$ Alcohols | 73.9 | 71.1 | 72.5 | 62.0 | 23.7 | 14.3 |
| Phenols | 76.5 | 84.1 | 80.3 | 68.0 | 17.3 | 14.7 |
| $1^{\circ}$ Amines | 72.2 | 75.0 | 73.6 | 63.0 | 23.0 | 14.0 |
| Ethers and acetals | 63.5 | 64.9 | 64.2 | 57.3 | 32.0 | 10.7 |
| Nitro and nitroso | 88.5 | 85.5 | 87.0 | 73.7 | 11.0 | 15.3 |

random guessing of the 15.3% in the no-decision region would reduce the success to the 81-82% of the single vector method. In other words, it is seen and was true in virtually every case tested that no real gain is made by classifying patterns in the no-decision region.

The logical extension of this approach would be to develop as many different weight vectors as convenient from different initial vectors and use a multiple-decision process in which the confidence in the result would be a function of the number of vectors that give agreeing classifications.

Various methods have been demonstrated to improve each of the four features. Undoubtedly the choice of method, particularly in preprocessing and feature selection, will be strongly dictated by the type of data. Combining of data from various sources has been demonstrated to be feasible [21]. In order to evaluate the effect of combining data from diverse sources, compounds were classified on the presence or absence of one or more double bonds. Attempts to answer this question using infrared patterns or mass spectrometry patterns met with limited success as shown in the first two sections of Table 8. In each case parameters are discarded from the 125 starting parameters by the previously described criterion [18]. Prediction is around 82% for the infrared case and falls off rapidly after the number of parameters is below 50. The mass spectrometry gives prediction in the region of 87% which slowly decreases until the number of parameters is reduced below 20 when it drops off rapidly toward random success. The

TABLE 8

Double Bond Presence

| Section | Para-meters | No. of feed-backs | Recog-nition | Per cent prediction | MS/ir |
|---|---|---|---|---|---|
| 1. Infrared | 262 | | | | |
| | 162 | | | | |
| | 125 | 156 | x | 79 | |
| | 100 | 144 | x | 83 | |
| | 70 | 177 | x | 82 | |
| | 50 | 261 | x | 77 | |
| | 30 | >5000 | 168 | 62 | |
| | 20 | >5000 | 136 | 47 | |
| | 10 | >5000 | 129 | 50 | |
| | 5 | >5000 | 129 | 69 | |
| 2. Mass spectrometry | 262 | | | | |
| | 162 | | | | |
| | 125 | 1105 | x | 87 | |
| | 100 | 958 | x | 88 | |
| | 70 | 1529 | x | 85 | |
| | 50 | 1677 | x | 84 | |
| | 30 | >5000 | 179 | 81 | |
| | 20 | >5000 | 178 | 81 | |
| | 10 | >5000 | 141 | 69 | |
| | 5 | >5000 | 131 | 56 | |
| 3. Combined patterns | 262 | 1182 | x | 86 | 136/126 |
| | 162 | 1024 | x | 84 | 99/6 |
| | 125 | 1005 | x | 87 | 92/33 |
| | 100 | 1056 | x | 85 | 88/12 |
| | 70 | 1176 | x | 85 | 69/1 |
| | 50 | 1845 | x | 85 | 50/0 |
| | 30 | >5000 | 182 | 79 | 30/0 |
| | 20 | >5000 | 179 | 82 | 20/0 |
| | 10 | >5000 | 135 | 61 | 10/0 |
| | 5 | >5000 | 142 | 61 | 5/0 |

Table 8 (continued)

| | Para-meters | No. of feed-backs | Recog-nition | Per cent prediction | MS/ir |
|---|---|---|---|---|---|
| 4. Combined patterns | 262 | 141 | x | 78 | 136/126 |
| | 162 | 149 | x | 80 | 52/110 |
| | 125 | 145 | x | 83 | 24/101 |
| | 100 | 133 | x | 80 | 11/89 |
| | 70 | 121 | x | 81 | 1/69 |
| | 50 | 128 | x | 78 | 0/50 |
| | 30 | >5000 | 186 | 69 | 0/30 |
| | 20 | >5000 | 140 | 52 | 0/20 |
| | 10 | >5000 | 136 | 52 | 0/10 |
| | 5 | >5000 | 139 | 55 | 0/5 |
| 5. Combined patterns | 262 | 105 | x | 89 | 136/126 |
| | 162 | 82 | x | 88 | 76/86 |
| | 125 | 82 | x | 89 | 61/64 |
| | 100 | 87 | x | 90 | 46/54 |
| | 70 | 93 | x | 89 | 30/40 |
| | 50 | 169 | x | 89 | 23/27 |
| | 30 | 146 | x | 92 | 13/17 |
| | 20 | >5000 | 174 | 88 | 9/11 |
| | 10 | >5000 | 164 | 75 | 6/4 |
| | 5 | >5000 | 137 | 62 | 4/1 |
| 6. Combined patterns | 262 | 87 | x | 87 | 136/126/2 [a] |
| | 162 | 83 | x | 88 | 76/76/2 |
| | 125 | 90 | x | 89 | 55/70/2 |
| | 100 | 88 | x | 89 | 44/56/2 |
| | 70 | 121 | x | 91 | 31/39/2 |
| | 50 | 129 | x | 90 | 25/25/2 |
| | 30 | 468 | x | 89 | 15/15/2 |
| | 20 | 1520 | x | 92 | 12/8/2 |
| | 10 | >5000 | 176 | 84 | 8/2/2 |
| | 5 | >5000 | 129 | 64 | 5/1/1 |

[a] MS/ir/other.

mass spectrometry also fails to converge within the allotted number of feedbacks below 50 parameters.

Sections 3, 4, and 5 of Table 8 show the results of training with patterns formed by combining the mass spectrum and infrared spectrum. In Section 3, the patterns have been normalized such that the mass spectral intensities are much greater than the infrared components. The mass spectral peak intensities were on the range 10 to 100 and the ir intensities were set to 0, 1.0, 2.0, 3.0. Thus the contribution from the ir data to the overall length of any pattern vector is relatively small compared to that from the mass spectrum portion of the pattern. As is expected the results are almost identical to those of the mass spectra alone, and by the time the number of parameters is 50 there remain only components of the mass spectral patterns. In Section 4, the results of normalizing the patterns so that the infrared data predominate is shown. Here the ir intensities are set to 0, 1000, 2000, 3000, so most of the pattern vector's length is due to the ir contribution. As expected the results are quite like the independent infrared spectra, and when only 50 parameters are left they are all from infrared patterns.

Section 5 shows the results of normalizing the patterns so that each data source contributes equally to the total amplitude of the pattern set. In this case the prediction starts out around 90% and remains very high until the number of pattern components is reduced below 20. Even with only 10 components left, prediction is 75% and recognition is 82% showing better than random behavior. It is also interesting to note that components of both the mass spectra and the infrared spectra have been retained throughout the parameter reduction process.

Section 6 of Table 8 shows the further improvement gained by adding two more data points to the patterns, the melting point and boiling point for each compound. (Note that in every case where the boiling and melting points are included, the total number of parameters is 2 more than in the combined MS/ir patterns or the uncombined patterns.) In comparison to Section 5 no particular improvement is noted down to 30 parameters, both the predictive abilities and convergence rates being about the same. However, with addition of the boiling and melting points, the learning machine still converges within the limit with only 20 parameters and also retains approximately a 90% predictive ability at that level. Furthermore, the prediction at 10 parameters is still notably higher than the other cases. The decision process used to reduce the number of parameters contributing little to the classifications retains the melting and boiling point information almost until the end of the calculation.

There is thus no restriction by the pattern recognition method that the data must come all from one source. Indeed, one might go so far as to envision the "automated laboratory data analyzer" as a simple computer with a minimal central processor capable of calling on a mass storage device containing a variety of weight vectors and carrying out the necessary vector multiplications and binary branches to arrive at the desired results. Such a system could be instructed by the decision vectors developed at a large central system.

## IV. IMPROVEMENT OF LEARNING MACHINE CAPABILITIES

The previous section has described work done with linear TLU's and relatively simple transformations. This section describes some more advanced methods that have been investigated.

### A. Preprocessing, Feature Selection, Transformation

#### 1. Feature Selection

In investigating the classification powers of TLU's, it is extremely desirable to perform feature extraction on the original data patterns before classification, that is, reduce the number of features per pattern as much as possible. Two reasons are as follows: (1) The length of the dot product operation depends on the number of features per pattern. In terms of the computer program implementing the learning machine procedure, the number of multiplications in forming the dot product $\underline{W} \cdot \underline{X} = s$ is a monotonically increasing function of the number of dimensions of $\underline{W}$ and $\underline{X}$. Thus, reducing the dimensionality of $\underline{W}$ and $\underline{X}$ decreases the dot product computation time. A substantial decrease in the dimensionality of the system makes it possible to calculate these dot products on a desk calculator instead of with a computer. (2) By reducing the number of features per pattern, one can investigate which portions of the original data are most important with regard to the classification being made. That is, if the number of features per pattern can be substantially reduced with no concurrent loss in recognition ability or predictive ability of the lower dimensional pattern classifiers, then one has narrowed the locus of information placement within the original pattern. Additionally, if the pattern components have physical meaning, one can then look at the results in terms of the physical interpretation of this reduction.

Many methods have been proposed in the literature for the extraction of important features from data destined for subsequent use in pattern classification. Most of them rest on statistical treatments of the data, and they often involve calculating eigenvalues and eigenvectors of covariance matrices (e.g., [34]). These methods are not directly applicable to the type of physically meaningful data being considered here for several reasons. The basic reason is that mass spectra peak intensities do not obviously follow simplified probability density distributions, the situation for which most of the statistical methods of feature extraction are designed. Other, nonstatistical methods must evidently be used to extract features from such complicated data sources as mass spectra.

Mass spectra data consisting of 119 features per pattern were investigated by a feature selection program based on a linear learning machine method. The sequence of operations is as follows. Two weight vectors are to be trained to detect oxygen presence/absence. One of the weight vectors is initialized with all components set equal to +1, and the other one is initialized with all components set to -1. The two weight vectors are then trained to classify the members of the training set. Their predictive abilities are tested, and then the m/e positions not contributing to the classification are discarded by the following scheme. The components of the two weight vectors which correspond to the same m/e position are compared; if both

TABLE 9

Training for Oxygen Presence

| m/e positions | +1 weight vector | | -1 weight vector | | Av. % prediction |
|---|---|---|---|---|---|
| | Feedbacks | Prediction, % | Feedbacks | Prediction, % | |
| 119 | 236 | 92.1 | 219 | 93.3 | 92.7 |
| 69 | 179 | 93.6 | 221 | 93.3 | 93.4 |
| 51 | 208 | 94.9 | 223 | 94.9 | 94.9 |
| 43 | 202 | 94.2 | 256 | 94.9 | 94.5 |
| 40 | 217 | 92.7 | 217 | 94.2 | 93.4 |
| 38 | 184 | 93.9 | 203 | 92.7 | 93.3 |
| 37 | 199 | 94.6 | 213 | 93.9 | 94.2 |
| 31 | 235 | 94.6 | 202 | 93.3 | 93.9 |
| | | | | Ave. | 93.8 |

| | + | - | Total |
|---|---|---|---|
| Training set | 82 | 218 | 300 |
| Prediction set | 92 | 238 | 330 |

| m/e positions | +1 weight vector | | -1 weight vector | | Av. % prediction |
|---|---|---|---|---|---|
| | Feedbacks | Prediction, % | Feedbacks | Prediction, % | |
| 119 | 219 | 90.0 | 208 | 93.3 | 91.7 |
| 74 | 187 | 93.3 | 174 | 92.7 | 93.0 |
| 53 | 179 | 93.3 | 187 | 92.7 | 93.0 |
| 45 | 165 | 92.7 | 281 | 92.1 | 92.4 |
| 42 | 161 | 92.1 | 221 | 93.0 | 92.5 |
| 38 | 158 | 92.1 | 223 | 93.3 | 92.7 |
| 31 | 210 | 93.0 | 192 | 93.9 | 93.5 |
| | | | | Ave. | 92.7 |

components have the same sign, the m/e position is said to correlate well with oxygen absence if they are both negative and oxygen presence if they are both positive, and that m/e position is retained. But m/e positions for which the signs are different are considered ambiguous and are discarded. After all the m/e positions have been checked, the entire cycle begins anew with the spectra of reduced dimensionality. The process is repeated as long as the feature extraction routine can find m/e positions which are ambiguous. Table 9 shows the results of applying the feature selection procedure to oxygen presence/absence classification. The pattern classifiers are trained to detect oxygen presence in the compound yielding each spectrum regardless of the type of oxygen group present. The populations of the training set and prediction set are given at the bottom of the table. The training set contained 82 spectra of compounds that contain oxygen and 218 which do not. The first column in Table 9 gives the number of m/e positions being considered at each stage of the feature selection process. Column two gives the number of feedbacks performed while training to complete recognition of the members of the training set by the weight vector initiated with +1's. Column four is the same parameter for the weight vector initiated with -1's. Columns three and five give the predictive percentages exhibited by the two weight vectors on the 330 members of the prediction set for each stage of the feature selection process.

Through seven iterations, the number of features per pattern was reduced from 119 to 37 m/e positions. Despite this decrease in dimensionality, the number of feedbacks performed during training remains approximately constant; the total computer time used in training thus falls because each classification involves a shorter calculation in a lower dimensional space. The average predictive ability remains high even with only 37 out of the original 195 m/e positions being considered.

The lower part of Table 9 shows the results of another test conducted identically to that of the top of the table except that a somewhat different training procedure was used. The difference in the training procedure is that the members of the training set are presented in a different sequence, thus yielding different decision surfaces. The results are comparable. After the two feature selection routines had been allowed to reduce the number of features per pattern to 37 and 38 m/e positions, respectively, a list of features common to both lists was compiled. Thirty-one m/e positions were selected by both feature selection routines. Then both routines were allowed to train using these 31 m/e positions. Neither routine could find any more ambiguous m/e positions, and they both terminated with the results shown at the bottom of the two sections of the table. It is interesting to note that in both cases the ability of the classifiers to correctly categorize complete unknowns was nearly the highest observed for any training sequence, 93.9% and 93.5%, respectively. It appears that removal of the ambiguous m/e positions from the problem does not degrade the classifier's performance.

TABLE 10

Probable Fragments of the 14 m/e Positions Correlating with Oxygen Presence

| m/e | Fragments |
|---|---|
| 14 | $CH_2$ |
| 27 | $C_2H_3$ |
| 31 | $CH_3O$ |
| 37 | $C_3H$ |
| 38 | $C_3H_2$, $C_2N$ |
| 43 | $C_2H_3O$, $C_3H_7$, $CH_3N_2$, $C_2H_5N$ |
| 45 | $C_2H_5O$, $C_2H_7N$(r) |
| 46 | $C_2H_6O$ |
| 59 | $C_3H_7O$, $C_2H_3O_2$, $C_2H_7N_2$, $C_2H_5NO$(r) |
| 69 | $C_4H_5O$, $C_5H_9$ |
| 73 | $C_4H_9O$, $C_4H_{11}N$(r) |
| 83 | $C_5H_7O$, $C_6H_{11}$ |
| 100 | $C_5H_{10}NO$, $C_6H_{14}N$, $C_6H_{12}O$(r) |
| 135 | $C_8H_7O_2$, $C_9H_{11}O$, $C_8H_9NO$ |

Of the 31 m/e positions selected by the feature selection process, 14 have positive weight vector components — i. e., they correlate with oxygen presence. Table 10 lists these 14 m/e positions along with the fragments that correspond to each position, taken from a published compilation of mass spectral fragments [38]. Rearrangement reactions are denoted by (r).

Many of the m/e positions in the list are those that would be expected, especially the series 31, 45, 59, 73. Most of the m/e positions correspond to fragments containing oxygen and are therefore not surprising. However, other m/e ratios such as 14, 27, 37, and 38 have no oxygen-containing fragments. These peaks appear to arise preferentially from the oxygen-containing compounds in the training set. They apparently are used by the learning machine to classify the compounds of the training set which cannot be classified on the basis of their oxygen-containing fragments alone.

## 2. Fourier Transform Representation of Mass Spectra

The advent of the fast Fourier transform [39] has been of great significance in spectroscopy because it has made it economical to perform finite Fourier transformation on a digital computer.

The FT is most often thought of as relating the time domain to the frequency domain, e. g., the positive part of the FT of a finite cosine wave is just a peak centered at the frequency of the wave. More generally the FT is just frequency per unit of X where f(X) is the function being transformed. (A discussion of the general theory and some applications of the Fourier transform can be found in the book by Bracewell [40].) In interferometry X represents distance while in mass spectrometry X corresponds to mass/charge ratio. That is, the transform may be thought of as a frequency analysis of the original spectrum.

A second method of interpreting the transform is helpful in appreciating data transmission advantages in the Fourier domain: each point in the Fourier domain is a weighted sum of all the points in the original spectrum, that is, the data at a given point (e.g., a mass position) in the original domain is spread out over the entire spectrum in the Fourier domain. This is referred to as the averaging property of the transform. Hence when errors occur or bits are lost during transmission of the data in the Fourier domain there will be a minimal effect on the original spectrum, whereas had the data been transmitted in the mass domain loss of one bit could possibly result in a meaningless spectrum. It will be shown that this effect can be put to good advantage in pattern classification as well.

It is convenient to introduce some notation at this point to facilitate discussion. The FT of real data is generally complex, having a real part originating from the cosine transform and an imaginary part originating from the sine transform. The FT of a function f(X) is given by

$$G(\nu) = \int_{-\infty}^{\infty} f(X) e^{i2\pi\nu X} \, dX \qquad (15)$$

or equivalently

$$G(\nu) = \int_{-\infty}^{\infty} f(X) \cos 2\pi\nu X \, dX + i \int_{-\infty}^{\infty} f(X) \sin 2\pi\nu X \, dX \qquad (16)$$

$$= G_c(\nu) + iG_s(\nu)$$

In practice f(X) has finite limits and is set to 0 beyond these limits such that the integration is over a finite interval. Two additional means of representing the data in the Fourier domain are easily calculated from Eq. (16). These are the phase spectrum defined by

$$\Phi(\nu) = \arctan[G_s(\nu)/G_c(\nu)] \qquad (17)$$

and the intensity spectrum defined by

$$I(\nu) = [G_c(\nu)^2 + G_s(\nu)^2]^{1/2} \qquad (18)$$

With respect to pattern classification, the transformed data are a different representation of the same information which may make the implementation of a classifier simpler or more amenable to linear methods. The purpose here is to illustrate the use of the fast Fourier transform as an aid to pattern classification of real data (low resolution mass spectra) and to show how in practice the averaging property may be exploited for reduction of dimensionality.

Four different data sets were established, each consisting of 630 patterns. These are derived from (1) the cosine part of the transform (real part of G), (2) the sine part of the transform (imaginary part of G), (3) the phase spectra, and (4) the intensity spectra.

The FFT algorithm requires $2^N$ points, where N is a positive integer, for calculation of the transformation. Hence each mass spectra of 200 mass positions is augmented to 256 mass positions by addending 56 zeros. Upon application of the FFT there results a set of 512 points, 256 each from the cosine and sine parts of the transform. The practical significance of this with real data is that the cosine part is even whereas the sine part is odd, hence all information is contained in 128 points for either case. The other 128 points are easily generated by symmetry, and hence, are redundant. The data thus consist of the four sets of 630 patterns each, each pattern consisting of 128 components.

To determine whether the basic information is still present in the Fourier domain, pattern classifiers were developed on the transformed data and compared with those developed in the mass domain for the same questions. Some illustrative results are shown in Table 11. In each case the positive category consists of the compounds with the indicated carbon-hydrogen ratio with all other compounds making up the negative category. The data set

TABLE 11

Feasibility of Training and Prediction in the Fourier Domain Using C/H Questions[a]

| Positive category | Feedbacks to convergence | | | | | Prediction (% correct) | | | | |
|---|---|---|---|---|---|---|---|---|---|---|
| | MS | CT | ST | IS | PS | MS | CT | ST | IS | PS |
| $C_nH_{2n+2}$ | 30 | 40 | 145 | 286 | 161 | 97 | 93 | 96 | 96 | 79 |
| $C_nH_{2n}$ | 107 | 358 | 364 | 2100 | 94 | 95 | 95 | 94 | 94 | 83 |
| $C_nH_{2n-2}$ | 35 | 55 | 46 | 1380 | 115 | 96 | 95 | 96 | 88 | 89 |

[a]MS = mass spectra; CT = cosine transform; ST = sine transform; IS = intensity spectra; PS = phase spectra.

consisted of the 387 CH compounds, 200 of these randomly chosen as a training set while the remaining 187 were used as a prediction set. The most important result is that the basic information is still easily obtained from the Fourier data. Convergence rate (as measured by the number of feedbacks necessary to attain complete convergence) and predictive ability (classification of unknown patterns) in the Fourier domain are comparable to that in the mass domain.

The usefulness of the trained weight vectors is measured in two ways in this work. (1) Attaining complete convergence, and thereby proving linear separability, in a reasonable number of feedbacks is necessary if $\underline{W}$ is to be utilized in lieu of conventional data search and retrieval systems. (2) Accurate predictive performance of the trained weight vector is required if it is to be used as a pattern classifier for unknowns patterns. In work, described above, five out of forty-three categories of hydrocarbons, for which linear separability of the low resolution mass spectra was expected, were not found to be separable after allowing 2000 feedbacks for a training set of 200 compounds. The results of testing the Fourier domain for each of these categories are shown in Table 12 in comparison to the mass spectra treatment. The categories ethyl, n-propyl, and vinyl indicate compounds that have such groups, and a branch point carbon is one with three or more carbon-carbon bonds. Notice that, by using the phase spectra, convergence was accomplished in every case while no other form of the data was made to converge in any of the cases. Hence it is seen that the Fourier form of data may answer questions that the original data could not answer — at least within the convergence limit applied. This satisfies the first criterion for use as a searching device. However, prediction is poor for all questions, hence the second criterion is not satisfied.

TABLE 12

Fourier Domain Results on Questions Which Did Not Converge in Mass Domain[a]

| Positive category | Feedbacks to convergence | | | | | Prediction (% correct) | | | | |
|---|---|---|---|---|---|---|---|---|---|---|
| | MS | CT | ST | CST | PS | MS | CT | ST | CST | PS |
| c=c | >2000[b] | >10000 | >10000 | >10000 | 160 | 78 | 73 | 75 | 72 | 75 |
| Vinyl | >2000 | >10000 | >10000 | >10000 | 199 | 80 | 79 | 76 | 65 | 73 |
| Ethyl | >2000 | | | >10000 | 645 | 73 | | | 62 | 67 |
| n-Propyl | >2000 | | | >10000 | 422 | 72 | | | 60 | 81 |
| >2-Branch carbons | >2000 | | | >10000 | 217 | 62 | | | 62 | 66 |

[a]MS = mass spectra; CT = cosine transform; ST = sine transform; CST = cosine + sine transforms; PS = phase spectra.

[b]The symbol > signifies that training was discontinued at the indicated number of feedbacks without having attained convergence.

TABLE 13

Comparison of Dimension Reduction by Various Methods in Fourier Domain with the Mass Domain

| | Dimensions | Feedbacks to convergence | Prediction (% correct) | Feedbacks to convergence | Prediction (% correct) | Feedbacks to convergence | Prediction (% correct) |
|---|---|---|---|---|---|---|---|
| Fourier spectra cosine part | 128 | 75 | 95 | 75 | 95 | 75 | 95 |
| | 96 | 79 | 95 | 81 | 96 | 92 | 97 |
| | 64 | 162 | 98 | 103 | 95 | 147 | 98 |
| | 48 | 319 | 97 | 118 | 94 | 293 | 98 |
| | 32 | 2696 | 97 | 112 | 93 | 2088 | 97 |
| | 16 | >8000 | 80 | >8000 | 94 | >8000 | 97 |
| Mass spectra | 128 | 74 | 94 | 74 | 94 | 74 | 94 |
| | 96 | 116 | 96 | 142 | 98 | 74 | 94 |
| | 64 | 231 | 95 | 156 | 96 | 73 | 95 |
| | 48 | >2666 | 90 | 781 | 94 | 82 | 95 |
| | 32 | | | >4000 | 82 | 126 | 95 |
| | 16 | | | | | 575 | 94 |
| | | Highest dimensions omitted | | Dimensions omitted randomly | | Smallest average magnitude dimensions omitted | |

It is often desirable in problems of pattern classification and search and retrieval systems to reduce the dimensionality of the data. This greatly decreases both computation time and data storage requirements. The information in each mass position is spread over all dimensions in the Fourier domain by the averaging property mentioned. Therefore some components can be arbitrarily omitted in the Fourier domain and the basic spectrum is still obtained upon inverse transformation. (Therein lies one advantage of the transmission of images in the Fourier domain.) The problems of data collection and manipulation operations during which errors occur and the arbitrary omission of data are similar to the problems of noisy data transmission. Hence, due to averaging properties, the Fourier transformed patterns should be preferable to normal patterns such as mass spectra for the reduction of dimensionality.

Pattern classifiers were developed for mass spectra and the cosine (real) component of the Fourier transforms thereof, after reduction of dimensionality by several methods. The question was a C/H ratio of 1/2 versus all other CH compounds. The same training set of 200 compounds consisting of 81 compounds with a CH ratio of 1/2 was used in each case. The remaining 187 compounds consisting of 74 with a CH ratio of 1/2 were used in forming the prediction set. The results are summarized in Table 13. (Computation was terminated in each case when a set computation time was reached.) In every case the Fourier form of the data allows considerable reduction without serious increase in convergence time or decrease in predictive ability. The effect of data omission in the mass spectra varies, with that case in which dimensions are discarded on the basis of smallest average magnitude being the best method tried, at least as far as convergence rate to the final step to sixteen dimensions is considered. Two of the three methods using the Fourier data actually show slightly better predictive ability than the best mass spectra results. Hence it is seen that the Fourier data may be arbitrarily reduced in dimensionality to a great extent before degradation becomes noticeable, whereas the original mass spectra may only be so treated when a logical criterion, such as the smallest average magnitude peak, is used.

## B. Discriminant Training

### Committee Machine

The work discussed above utilizes a single threshold logic unit to classify patterns. A slight generalization involves using two levels of TLU's [5]. In this method the pattern to be classified is presented simultaneously to three (or five or seven, etc.) TLU's which operate in the normal manner. However, the outputs of the first layer of TLU's go to a second layer TLU which follows a majority rule law. Thus, the whole committee of classifiers sends its output to the vote-taker which classifies the original pattern into the category agreed upon by the majority of the first level TLU's.

TABLE 14

Comparison of Properties of Four Pattern Classification Methods. Oxygen Presence-Absence Determination

| | Randomization | Feedbacks +1WV/-1WV | % prediction +1WV/-1WV | Ave. % prediction | % recognition | |
|---|---|---|---|---|---|---|
| | | | | | $\sigma = 2\%$ | $\sigma = 5\%$ |
| Linear machine with threshold $\Delta = 50$ | 1 | 239/184 | 96.0/96.0 | 96.0 | 99.6 | 98.6 |
| | 2 | 286/211 | 96.3/95.7 | 95.8 | | |
| | 3 | 157/179 | 95.7/94.3 | 95.0 | | |
| Committee machine with threshold $\Delta = 50$ | 1 | 375 | | 98.0 | 100.0 | 99.8 |
| | 2 | 297 | | 96.3 | | |
| | 3 | 398 | | 95.7 | | |
| | Randomization | Training set, +/- | Prediction set, +/- | | | |
| | 1 | 84/216 | 89/211 | | | |
| | 2 | 92/208 | 81/219 | | | |
| | 3 | 78/222 | 95/205 | | | |

TABLE 15

Properties of Layered Pattern Classifiers as a Function of Threshold, $\Delta$

| $\Delta$ | Feedbacks | % prediction | % recognition | | % prediction | |
|---|---|---|---|---|---|---|
| | | | $\sigma = 2\%$ | $\sigma = 5\%$ | $\sigma = 2\%$ | % fall |
| 0 | 180 | 96.0 | 95.0 | 90.7 | 92.8 | 3.3 |
| 25 | 397 | 97.0 | 99.8 | 98.5 | 96.7 | 0.3 |
| 50 | 375 | 98.0 | 100.0 | 99.8 | 97.7 | 0.3 |
| 75 | 359 | 97.3 | 99.8 | 100.0 | 96.5 | 0.8 |

A useful and simple training method is to make only the minimum necessary error correction feedback at each step in the training. Thus, when an incorrect classification is made during training, the weights of the TLU's that missed being correct by the smallest amount are altered. Only enough TLU's are changed to ensure correct classification after the feedback. Table 14 shows the results of training a committee machine with three members with the same three sets of data used by the other pattern classification implementations. Convergence to 100% recognition is seen to be fast, and the predictive ability is high. This pattern classifier could correctly classify 95.7, 96.3, and 98.0% of the members of the prediction set, which are complete unknowns. The percentage recognition is seen to be very high also; this implementation is nearly impervious to errors in the patterns of this size.

A study was made of some of the properties of committee machines as a function of threshold size. The results are shown in Table 15. The number of feedbacks necessary for convergence jumps dramatically for t = 25 and then remains at approximately the same level. The predictive ability goes through a maximum for t = 50. The per cent recognition is very high for all nonzero thresholds. The sixth column gives the results of testing the predictive ability with random errors imposed on the prediction set members, and the last column gives the per cent drop predictive ability with these errors compared to the predictive ability on the prediction set members themselves. Thus, a committee of machines each operating with nonzero threshold is seen to be superior in terms of predictive ability and reliability.

## C. Piecewise Linear Learning Machine

Table 3 lists a number of examples of molecular features which are directly evident upon subjecting low resolution mass spectra to treatment by trained binary weight vectors. In general, the weight vectors associated with the features listed in Table 3 are sufficiently reliable for library purposes and perform well even on spectra not included in the training set, i.e., the spectra of "unknown" compounds. However, in a few cases, performance is less than satisfactory, as manifest by the failure of the weight vector to converge and/or by poor predictive ability (generally 75% or less). It should be noted here that classification errors are particularly damaging since they may occur conjointly with the possibility of structural isomers.

A number of explanations may be advanced to account for the difficulty in training a reliable weight vector in such cases. First, the information sought may simply not be contained in the spectrum, either because the actual physical mass spectroscopic measurement does not reflect that particular property or because the mode of presentation or data format effectively suppresses this information. (Indeed, the formulation of the question itself may be at fault.) Alternatively, the structure of the data may be such that the data are intractable to dichotomization by a single decision surface, or, in other words, the data are linearly inseparable; the distribution of the individual classes may exhibit an unknown number of prominent local maxima or modes. In such cases, the data become separable only through the use of more complex discriminant functions which are able to define nonconvex decision regions in the pattern space. The research described in this section demonstrates that, to a considerable extent, the latter situation, namely linear inseparability, exists in the mass spectral data examined, and that the use of a more sophisticated pattern classifier leads to improved performance of trained weight vectors both in library applications and in prediction on mass spectra of unknown compounds.

Assuming a need for a pattern classifier capable of implementing non-linear decision surfaces, one is faced with a variety of techniques, ranging from polynomial discriminant functions to piecewise-linear discriminant functions, to the various clustering techniques. Using such an arrangement, a pattern is assigned to category k if $S_k$ has the largest value of all the discriminant functions, $S_i$, $i = 1, \ldots, R$.

An iterative training process using an error-correction method adjusts the individual weights comprising the weight $W_i(j)$ until all patterns are correctly classified. Suppose a pattern belonging to the kth category is presented to the classifier and that $S_l$ has the largest value among the discriminant functions $S_i$. The generalized error-correction procedure is then

$$\underline{W}'_k = \underline{W}_k + c \cdot \underline{Y}$$

$$\underline{W}'_l = \underline{W}_l - c \cdot \underline{Y}$$

where c is some positive correction increment. Several rules for choosing c are in use. Duda and Fossum, for example, suggest changing the value of c regularly over the range $0 < c_{min} \leq c \leq c_{max}$, after each presentation of a block of N patterns. While this procedure is probably the most satisfactory, it has the disadvantage of introducing the adjustable parameters, $c_{min}$, $c_{max}$, and N, which must be set experimentally; in diference to the design criterion stated earlier, a simpler rule, a variant of the fractional correction rule introduced by Matzkin and Schoenberg [45] was chosen.

Under the fractional correction rule, the increment c is always chosen such that the decision surface defined by the weight vectors $\underline{W}_k$ and $\underline{W}_l$ is moved a fixed fraction $\lambda$ of its normal distance to the pattern point $\underline{Y}$; thus, for $\lambda = 2$, the new decision surface defined by $\underline{W}'_k$ and $\underline{W}'_l$ would lie an equal distance on the opposite side of the pattern point $\underline{Y}$. An alternative expression for this is given by the following:

$$S'_k = \underline{Y} \cdot \underline{W}'_k = S_l \tag{19a}$$

and

$$S'_l = \underline{Y} \cdot \underline{W}_l = S_k \tag{19b}$$

In cases where $R > 2$, a complication arises due to the varying magnitudes of the weight vectors $\underline{W}_i^{(j)}$; since the lengths of the $\underline{W}_i^{(j)}$ enter into the evaluation of the $S_i^{(j)}$ the decision process will be biased toward the longer weight vectors. Such a bias is avoided if the weight vectors are normalized to unit length; normalization, however, adds the restrictions that

$$\underline{W}'_k \quad \frac{\underline{W}_k + c_k \cdot \underline{Y}}{|\underline{W}_k + c_k \cdot \underline{Y}|} \tag{20a}$$

and

$$W'_l = \frac{\underline{W}_l - c_l \cdot \underline{Y}}{|\underline{W}_l - c_l \cdot \underline{Y}|} \tag{20b}$$

Substitution of Eqs. (20a) and (20b) into (19a) and (19b), respectively, leads to quadratics in $c_k$ and $c_l$:

$$c_k^2 [(\underline{YY})^2 - (\underline{YY})S_l^2] + c_k[2S_k[(YY) - S_l^2]] + S_k^2 - S_l^2 = 0$$

$$c_l^2 [(YY)^2 - (YY)S_k^2] + c_l[2S_l[(YY) - S_k^2]] + S_l^2 - S_k^2 = 0$$

The values of $c_k$ and $c_l$ must be obtained from these equations for each feedback correction.

While the number of categories R is usually known when formulating the classification problem, the number of subcategories, $L_i$, is, in general, unknown. The conventional method for dealing with this uncertainty is to assign an adequate but fixed number of weight vectors to each category in advance of the actual training process. Although such a procedure leads to a solution, it allows for the possibility of introducing unnecessary calculation both in the training process and in classification tasks after training, if the number of vectors exceeds the number actually required; in addition, the use of an excessive number of weight vectors can produce "overfitting" of the data distribution of the training set, which leads subsequently to poor performance on "unknown" patterns.

In designing the present classifier, the intention was to provide a simple means for internal generation of new weight vectors as dictated by the training process itself; thus, hopefully, the pattern classifier would evolve to that level of complexity just sufficient to provide the desired solution. Any pattern classifier of a given complexity, which uses an error-correction training procedure, will exhibit an oscillatory behavior when presented with an insoluble problem; a classifier using a single linear discriminant function, for example, would never converge to a solution for the problem illustrated in

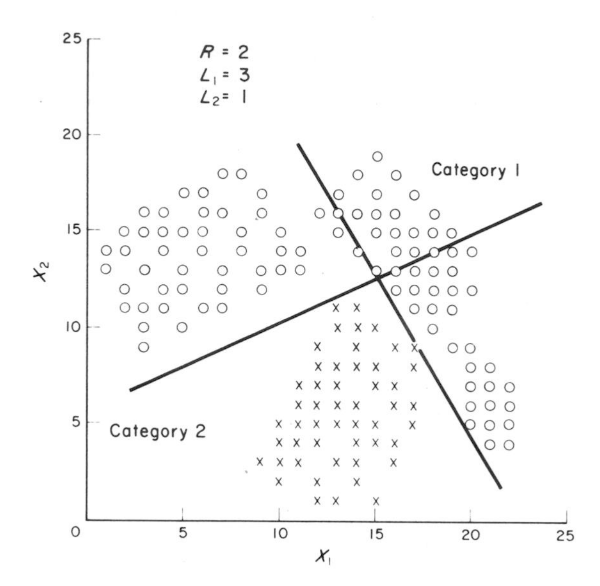


Fig. 5. Two-dimensional classification problem with R = 2, $L_1$ = 3, and $L_2$ = 1.

Fig. 5; instead the orientation of the single decision surface would undergo major oscillations for an indefinite period. It is logical, then, to look for means of detecting such oscillations, which would indicate the need for a more complex decision surface.

The method proposed here involves periodic evaluation of a function based on the following quantity:

$$\sum_{m=1}^{M} \frac{(S_k - S_l)}{M}$$

Here, M is the total of patterns in the training set; it is understood that k represents the category of the pattern m and that $S_l$ is the largest discriminant function among the $S_i$, $i \neq k$. Thus, if the pattern m is correctly classified, it will contribute a positive term to the sum, whereas, if the pattern is misclassified, that term will be negative. Although no rigorous argument can be offered for the use of the above quantity, particularly as to how its absolute magnitude should vary during training, the relative changes in its value have been found empirically to reflect oscillations in the decision surface, and were deemed worthy of investigation.

The actual function evaluated has the following form:

$$P_t = \frac{1}{\tau} P_{t-1} + \sum_{m-1}^{M} \frac{(S_k - S_l)}{M}$$

TABLE 16

Sample Evaluation of the Function

$$P_t = \frac{1}{\tau} [P_{t-1} + I_t] \text{ for } \tau = 10.0$$

| Iteration (t) | Increment ($I_t$) | $P_{t-1}$ | $P_t$ |
|---|---|---|---|
| 1 | 10.0 | 0 | 1.00000 |
| 2 | 10.0 | 1.00000 | 1.10000 |
| 3 | 10.0 | 1.10000 | 1.11000 |
| 4 | 10.0 | 1.11000 | 1.11100 |
| 5 | 10.0 | 1.11100 | 1.11110 |
| 6 | 10.0 | 1.11110 | 1.11111 |
| 7 | 5.0 | 1.11111 | 0.61111 |

The function P is therefore a "running weighted integral" of the past values of the summation, with a "time constant" $\tau$; Table 16 shows the smoothing effect of a function of this general form and the decreasing contribution of earlier increments I with time, as governed by the time constant. A simple criterion for addition of another subsidiary discriminant function is then,

$$P_t \geq \alpha P_{t-1} \quad \text{no addition}$$

$$P_t < \alpha P_{t-1} \quad \text{addition}$$

where $\alpha$ is a positive fraction, $0 < \alpha \leq 1$. The parameters $\tau$ and $\alpha$ determine the sensitivity of the function P to changes in decision surface orientation and are adjusted experimentally.

Figure 6 illustrates the behavior of P during training on the two-dimensional problem of Fig. 5. The pattern classifier begins with a single hyperplane decision surface, but since a single plane is insufficient, the value of P eventually falls at the interval t = 4; at this point $P_4 < P_3$, and another subsidiary discriminant function is introduced, which allows the training process to converge to a solution. The dotted lines in Fig. 6 indicate the result of suppressing the addition of a new subsidiary function.

The overall organization of the pattern classifier (SUBROUTINE LEARN) is summarized in Fig. 7. LEARN uses a collapsing subset procedure described previously [22] in which patterns misclassified on a given iteration are placed in a new subset of all such patterns to be used in the next iteration. If no errors are made, the program proceeds through previous subsets until all patterns in the training set have been correctly classified. At set intervals, the program breaks out of the subsetting procedure and the function P

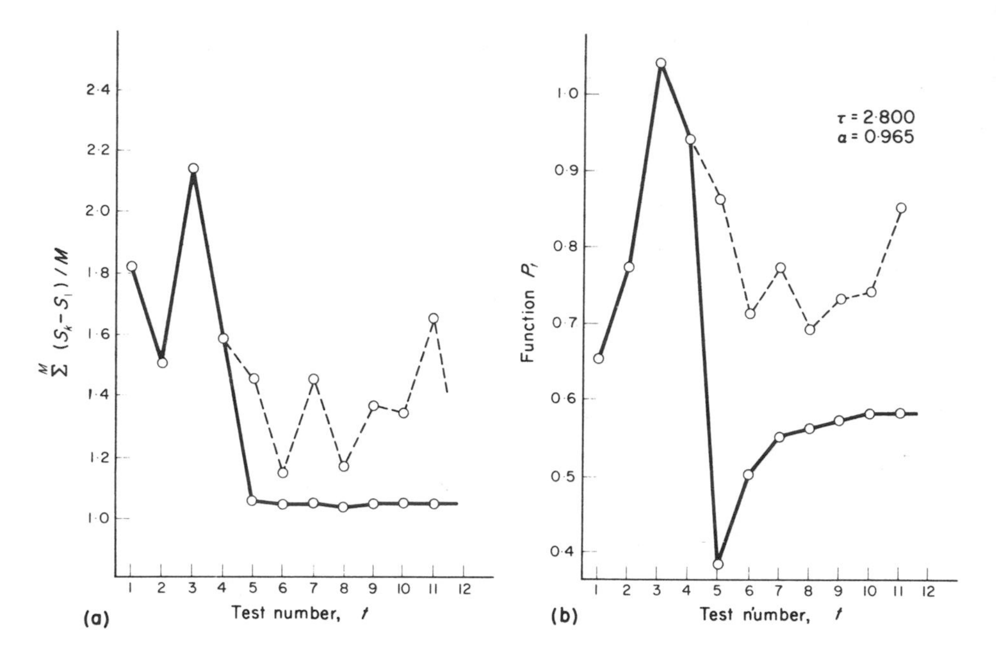


Fig. 6. (a) Plot of $\Sigma^m(S_k - S_l)/M$ vs test number for classification problem of Fig. 5. (b) Plot of function P vs test number for classification problem of Fig. 5.

is evaluated. If the criterion for smooth convergence is satisfied, the program returns to the subsetting operation; otherwise, the program determines the category of the discriminant function (weight vector) having the highest error rate and assigns a new weight vector to that category. This new vector is initilized as the last pattern misclassified by the parent vector. The program then resumes its normal flow.

Generally, sets of 200 spectra were randomly chosen for training and the remaining 187 spectra were used to test predictive ability of the trained classifier. The storage for subroutine LEARN and its associated subroutines is approximately 12 K bytes, excluding storage for the data and weight vectors. A typical run including training on 200 spectra and prediction of 187 spectra requires approximately 30-40 sec of CPU time, using from three to five 200-dimensional weight vectors.

The piecewise-linear pattern classifier has been applied to selected problems involving the identification of certain structural features in molecules from their mass spectra: the results discussed here indicate varying success and are compared with results obtained from binary classifiers.

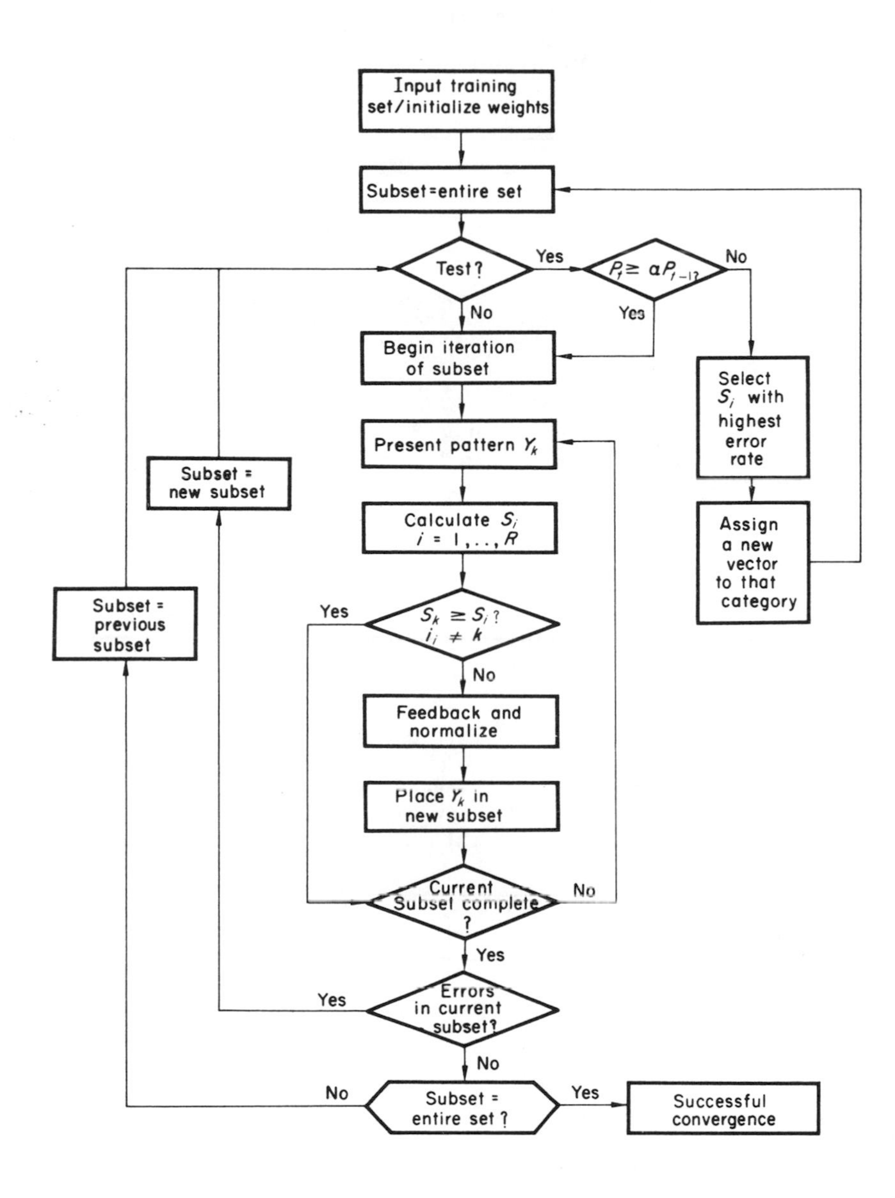


Fig. 7. Flow diagram of subroutine LEARN.

TABLE 17

Double Bonds in Various Hydrocarbons

| Molecular formula | Structural formula | | |
|---|---|---|---|
| $C_nH_{2n}$ | | $H_2C{=}CH{-}CH_2{-}C(CH_3)(CH_2){-}CH_3$ | |
| $C_nH_{2n-2}$ | $H_2C{=}C{=}CH_2$ | $(H_2C)_2{>}CH{-}C(CH_3){=}CH_2$ | $H_2C$, $CH_2$, $CH_2$, $C{=}CH$, $H_3C$ (ring) |
| $C_nH_{2n-4}$ | | $H_3C$, $C{=}CH_2$, $CH_2$, $H_2C$, $CH{-}C(CH_3){=}CH_2$, $CH_2$ (ring) | $H{\equiv}C{-}CH{=}CH_2$ |
| $C_nH_{2n-6}$ | a | $HC$, $CH$, $C{-}CH_3$, $C{-}CH_3$, $CH$, $HC$ (ring) | |
| $C_nH_{2n-8}$ | | | |
| $C_nH_{2n-10}$ | | | |

[a]For this work three double bonds were assigned to an aromatic ring.

TABLE 18

Performance of Self-Generating Piecewise Classifier on Carbon-Carbon Double Bond Problem

| Trial | $\tau$ | $\alpha$ | Classification | Feedbacks | Weight vectors | Prediction | | | |
|---|---|---|---|---|---|---|---|---|---|
| | | | | | | Range | Average | | |
| | | | | | | | Overall | Category 1 | Category 2 |
| 1 | 2.80 | 0.985 | 6925 | 930 | 3-4 | 78.6-86.1 | 81.9 | 82.3 | 81.5 |
| 2 | 3.10 | 0.985 | 6680 | 895 | 4 | 78.1-85.0 | 81.0 | 82.5 | 79.6 |
| 3 | 3.50 | 0.985 | 7623 | 1085 | 4-5 | 77.5-86.6 | 81.4 | 83.1 | 79.9 |
| 4 | 4.50 | 0.985 | 6159 | 914 | 5 | 79.1-86.6 | 81.7 | 84.8 | 79.2 |
| 5 | 2.00 | 0.990 | 8898 | 1216 | 6-10 | 78.6-82.9 | 81.5 | 84.1 | 78.9 |

An interesting test of the piecewise classifier and its ability to evolve in complexity is presented by the problem of detecting the presence or absence of the carbon-carbon double bond, an important structural feature. The double bond may occur in one of several interesting ways in a given molecule; a few representative examples and the corresponding carbon/hydrogen ratios are listed in Table 17. As indicated in Table 3, a simple binary classifier fails to converge in less than several thousand feedbacks during training on this problem, and prediction with the resulting weight vector is approximately 75%. The results of training with the piecewise-linear classifier are shown in Table 18, which lists the values of $\tau$ and $\alpha$, the number of classifications during training, the number of feedbacks, the number of weight vectors arrived at, and finally, the predictive ability of the trained classifier. The data listed for each trial in Table 18 represent averages for five separate runs, each using different weight initializations and order of pattern presentation. Convergence was obtained in all runs, and the number of classifications may be taken as roughly proportional to length of training.

In each trial, the classifier begins with only two weight vectors and subsequently adds vectors as necessary; the effect of varying the time constant $\tau$ on the final number of weight vectors is evident: larger values of $\tau$ increase the sensitivity of P to oscillations and, therefore, yield more weight vectors. (The seemingly large numbers of vectors in trial 5 results from pairwise addition of weight vectors, one from each category.) The results show predictive abilities ranging from roughly 78-87%, an improvement of 3-12% over the binary classifier.

## V. BLUE SKIES—THE FUTURE

The above suggestion of the centralization of such calculations so that small machines in individual laboratories could make decisions on data is a real possibility. There are other considerations, both likely and bizarre, that remain possibilities in the realm of artificial intelligence.

The basic speed of operations within present electronic computers already far exceeds that of the human brain. That is, the rate of flip-flop operations is several orders of magnitude faster than the propagation of signals through the neural network of the human brain. However, these systems appear to make decisions on different bases. For example, the computations involved in catching a baseball thrown with a curve on it seem far too complex for neural speeds to be of any use. Several sightings of the ball would have to be made, and then the optical data fit to some sort of polynomial or differential equation and finally the equation solved for the time and coordinates of the ball reaching the catcher. (Additional problems such as wind, eyeball error, etc., would make the problem much more complicated.) However, it is observed that baseballs are caught by human beings with great regularity. Obviously some sort of learning process employing feedback corrections is involved. It is also likely that the human brain solves problems by some sort of parallel calculations that involve various estimations and extrapolations and even voting machines to make a final decision. The same sort of parallel logic might be introduced to digital computer pattern recognition systems and give manyfold increases in speed and complexity of problem to be handled.

While digital computers based on electronic phenomena seem to be rapidly approaching certain fundamental limits of speed, these limitations may be overcome by higher component density and even other phenomenological approaches such as using light rather than electronic processes.

Software advances seem, theoretically, almost unlimited. There are parallels to biological evolution. Biological systems can be placed under stress to encourage evolution toward given goals. The question is, can machines be made to evolve purposefully? If the answer is "yes" then we may be looking toward an era of cybernetic chemistry.

In conclusion it may be said that the problem is "how to solve problems." Perhaps computer systems can eventually answer that question with respect to specific issues, and more excitingly with respect to general issues.

## ACKNOWLEDGMENT

The research reviewed in this chapter was supported in part by the National Science Foundation.

REFERENCES

[1]. J. Von Neumann, The Computer and the Brain, Yale Univ. Press, New Haven, Conn., 1958.

[2]. G. S. Sebestyen, Decision-Making Processes in Pattern Recognition, Macmillan, New York, 1962.

[3]. E. Feigenbaum and J. Feldman, eds., Computers and Thought, McGraw-Hill, New York, 1963.

[4]. J. T. Tou and R. H. Wilcox, eds., Computer and Information Sciences, Spartan Books, Washington, D.C., 1964.

[5]. N. J. Nilsson, Learning Machines, McGraw-Hill, New York, 1965.

[6]. J. T. Tou, ed., Computer and Information Sciences -- II, Academic Press, New York, 1967.

[7]. S. Watababe, ed., Methodologies of Pattern Recognition, Academic Press, New York, 1969.

[8]. M. Minsky and S. Papert, Perceptrons, M.I.T. Press, Cambridge, Mass., 1969.

[9]. J. M. Mendel and K. S. Fu, eds., Adaptive, Learning, and Pattern Recognition Systems, Academic Press, New York, 1970.

[10]. K. S. Fu, Sequential Methods in Pattern Recognition and Machine Learning, Academic Press, New York, 1968.

[11]. B. Meltzer and D. Michie, eds., Machine Intelligence, Vol. 5, American Elsevier, New York, 1970. See earlier volumes also.

[12]. M. Minsky, Proc. IRE, 49, 8 (1961).

[13]. R. J. Solomonoff, Proc. IEEE, 54, 1687 (1966).

[14]. C. A. Rosen, Science, 156, 38 (1967).

[15]. R. G. Casey and G. Nagy, Scientific American, 224, 56 (1971).

[16]. G. Nagy, Proc. IEEE, 56, 836 (1968).

[17]. P. C. Jurs, B. R. Kowalski, and T. L. Isenhour, Anal. Chem., 41, 21 (1969).

[18]. P. C. Jurs, B. R. Kowalski, T. L. Isenhour, and C. N. Reilley, Anal. Chem., 41, 690 (1969).

[19]. B. R. Kowalski, P. C. Jurs, T. L. Isenhour, and C. N. Reilley, Anal. Chem., 41, 695 (1969).

[20]. B. R. Kowalski, P. C. Jurs, T. L. Isenhour, and C. N. Reilley, Anal. Chem., 41, 1945 (1969).

[21]. P. C. Jurs, B. R. Kowalski, T. L. Isenhour, and C. N. Reilley, Anal. Chem., 41, 1949 (1969).

[22]. L. E. Wangen and T. L. Isenhour, Anal. Chem., 42, 737 (1970).

[23]. P. C. Jurs, B. R. Kowalski, T. L. Isenhour, and C. N. Reilley, Anal. Chem., 43, 3187 (1970).

[24]. P. C. Jurs, Anal. Chem., 42, 1633 (1970).

[15]. P. C. Jurs, Anal. Chem., 43, 22 (1971).

[26]. L. B. Sybrandt and S. P. Perone, Anal. Chem., 43, 382 (1971).

[27]. L. E. Wangen, N. M. Frew, T. L. Isenhour, and P. C. Jurs, Appl. Spec., 25, 203 (1971).

[28]. P. C. Jurs, Appl. Spec., 25, 483 (1971).

[29]. L. E. Wangen, N. M. Frew, and T. L. Isenhour, Anal. Chem., 43, 845 (1971).

[30]. N. M. Frew, L. W. Wangen, and T. L. Isenhour, Pattern Recognition, 3, 281 (1971).

[31]. B. R. Kowalski and C. A. Reilly, J. Phys. Chem., 75, 1402 (1971).

[32]. T. L. Isenhour and P. C. Jurs, Anal. Chem., 43 (10), 20A (1971).

[33]. P. C. Jurs, Anal. Chem., 43, 1812 (1971).

[34]. J. T. Tou, Pattern Recognition, 1, 8 (1968).

[35]. M. D. Levine, Proc. IEEE, 57, 1391 (1969).

[36]. Y. Uesak, IEEE Trans., SMC-1, 194 (1971).

[37]. A. L. Crittenden, University of Washington, personal communication, 1968.

[38]. F. W. MacLafferty, Mass Spectral Correlations, Advan. Chem. Ser., Vol. 40, American Chemical Society, Washington, D.C., 1963.

[39]. R. Bracewell, The Fourier Transform and Its Applications, McGraw-Hill, New York, 1965.

[41]. N. J. Nilsson, Ann. N. Y. Acad. Sci., 161, 380 (1969).

[42]. G. H. Ball, Proceedings-Fall Joint Computer Conference, 533 (1965).

[43]. D. F. Specht, IEEE Trans. Electronic Computers, EC-16, 308 (1967).

[44]. R. O. Duda and H. Fossum, IEEE Trans. Electronic Computers, EC-15, 222 (1966).

[45]. T. S. Matzkin and I. J. Schoenberg, Can. J. Math., 6, 393 (1954).

# AUTHOR INDEX

Numbers in parentheses are reference numbers and indicate that an author's work is referred to although his name is not cited in the text. Underlined numbers give the page on which the complete reference is listed.

SUBJECT INDEX

B



C

T

# Other books of interest to you...

Because of your interest in our books, we have included the following catalog of books for your convenience.

Any of these books are available on an approval basis. This section has been reprinted in full from our ***analytical chemistry/ spectroscopy*** catalog.

If you wish to receive a complete catalog of MDI books, journals and encyclopedias, please write to us and we will be happy to send you one.

# analytical chemistry spectroscopy

## ADAMS *Electrochemistry at Solid Electrodes*

*(Monographs in Electroanalytical Chemistry and Electrochemistry Series)*

by RALPH N. ADAMS, *Department of Chemistry, University of Kansas, Lawrence*

416 pages, illustrated. 1969

The first book specifically treating solid electrode electrochemistry. Covers basic theory of recently developed techniques and relies heavily on the experimental side of organic electrochemistry. For graduate students and researchers specializing in solid electrochemistry.

CONTENTS: Introduction • Scope and limitations of solid electrodes • Mass transfer to stationary electrodes in quiet solutions • Mass transfer by forced convection • Current-potential curves • Electrochemical methods employing controlled current • Electrode surface conditions • Investigation of electrode processes • Fabrication of electrode systems • Applications to organic compounds.

## AMAT, NIELSEN, and TARRAGO *Rotation-Vibration of Polyatomic Molecules: Higher Order Energies and Frequencies of Spectral Transitions*

by GILBERT AMAT, *Faculté des Sciences de Paris, France,* HARALD H. NIELSEN, *The Ohio State University, Columbus,* and GINETTE TARRAGO, *Faculté des Sciences de Paris, France*

448 pages, illustrated. 1971

Develops the theory involved in interpreting high precision spectra obtained from polyatomic molecules both in the infrared and the microwave regions. Provides explicit formulae for the frequencies of the spectral lines expressed in terms of quantum numbers and molecular parameters. Of significant value to all physicists and chemists interested in molecular spectroscopy.

CONTENTS: Expansion of the Hamiltonian • First contact transformation (to third order) • Second contact transformation (to third order) • First and second contact transformations (fourth order terms) • Contact transformations in the case of a Fermi resonance • Non-vanishing matrix elements for a molecule of given symmetry • Orders of magnitude of matrix elements and of their contributions to the energy • Matrix elements of the transformed Hamiltonian in terms of quantum numbers (to third order) • Matrix elements of the transformed Hamiltonian in terms of quantum numbers (fourth order) • Molecules with a threefold symmetry • Molecules with a fourfold symmetry • Linear molecules.

## BARD *Electroanalytical Chemistry: A Series of Advances*

edited by ALLEN J. BARD, *Department of Chemistry, University of Texas, Austin*

**Vol. 1** 440 pages, illustrated. 1966
**Vol. 2** 288 pages, illustrated. 1967
**Vol. 3** 328 pages, illustrated. 1969
**Vol. 4** 344 pages, illustrated. 1970
**Vol. 5** 400 pages, illustrated. 1971
**Vol. 6** 384 pages, illustrated. 1973
**Vol. 7** in preparation. 1974

"... Professor Bard, by stressing the electroanalyst's viewpoint, has introduced a complementary series which . . . will be welcome and a useful addition to the library of those interested in any aspect of electrochemistry."—R. A. Osteryoung, *Analytical Chemistry*.

The series provides a background and a starting point for graduate students undertaking research and is valuable to practicing analytical chemists interested in electroanalytical techniques.

CONTENTS:

**Volume 1:** AC Polarography and related techniques: Theory and practice, *D. E. Smith.* Applications of chronopotentiometry problems in analytical chemistry, *D. G. Davis.* Photoelectrochemistry and electroluminescence, *T. Kuwana.* Electrical double layer, Part I: Elements of double-layer theory, *D. M. Mohilner.*

*(continued)*

**BARD** *(continued)*

**Volume 2:** Electrochemistry of aromatic hydrocarbons and related substances, *M. E. Peover.* Stripping voltammetry, *E. Barendrecht.* Anodic film on platinum electrodes, *S. Gilman.* Oscillographic polarography at controlled alternating current, *M. Heyrovsky and K. Micka.*

**Volume 3:** Application of controlled-current coulometry to reaction kinetics, *J. Janata and H. B. Mark, Jr.* Nonaqueous solvents for electrochemical use, *C. K. Mann.* Use of radioactive-tracer method for investigation of electric double-layer structure, *N. A. Balashova and V. E. Kazarinov.* Digital simulation: A general method for solving electrochemical diffusion-kinetic problems, *S. W. Feldberg.*

**Volume 4:** Sine wave methods in the study of electrode processes, *M. Sluyters-Rehbach and J. H. Sluyters.* The theory and practice of electrochemistry with thin layer cells, *A. T. Hubbard and F. C. Anson.* Application of controlled potential coulometry to the study of electrode reactions, *A. J. Bard and K. S. V. Santhanam.*

**Volume 5:** Hydrated electrons and electrochemistry, *G. A. Kenny and D. C. Walker.* The fundamentals of metal deposition, *J. A. Harrison and H. R. Thirsk.* Chemical reactions in polarography, *R. Guidelli.*

**Volume 6:** Electrochemistry of biological compounds, *A. L. Harwood and R. W. Burnett.* Electrode processes in solid-electrolyte systems, *D. O. Raleigh.* The fundamental principles of current distribution and mass transport in electrochemical cells, *J. Newman.*

## BLACKBURN *Spectral Analysis: Methods and Techniques*

edited by JAMES A. BLACKBURN, *NASA Electronics Research Center, Cambridge, Massachusetts*

304 pages, illustrated. 1970

A broad sampling of current activity in the field of spectrum analysis. Of interest to all scientists faced with the problem of reducing and interpreting spectral data.

CONTENTS: Information and spectra, *E. J. Gauss.* Spectrum analysis, *J. A. Blackburn.* Numerical filtering, *J. F. Ormsby.* A numerical least-square method for resolving complex pulse-height spectra, *J. I. Trombka and R. L. Schmadebeck.* Biological applications, *R. D. Fraser and E. Suzuki.* Activation analysis, *F. J. Kerrigan.* Mass spectrometry, *J. I. Brauman.* Gamma-ray spectroscopy, *J. I. Trombka.*

## BRAME *Applied Spectroscopy Reviews*

*(Book Edition)*

a series edited by EDWARD G. BRAME, JR., *E. I. duPont de Nemours & Co., Wilmington, Delaware*

**Vol. 1** *out of print*
**Vol. 2** 384 pages, illustrated. 1969
**Vol. 3** 358 pages, illustrated. 1970
**Vol. 4** 396 pages, illustrated. 1971
**Vol. 5** 372 pages, illustrated. 1972
**Vol. 6** 398 pages, illustrated. 1973

These *Reviews* relate physical concepts to chemical applications. They are written by noted spectroscopists and cover the entire field for chemists, physicists, biochemists, and other scientists who use spectroscopy.

CONTENTS:

**Volume 1:** Atomic fluorescence flame spectrometry, *J. D. Winefordner and J. M. Mansfield, Jr.* Integrated intensities of absorption bands in infrared spectroscopy, *A. S. Wexler.* Internal reflection spectroscopy, *P. A. Wilks, Jr. and T. Hirschfeld.* Methods and applications in examination of small samples by high-resolution NMR, *R. E. Lundin, R. H. Elsken, R. A. Flath, and R. Teranishi.* Chemical far infrared spectroscopy, *J. W. Brasch, Y. Mikawa, and R. J. Jakobsen.* Infrared spectra of adsorbed molecules. *M. R. Basila.* Examination of polymers by high-resolution NMR, *H. A. Willis and M. E. A. Cudby.* Instrumentation, spectral characteristics, and applications of soft x-ray spectroscopy, *W. L. Baun.*

**Volume 2:** Near-infrared spectrophotometry, *K. B. Whetsel.* Spectroscopic studies of the hydrogen bond, *A. S. Murphy and C. N. Rao.* Molecular vibrations of high polymers, *G. Zerbi.* Analysis of ABX spectra in NMR spectroscopy, *G. Slomp.*

**Volume 3:** Recent advances in analytical emission spectrometry, *A. Yoakum.* Application of infrared spectroscopy to structure studies of nucleic acids, *M. Tsuboi.* Infrared studies of hydrogen-deuterium exchange in biological molecules, *F. Parker and K. Bhaskar.* Application of x-ray spectroscopy to clinical analyses and biological research, *S. Natelson.* Electronic spectra of radical ions, *C. N. Rao, V. Kalyanaraman, and M. V. George.* Quantitative analysis of infrared spectrophotometry, *J. A. Perry.* The combination of gas chromatography with mass spectrometry, *C. Merritt, Jr.*

**Volume 4:** Aluminum-27 and proton NMR of organoaluminums, *L. Petrakis and F. E. Dickson.* Electronic relaxation and molecular vibrations, *J. K. Burdett.* Application of spectroscopy in the study of glassy solids, Part I. Introduction, $\gamma$-ray spectroscopy (Mössbauer effect), x-ray absorption fine structure, UV absorption, and visible spectra, *J. Wong and C. A. Angell.* Application of spectroscopy in the study of glassy solids, Part II. Infrared, Raman, EPR, and MNR spectral studies, *J. Wong and C. A. Angell.* Raman scattering of synthetic polymers —a review, *J. L. Koenig.* Chemical applications of nuclear magnetic resonance at high fields, *J. K. Becconsall, P. A. Curnuck, and M. C. McIvor.*

**Volume 5:** Spectroscopy of electron donor-acceptor systems, *C. R. Rao, S. N. Bhat, and*

*P. C. Dwivedi*. Recording infrared spectra at low signal levels, *M. J. Low, J. C. McManus, and L. Abrams*. The application of spectroscopic techniques to the structural analysis of coal and petroleum, *J. G. Speight*. Nuclear magnetic resonance in ferromagnets and antiferromagnets, *M. P. Petrov and E. A. Turov*.

**Volume 6:** Dichroic spectra of biopolymers oriented by flow, *A. Wada*. Rapid-scan infrared Fourier transform spectroscopy, *P. Griffiths, C. Foskett, and R. Curbelo*. Neutron scattering and vibrational spectra of molecular crystals, *T. Kitagawa*. Diagnostics of high temperature high density plasma by radiation analysis, *H. Conrads*. Spectroscopy of liquid crystals, *S. Chandrasekhar and N. Madhusudana*. Metals analysis in particulate pollutants emission spectroscopy, *R. Sacks and S. Brewster, Jr*. Communication, *V. Rajan*.

## CHATTEN *Pharmaceutical Chemistry*

**In 2 Volumes**

edited by LESLIE G. CHATTEN, *Faculty of Pharmacy, University of Alberta, Edmonton*

**Vol. 1** ***Theory and Application***
520 pages, illustrated. 1966

**Vol. 2** ***Instrumental Techniques***
792 pages, illustrated. 1969

This comprehensive two-volume textbook in pharmaceutical chemistry is for use of the senior undergraduate and graduate student studying analytical chemistry and instrumental techniques as applied to chemistry.

"It will be welcomed by workers engaged in pharmaceutical analysis and should find a place in libraries of all colleges of pharmacy and medicine."—F. Kurzer, *Chemistry in Britain*. April, 1967

**Volume 1:** Introduction and technique, *M. Pernarowski*. Gravimetric analysis, *R. T. Coutts*. Acid-base titrations and pH, *J. W. Steele*. Precipitation, complex formation, and oxidation–reduction methods, *J. A. Zapotocky*. Acidimetry and alkalimetry, *M. I. Blake*. Nonaqueous titrimetry, *L. G. Chatten*. Complexometric titrations, *J. P. Leyda*. Alkaloidal assay and crude drug analysis, *W. C. Evans*. Miscellaneous methods, *J. E. Sinsheimer*. Ion exchange, *M. C. Vincent*. Column, thin-layer, and paper chromatography, *J. C. Morrison*. Analysis of fixed oils, fats, and waxes, *A. C. Glasser*. Analyses of volatile oils, *I. C. Nigam*.

**Volume 2:** Absorption spectrophotometry, *M. Pernarowski*. Infrared spectroscopy, *R. Coutts*. Raman spectroscopy, *E. A. Robinson and D. S. Lavery*. Fluorometry, *D. E. Guttman*. Atomic absorption spectroscopy, *J. W. Robinson*. Mass spectrometry, *A. Chisholm*. Nuclear magnetic resonance spectroscopy, *M. P. Mertes*. Turbidimetry: Nephelometry; colloidimetry, *F. T. Semeniuk*. Optical crystallography, *J. A. Biles*. X-ray analysis, *J. W. Shell*. Refractometry, *R. A. Locock*. Polarimetry, *R. A. Locock*. Potentiometric titrations, *J. G. Jeffrey*. Current flow methods, *S. Eriksen*. Coulometric methods and chronopotentiometry, *P. Kabasakalian*. Polarography, *F. W. Teare*. Amperometric titrations, *F. W. Teare*. Gas chromatography, *C. A. Bliss*. Radiochemical techniques, *A. Noujaim*.

## DEAN and RAINS *Flame Emission and Atomic Absorption Spectrometry*

**In 3 Volumes**

edited by JOHN A. DEAN, *University of Tennessee, Knoxville*, and THEODORE C. RAINS, *National Bureau of Standards, Washington, D.C.*

**Vol. 1** ***Theory***
456 pages, illustrated. 1969

**Vol. 2** ***Components and Techniques***
384 pages, illustrated. 1971

**Vol. 3** ***Applications***
in preparation. 1974

A three volume set which considers the various aspects of flame emission, atomic absorption, and atomic fluorescence spectrometric methods. Provides a source of theoretical and practical analytical information for all persons who are either engaged in, or considering the use of these flame spectrometric methods.

CONTENTS:

**Volume 1:** Flame methods: Their development and application, *R. L. Mitchell*. Basic principles of flame emission, atomic absorption and fluorescence methods, *J. Ramirez-Muñoz*. Emission problems of unsalted flames, *E. Pungor and I. Cornides*. Fundamental aspects of decomposition, atomization, and excitation of the sample in the flame, *C. Th. J. Alkemade*. Radicals and molecules in flame gases, *D. R. Jenkins and T. M. Sugden*. Flames for atomic absorption and emission spectrometry, *R. N. Kniseley*. Measurement and calculation of flame temperatures, *W. Snelleman*. Distribution of atomic concentration in flames, *A. N. Hambly and C. S. Rann*. Spectral interferences, *B. E. Buell*. Physical interferences in flame emission and absorption methods, *S. R. Koirtyohann*. Chemical interferences in the vapor phase, *I. Rubeška*. Chemical interferences in condensed phase, *T. C. Rains*. Accuracy and precision, *R. K. Skogerboe*.

**Volume 2:** Introduction to the practice and technique of flame spectrometry, *J. Dean and B. Bailey*. Light sources for atomic absorption and atomic fluorescence spectrometry, *L. Butler and J. Brink*. Nebulizers and burners, *R. Herrmann*. Nonflame absorption devices in atomic absorption spectrometry, *H. Massmann*. The optical train, *J. Dean*. Electronics, *C. Veillon*. Instrument operation, *C. Veillon*. Atomic fluorescence spectrometry, *A. Syty*. Commercial instruments, *J. Dean*. Sample preparation and separation methods, *D. Ellis*. Trace analysis and micromethods, *R. Herrmann*. Evaluation of data, *W. Schrenk*. Standard solutions for flame spectrometry, *J. Dean and T. Rains*.

**DOMSKY and PERRY** ***Recent Advances in Gas Chromatography***

edited by IRVING I. DOMSKY, *Armour-Dial, Inc., Chicago, Illinois,* and JOHN A. PERRY, *Consultant, Instrumental Analytical Chemistry, Chicago, Illinois*

432 pages, illustrated. 1971

Enables the gas chromatographic worker to increase his understanding of, and to consider in detail, selected areas of the frontiers of instrumental chromatography. Of prime importance to all those engaged in the field of instrumental chromatography.

CONTENTS: Column technology, *J. Perry.* Micro- and macro-flow patterns in gas chromatography, *S. J. Hawkes.* Advances in, and status of glass as a gas chromatographic support material, *A. Filbert.* The use of Rohrschneider constants, *W. R. Supina.* Dependence of electron capture response on pulse interval, *W. E. Wentworth.* Detectors–a critical review, *L. Giuffrida.* Gel permeation chromatography –calibration, quantitation, and the effect of operational variables, *R. E. Thompson, D. C. Ford, and E. G. Sweeney.* Little things to do for troubled GPCs, *S. Chechakli and J. B. Himes.* The development of a reagent for gas chromatography, *G. D. Brittain.* Silylation in the presence of water–the development of a commercial reagent for silylating aqueous solutions of hydroxy and polyhydroxy compounds, *G. D. Brittain, J. E. Sullivan, and L. R. Schewe.* Some practical considerations of small scale preparative gas chromatography, *F. J. Debbracht.* Application of gas chromatography in food packaging, *M. Kulisz, E. S. Derzko, and J. B. Himes.* Lipid analysis by gas chromatography, *L. D. Metcalfe.* Gas chromatography of pesticides, *G. Zweig and J. M. Devine.* The application of electron capture detection for the determination of halogenated derivatives of organic compounds, *L. M. Cummins.* Metabolic profiles: The study of human metabolites by gas chromatography and gas chromatography-mass spectrometry, *E. C. Horning and M. G. Horning.* The use of computers in the gas chromatographic laboratory, *D. Ford and K. Weihman.*

**DURIG** ***Vibrational Spectra and Structure:*** *A Series of Advances*

a series edited by JAMES R. DURIG, *Department of Chemistry, University of South Carolina, Columbia*

**Vol. 1** 224 pages, illustated. 1972

**Vol. 2** in preparation. 1973

Contains critical summaries of recent work in vibrational spectroscopy and reviews current research. This series is of special interest to all physical chemists, spectroscopists, physicists, and other research scientists who use vibrational spectroscopy in their work.

CONTENTS:

**Volume 1:** Far-infrared spectra of four-membered-ring compounds, *C. S. Blackwell and R. C. Lord.* Pseudorotation of five-membered rings, *J. Laane.* Determination of torsional barriers from far-infrared spectra, *J. R. Durig, S. M. Craven, and W. C. Harris.* High pressure vibrational spectroscopy, *A. J. Melveger, J. W. Brasch, and E. R. Lippincott.*

**FITZGERALD** ***Analytical Photochemistry and Photochemical Analysis:*** *Solids, Solutions, and Polymers*

edited by JERRY M. FITZGERALD, *University of Houston, Texas*

376 pages, illustrated. 1971

Presents photochemical techniques as applied in analytical chemistry, and also analytical methods used in photochemical reactions. Essential reading for photochemists; for analytical chemists who can be aided by photochemical techniques or who are engaged in the study of photochemistry; and for polymer chemists seeking a review of analytical techniques used in the photodegradation of polymeric materials.

CONTENTS: Apparatus for photochemical experimentation, *E. Smith.* Lasers in analytical chemistry, *F. Fry.* Analytical methods and techniques for actinometry, *H. Taylor.* Analytical methods employing photochemical pretreatment, *W. Riggs.* Photochemical titrations, *J. Fitzgerald.* Analytical problems in the study of photochemical reactions, *E. Wehry.* Electrochemical and other analytical methods for flash-photolysis studies, *S. Perone* and *H. Drew.* Analytical techniques for the solid-state photolysis of coordination compounds, *W. Wendlandt.* Analytical methods for the study of photodegradation of polymers, *N. Searle.*

**FLASCHKA and BARNARD** ***Chelates in Analytical Chemistry:*** *A Collection of Monographs*

edited by HERMANN A. FLASCHKA, *School of Chemistry, Georgia Institute of Technology, Atlanta,* and A. J. BARNARD, JR., *J. T. Baker Chemical Company, Phillipsburg, New Jersey*

**Vol. 1** 432 pages, illustrated. 1967

**Vol. 2** 412 pages, illustrated. 1969

**Vol. 3** 248 pages, illustrated. 1972

**Vol. 4** 328 pages, illustrated. 1972

Employs chelation as a unifying concept for a large area of analytical chemistry.

Most of the monographs are of direct interest to the analyst; many include representative procedures written for immediate laboratory use; a few are largely theoretical. Valuable for research and applied chemists and students of analytical chemistry.

CONTENTS:

**Volume 1:** Outline of the history of analytical methods based on complex formation, *F. Szabadváry and M. T. Beck.* Xylenol orange and methylthymo blue as chromogenic reagents, *B. Budĕšínsky.* Chelating ion-exchange resins, *E. Blasius and B. Brozio.* Conductometric and high-frequency impedimetric titrations involving chelates and chelating agents, *F. Vydra and K. Štulík.* Thermal dissociation of chelating agents and chelates of analytical interest, *W. W. Wendlandt.* Chelates in organic polarographic analysis: Fundamentals, *M. Kopanica, J. Doležal, and J. Zýka.* Chelates in inorganic polarographic analysis: Applications, *M. Kopanica, J. Doležal, and J. Zýka.* Chelates and chelating agents in analytical chemistry of molybdenum and tungsten, *R. Püschel and E. Lassner.*

**Volume 2:** Monoarylazo and bis(arylazo) derivatives of chromotropic acid as photometric reagents, *B. Budĕšínsky.* Electrometric titrations with two polarizable electrodes involving chelating agents, *K. Štulík and F. Vydra.* Chelating agents as metal precipitants: Advances, 1960-1965, *F. H. Firsching.* Analytical applications of Schiff bases, *E. Jungreis and S. Thabet.* Selectivity and analytical application of dimethylglyoxime and related dioximes, *K. Burger.* Chelates and chelating agents in the analytical chemistry of niobium and tantalum, *E. Lassner and R. Püschel.*

**Volume 3:** Extractive titrations involving chelates and chelating agents, *A. Galík.* Electroanalytical methods in the study of chelation reactions, *M. Kopanica and J. Zýka.* Amperometric titrations involving chelates and chelating agents, *J. Doležal, K. Štulík, and J. Zýka.*

**Volume 4:** 2-Pyridylazo compounds in analytical chemistry, *S. Shibata.* The application of chelates to flame analytical techniques, *J. W. Robinson, P. F. Lott, and A. J. Barnard, Jr.*

## GIDDINGS *Dynamics of Chromatography: Principles and Theory*

*(Chromatographic Science Series, Volume 1)*

by J. Calvin Giddings, *University of Utah, Salt Lake City*

336 pages, illustrated. 1965

CONTENTS: Introduction and dynamics of zone migration • Dynamics of zone spreading • Nonequilibrium and the simple mass transfer terms • The generalized nonequilibrium theory • Packing structure and flow dynamics • Diffusion and kinetics in chromatography • The achievement of separation.

## GIDDINGS and KELLER *Advances in Chromatography*

a series edited by J. Calvin Giddings, *University of Utah, Salt Lake City,* and Roy A. Keller, *State University of New York, Fredonia*

**Vol. 1** *out of print*
**Vol. 2** 400 pages, illustrated. 1966
**Vol. 3** *out of print*
**Vol. 4** 400 pages, illustrated. 1967
**Vol. 5** 336 pages, illustrated. 1968
**Vol. 6** 360 pages, illustrated. 1968
**Vol. 7** 336 pages, illustrated. 1968
**Vol. 8** 416 pages, illustrated. 1969
**Vol. 9** 376 pages, illustrated. 1970
**Vol. 10** in preparation. 1974

Presents critical, current surveys of the most important advances in the respective areas of chromatographic science. The authors have summarized and developed their own perspective in single papers.

"For the expert, each volume should be required reading." — Barry L. Karger, Northeastern University

CONTENTS:

**Volume 1:** Ion-exchange chromatography, *F. Helfferich.* Chromatography and electrophoresis on paper and thin layers: A teachers guide, *I. Smith.* Stationary phase in paper chromatography, *G. H. Stewart.* Techniques of laminar chromatography, *E. V. Truter.* Qualitative and quantitative aspects of separation of steroids, *E. C. Horning and W. J. A. Vandenheuvel.* Capillary columns: Trials, tribulations, and triumphs, *D. H. Desty.* Gas chromatographic characterization of organic substances in the retention index system, *E. Kováts.* Inorganic gas chromatography, *R. S. Juvet, Jr., and F. Zado.* Lightly loaded columns, *B. L. Karger and W. D. Cooke.* Interactions of solute with liquid phase, *D. E. Martire and L. Z. Pollara.*

**Volume 2:** Ion exchange chromatography of amino acids: Recent advances in analytical determinations, *P. B. Hamilton.* Ion mobilities in electrochromatography, *J. T. Edward.* Partition paper chromatography and chemical structure, *J. Green and D. McHale.* Gradient techniques in thin-layer chromatography, *A. Niederwieser and C. C. Honegger.* Geology — an inviting field to chromatographers, *A. S. Ritchie.* Extra-column contributions to chromatographic band broadening, *J. C. Sternberg.* Gas chromatography of carbohydrates, *J. W. Berry.* Ionization detectors for gas chromatography, *A. Karmen.* Advances in programmed temperature gas chromatography, *L. Mikkelsen.*

**Volume 3:** Occurrence and significance of isotope fractionation during analytical separations of large molecules, *P. D. Klein.* Adsorption chromatography, *C. H. Giles and I. A. Easton.* History of thin-layer chromatography, *N. Pelick, H. R. Bolliger, and H. K. Mangold.* Chroma-

*(continued)*

**GIDDINGS and KELLER** *(continued)*

tography as a natural process in geology, *A. S. Ritchie.* Chromatographic support, *D. M. Ottenstein.* Electrolytic conductivity detection in gas chromatography, *D. M. Coulson.* Preparative-scale gas chromatography, *G. W. A. Rijnders.*

**Volume 4:** $R_F$ values in thin-layer chromatography on alumina and silica, *L. R. Snyder.* Steroid separation and analysis: Techniques appropriate to the goal, *R. Neher.* Some fundamentals of ion-exchange-cellulose design and usage in biochemistry, *C. S. Knight.* Adsorbents in gas chromatography, *A. V. Kiselev.* Packed capillary columns in gas chromatography, *I. Halász and E. Heine.* Mass-spectrometric analysis of gas-chromatographic eluents, *W. H. McFadden.* Polarity of stationary liquid phases in gas chromatography, *L. Rohrschneider.*

**Volume 5:** Prediction and control of zone migration rates in ideal liquid – liquid partition chromatography, *E. Soczewiński.* Chromatographic advances in toxicology, *P. L. Kirk.* Inorganic chromatography of natural and substituted celluloses, *R. A. A. Muzzarelli.* Quantitative interpretation of gas chromatographic data, *H. W. Johnson, Jr.* Atmospheric analysis by gas chromatography, *A. P. Altshuller.* Non-ionization detectors and their use in gas chromatography, *J. D. Winefordner and T. H. Glenn.*

**Volume 6:** Systematic use of chromatography in structure elucidation of organic compounds by chemical methods, *J. Gasparič.* Polar solvents, supports, and separation, *J. A. Thoma.* Liquid chromatography on lipophilic sephadex: Column and detection techniques, *J. Sjövall, E. Nyström, and E. Haahti.* Statistical moments theory of gas-solid chromatography: Diffusion-controlled kinetics, *O. Grubner.* Identification by retention and response values, *G. Schomburg.* Use of liquid crystals in gas chromatography, *H. Kelker and E. von Schivizhoffen.* Support effects on retention volumes in gas-liquid chromatography, *P. Urone and J. F. Parcher.*

**Volume 7:** Theory and mechanics of gel permeation chromatography, *K. H. Altgelt.* Thin-layer chromatography of nucleic acids, bases, nucleosides, nucleotides, and related compounds, *G. Pataki.* Review of current and future trends in paper chromatography, *V. C. Weaver.* Chromatography of inorganic ions, *G. Nickless.* Process control by gas chromatography, *I. G. McWilliam.* Pyrolysis gas chromatography of involatile substances, *S. G. Perry.* Labeling by exchange on chromatographic columns, *H. Elias.*

**Volume 8:** Principles of gel chromatography, *H. Determann.* Thermodynamics of liquid-liquid partition chromatography, *D. Locke.* Determination of optimum solvent systems for countercurrent distribution and column partition chromatography from paper chromatographic data, *E. Soczewiński.* Some procedures for the chromatography of the fat-soluble chloroplast pigments, *H. Strain and W. Svec.* Comparison of the performances of the various column types used in gas chromatography, *G. Guiochon.* Pressure (flow) programming in gas chromatography, *L. Ettre, L. Mázor, and J. Takács.* Gas chromatographic analysis of vehicular exhaust emissions, *B. Dimitriades, C. Ellis, and D. Seizinger.* The study of reaction kinetics by the distortion of chromatographic elution peaks, *M. van Swaay.*

**Volume 9:** Reversed-phase extraction chromatography in inorganic chemistry, *E. Cerrai and G. Ghersini.* Determination of the optimum conditions to effect a separation by gas chromatography, *R. Scott.* Advances in the technology of lightly-loaded glass bead columns, *C. Hishta, J. Bomstein and W. D. Cooke.* Radiochemical separations and analyses by gas chromatography, *S. P. Cram.* Analysis of volatile flavor components of foods, *P. Issenberg and I. Hornstein.*

**Volume 10:** Quantitative analysis by gas chromatography, *J. Novak.* Porous-layer open tubular columns—theory, practice, and applications, *L. Ettre and J. Purcell.* Resolution of optical isomers by gas-liquid chromatography of diastereoisomers, *E. Gil–Av and D. Nurok.* Gas liquid chromatography of terpenes, *E. Von Rudloff.* Polyamide-layer chromatography, *K. Wang, Y. Lin and I. Wang.* Specifically adsorbing silica gels, *H. Bartels and B. Prijs.* Non-destructive detection methods in paper and thin-layer chromatography, *G. Barrett.*

## GROB *Chromatographic Analysis of the Environment*

edited by ROBERT L. GROB, *Department of Chemistry, Villanova University, Pennsylvania*

in preparation. 1973

Consists of a comprehensive series of methods for determining the various toxic substances in the atmosphere, water, waste effluents, and soil. An indispensable aid in establishing the criteria for choosing one technique in preference to another with regard to specific sample type and analysis. Of practical value to all environmentalists including engineers, chemists, technicians, students, and teachers.

CONTENTS: **Plenary Section:** Theory and practice of chromatography, *T. Bunting.* **Air (Pollution) Section:** Gas chromatographic analysis, *R. Braman.* Liquid chromatographic analysis, *G. Chesters and D. Graetz.* Thin-layer chromatographic analysis, *D. Bender and W. Elbert.* **Soil Chemistry Section:** Gas chromatographic analysis, *S. Cram.* Liquid chromatographic analysis, *G. Chesters and D. Graetz.* Paper chromatographic analysis, *G. Chesters and D. Graetz.* Thin-layer chromatographic analysis, *W. Thornburg.* Ion exchange analysis, *H. Walton.* **Water (Pollution) Section:** Gas chromatographic analysis, *S. Cram.* Liquid chromatographic analysis, *I. Suffet and E. Sowinski, Jr.* Paper chromatographic analysis, *O. Aly and S. Faust.* Thin-layer chromatographic analysis, *O. Aly and S. Faust.* Ion exchange analysis, *H. Walton.* **Waste Chemistry Section:** Gas chromatographic analysis, *R. Dell'Aqua.* Liquid chromatographic analysis, *C. Hamilton.* Paper chromatographic analysis, *J. Hunter.*

Thin-layer chromatographic analysis, *E. McConigle.*

## GROVE *Analytical Emission Spectroscopy*

*(Analytical Spectroscopy Series, Volume 1)*

edited by E. GROVE, *IIT Research Institute, Chicago, Illinois*

**Part I** 416 pages, illustrated. 1971

**Part II** 584 pages, illustrated. 1972

**Part III** in preparation. 1974

A comprehensive treatise in the field of emission spectroscopy and its uses. Of interest to all persons working in the physical aspects of spectroscopy and especially valuable to those concerned with this field as an analytical tool.

CONTENTS:

**Part I:** Historical development, some uses and definitions, *E. Grove.* Origins of atomic spectra, *J. Devlin.* Prism systems, spectrographs, and spectrometers, *H. Faust.* Gratings and grating instruments, *R. Barnes and R. Jarrell.* Spectroradiometric principles, *H. Betz and G. Johnson.*

**Part II:** Excitation of spectra, *P. Boumans.* Flame spectrometry, *T. Vickers and J. Winefordner.* Qualitative and semiquantitative analysis, *M. Wang, W. Cave, and W. Coakley.* Quantitative analysis, *J. Rozsa.*

## GUDZINOWICZ *Gas Chromatographic Analysis of Drugs and Pesticides*

*(Chromatographic Science Series, Volume 2)*

by BENJAMIN J. GUDZINOWICZ, *Polaroid Corporation, Waltham, Massachusetts*

616 pages, illustrated. 1967

CONTENTS: Fundamentals of gas-liquid chromatography • Detectors: Operating principles and theory • Qualitative and quantitative methods of analysis • Phenothiazine drugs and barbiturates • Phenylethylamine-type and tryptamine-indole base alkaloids • Morphine-, nicotine-, and pyrrolizidine-related alkaloids and marihuana cannabinols • Antihistamines, high-boiling amine anesthetics, and vitamins • Miscellaneous drugs and pharmaceuticals • Pesticides, herbicides, and related compounds.

## GUILBAULT *Fluorescence: Theory, Instrumentation, and Practice*

edited by GEORGE G. GUILBAULT, *Department of Chemistry, Louisiana State University, New Orleans*

728 pages, illustrated. 1967

CONTENTS: Modern techniques of energy transfer (describing research of the late Jean T. Dubois), *F. Wilkinson.* Structural and environmental factors in fluorescence, *E. Wehry.* Laser excited vibrational fluorescence and its appliation to vibration→ vibration energy transfer in gases, *C. B. Moore.* Theory of molecular luminescence, *M. Kasha.* A corrected spectra instrument, *H. K. Howerton.* Fluorescence polarization by modulation techniques, *W. Kaye and D. West.* Fluorescent metal chelates in analytical chemistry, *C. E. White.* Kinetic methods of analysis, *G. G. Guilbault.* Use of fluorescence as a probe into mechanisms for concentrative transport of amino acids in living cells, *S. Udenfriend, G. Guroff, and P. Zaltzman-Nirenberg.* Phosphorimetry as a means of chemical analysis, *W. J. McCarthy and J. D. Winefordner.* Extrinsic and intrinsic fluorescence in the study of protein structure: A review, *R. F. Chen.* Chlorophyll fluorescence and photosynthesis, *Govindjee, G. Papageorgiou, and E. Rabinowitch.* Atomic fluorescence flame spectrometry, *J. D. Winefordner and J. M. Mansfield.* Electrogenerated chemiluminescence, *A. J. Bard, K. S. V. Santhanam, S. A. Cruser, and L. R. Faulkner.*

## GUILBAULT *Practical Fluorescence: Theory, Methods, and Techniques*

by GEORGE G. GUILBAULT, *Department of Chemistry, Louisiana State University, New Orleans*

680 pages, illustrated. 1973

A textbook which provides the reader with a practical introduction to the entire field of fluorescence spectroscopy. Concentrates on the fundamental principles of fluorescence, phosphorescence, chemiluminescence, and atomic fluorescence, and discusses actual analytical applications. Of utmost interest to students, researchers, and technicians in all areas of chemistry, biochemistry, immunology, physiology, environmental science, and spectroscopy, who are interested in the fundamental theories of luminescence.

CONTENTS: Introduction to luminescence • Instrumentation • Effects of molecular structure and molecular environment on fluorescence (authored by E. Wehry) • Practical aspects of measurement • Phosphorescence • Inorganic substances • Assay of organic compounds • Fluorescence in enzymology • Chemiluminescence • Atomic fluorescence flame spectrometry • Electrogenerated luminescence • Extrinsic and intrinsic fluorescence of proteins (authored by R. Chen) • Chlorophyll fluorescence and photosynthesis (authored by Govindjee, G. Papageorgiou, and E. Rabinowitch) • Analysis on solid surfaces • Fluorescent indicators • Forensic and environmental analysis.

## GUILBAULT and HARGIS
***Instrumental Analysis Manual:***
*Modern Experiments for the Laboratory*

by GEORGE G. GUILBAULT and LARRY G. HARGIS, *Louisiana State University, New Orleans*

456 pages, illustrated. 1970

Complements the available instrumental analysis texts, and provides experiments for the laboratory that demonstrate the principles discussed in the lecture series. Of great benefit to all teachers and students interested in instrumental analysis.

CONTENTS: Molecular absorption • Molecular emission • Atomic absorption and emission • Electron spin resonance • Nuclear magnetic resonance • Electronics and instrumentation • Potentiometry • Conductance • Amperometry • Coulometry • Polarography • Gas chromatography • Electrophoresis • Ion exchange • Liquid-solid chromatography • Radiochemistry • Mass spectrometry • Thermometric titrations.

## JORDAN *New Developments in Titrimetry*

*(Treatise on Titrimetry Series, Volume 2)*

edited by JOSEPH JORDAN, *Department of Chemistry, The Pennsylvania State University, University Park*

in preparation. 1974

Provides a concise and lucid presentation of important innovations in titrimetry which were previously scattered throughout the literature. Authoritatively reviews thermometric titrations and discusses several interesting and unconventional end point detection methods. Especially valuable reading for all practicing analytical chemists.

CONTENTS: Applications of thermometric titrimetry to analytical chemistry, *L. Hansen, R. Izatt, and J. Christensen.* Some unusual end point detection methods involving heterogeneous processes, *D. Curran.*

## KELLER *Separation Techniques in Chemistry and Biochemistry*

edited by ROY A. KELLER, *Department of Chemistry, University of Arizona, Tucson*

432 pages, illustrated. 1967

CONTENTS: Separation, identification, and estimation of human steroid hormones and their metabolites: Applications to adrenocortical steroids, *E. C. Horning, C. J. W. Brooks, L. Johnson, and W. L. Gardiner.* Quantitative analysis of the twenty natural protein amino acids by gas-liquid chromatography, *C. W. Gehrke and D. L. Stalling.* Continuous particle electrophoresis: a new analytical and preparative capability, *A. Strickler.* A critical evaluation of gel chromatography, *R. L. Pecsok and D. Saunders.* A comparison of mobile-phase peak dispersion in gas and liquid chromatography, *D. S. Horne, J. H. Knox, and L. McLaren.* Ultra-high-pressure gas chromatography in micro columns to 2000 atmospheres, *M. N. Myers and J. C. Giddings.* Structural effects and properties of oligomers of interest in separations, *H. H. Schmidt, R. K. Clark, and C. F. Gay.* Indeterminate errors in the measurement of chromatographic peaks, *D. L. Ball, W. E. Harris, and H. W. Habgood.* Criteria of identity and purity in chromatographic separations, *P. D. Klein.* Nonlinear distribution coefficients in gas chromatography, *P. Urone, J. F. Parcher, and E. N. Baylor.* Effect of carbon number, phase polarity, temperature, and flow rate on preparative scale gas-chromatographic separations of saturated methyl esters, *A. Rose, D. J. Royer, and R. S. Henly.* Support-coated open tubular columns, V. columns with various liquid-phase loadings, *L. S. Ettre, J. E. Purcell, and K. Billeb.* Gas-chromatographic behavior of pretreated silica gels, *E. K. Hurley, M. F. Burke, J. E. Heveran, and L. B. Rogers.* Introduction of gas-chromatographic samples to a mass spectrometer, *W. H. McFadden.* Observed plate height in TLC, *G. H. Stewart.* Thin-layer chromatography of the *N*-substituted maleimides on alumina: Stereochemical factors, *J. V. Dichiaro, R. A. Bate, and R. A. Keller.* An automated system for sample collection and computer analysis of thin-layer radiochromatograms, *F. Snyder and D. Smith.* Quantitative analysis by liquid chromatography, *A. Karmen.* Adduct formation with metal chelates involved in liquid–liquid extraction, *Q. Fernando.* Studies in solvent sublation: Extraction of methyl orange and rhodamine B, *B. L. Karger, A. B. Caragay, and S. B. Lee.*

## LODDING *Gas Effluent Analysis*

*(Thermal Analysis Series, Volume 1)*

edited by WILLIAM LODDING, *Rutgers—The State University, New Brunswick, New Jersey*

232 pages, illustrated. 1967

CONTENTS: Principles and general instrumentation, *W. Lodding.* Thermal conductivity detectors and their application to some thermodynamic and kinetic measurements, *T. R. Ingraham.* Mass spectrometric identification of gaseous products from thermal analysis, *H. G. Langer and R. S. Gohlke.* Pyrolysis–gas chromatography effluent analysis, *B. Groten.* Selective sorption and condensation of effluent gases, *W. Lodding.* Analysis of gas effluent streams by infrared absorption, *M. J. D. Low.* Thermoparticulate analysis, *C. B. Murphy.*

## MANN and BARNES *Electrochemical Reactions in Nonaqueous Systems*

*(Monographs in Electroanalytical Chemistry and Electrochemistry Series)*

by CHARLES K. MANN, *Florida State University, Tallahassee,* and KAREN K. BARNES, *St. Andrews College, Laurinberg, North Carolina*

576 pages, illustrated. 1970

A reference and guide to the literature for organic and inorganic nonaqueous electroanalytical reactions. Includes an introductory chapter on interpretation of electroanalytical measurements which facilitates the use of the book by chemists who do not have extensive backgrounds in electrochemistry or electroanalytical chemistry. Valuable to research chemists in electrochemistry, electroanalytical chemistry, physical, organic, and inorganic chemistry.

CONTENTS: Interpretation of electrochemical measurements • Reduction of hydrocarbons • Hydrocarbon oxidation • Anodic decarboxylation • Anodic substitution • Carbonyl compounds • Reactions of the carbon-halogen bond • The carbon-oxygen single bond • Amines, amides, and ammonium salts • Heterocyclic aromatic and other nitrogen compounds • Nitro and nitroso compounds • Organosulfur compounds • Organometallic compounds • Inorganic compounds.

## MATTSON, MARK, and MacDONALD *Computer Fundamentals for Chemists*

*(Computers in Chemistry and Instrumentation Series, Volume 1)*

edited by JAMES S. MATTSON, *Rosenstiel School of Marine and Atmospheric Sciences, University of Miami, Florida,* HARRY B. MARK, JR., *Department of Chemistry, University of Cincinnati, Ohio,* and HUBERT C. MACDONALD, JR., *Koppers Company, Inc., Monroeville, Pennsylvania*

368 pages, illustrated. 1973

Discusses the basic general principles and theories of electronic computers and their applications in the physical sciences. Individual chapters deal with specific applications to computation, data reduction, simulation, and instrumentation.

CONTENTS: Introduction to computers, *J. Kozak.* Basic principles of the electronic analog computer, *H. Mark, Jr.* Basic principles of digital circuitry, *R. Sacks.* Programming languages, *C. Thomas.* Simulation techniques, *V. LoDato.* Analog response by digital computers, *J. Sellers.* Interfacing experiments to computers, *D. Means.* Learning machines, *T. Isenhour and P. Jurs.*

## MATTSON, MARK, and MacDONALD *Electrochemistry: Calculations, Simulation, and Instrumentation*

*(Computers in Chemistry and Instrumentation Series, Volume 2)*

edited by JAMES S. MATTSON, *Rosenstiel School of Marine and Atmospheric Sciences, University of Miami, Florida,* HARRY B. MARK, JR., *University of Cincinnati, Ohio,* and HUBERT C. MACDONALD, JR., *Koppers Company, Inc., Monroeville, Pennsylvania*

488 pages, illustrated. 1972

Examines applications of computers to a wide variety of problems in electroanalytical chemistry and details how the computer and/or computer-based electronic circuitry has been applied to specific electrochemical problems. Of particular interest to analytical and electroanalytical chemists, this volume is of value to *all* graduate students, and industrial and academic research scientists who can use the techniques described in their own fields.

CONTENTS: Thermodynamic analysis of electrocapillary data, *P. R. Mohilner and D. M. Mohilner.* A program to calculate the relative surface excess of a substance adsorbed at the surface of a mercury electrode from differential capacitance data, *H. C. MacDonald, Jr.* Application of computers to solution of organic electrode reaction mechanisms, *H. B. Herman.* Computer analysis of data obtained by electrochemical transient perturbation techniques, *R. F. Martin and D. G. Davis.* Numerical solution of integral equations, *R. S. Nicholson and M. L. Olmstead.* Laplace plane analysis of electrode kinetics, *A. A. Pilla.* Digital simulation of electrochemical surface boundary phenomena: Multiple electron transfer and adsorption, *S. W. Feldberg.* Digital simulation of the rotating ring-disk electrode, *K. B. Prater.* Digital simulations of electrogenerated chemiluminescence at the rotating ring-disk electrode, *J. T. Maloy.* Operational amplifier instruments for electrochemistry, *R. R. Schroeder.* Introduction to the on-line use of computers in electrochemistry, *R. A. Osteryoung.* Applications of on-line digital computers in ac polarography and related techniques, *D. E. Smith.* Enhancement of electroanalytical measurement techniques by real-time computer interaction, *S. P. Perone.*

## MATTSON, MARK, and MacDONALD *Spectroscopy and Kinetics*

*(Computers in Chemistry and Instrumentation Series, Volume 3)*

edited by JAMES S. MATTSON, *Rosenstiel School of Marine and Atmospheric Sci-*

*(continued)*

**MATTSON, MARK, and MacDONALD** *(continued)*

*ences, University of Miami, Florida,* HARRY B. MARK, JR., *Department of Chemistry, University of Cincinnati, Ohio,* and HUBERT C. MACDONALD, JR., *Koppers Company, Inc., Monroeville, Pennsylvania*

352 pages, illustrated. 1973

The third volume in a series designed to cover the various computer applications to all areas of chemical research and quantitative measurement. Discusses the applications of computers to problems in spectroscopy and chemical kinetics. Of particular interest to analytical chemists, and also of value to all graduate students and industrial and academic researchers in physical chemistry, spectroscopy, and biochemistry.

CONTENTS: Application to magnetic resonance spectroscopy: The use of the large computer in chemical instrumentation, *L. Newman.* The calculation of the optical properties of thin metal films from internal reflectance data, *E. Randall and H. B. Mark, Jr.* Applications of computer circuitry and techniques to kinetic methods of analysis, *S. Crouch.* Analog computer simulation of kinetic models, *J. Janata.* The application of the Monte Carlo method to chemical kinetics, *J. Manock.* Integration of complex rate equations using infinite series, *N. Peterson and H. Butcher.*

## PERRY and VAN OSS *Separation and Purification Methods*

*(Book Edition)*

edited by EDMOND S. PERRY, *Eastman Kodak Company, Rochester, New York,* and CAREL J. VAN OSS, *Department of Microbiology, School of Medicine, State University of New York, Buffalo*

**Vol. 1** 512 pages, illustrated. 1973

Devoted to keeping the scientist informed on new developments in the field of separations. Covers all areas of separation and purification techniques and methods, ranging from new laboratory methods to chemical engineering practice, and dealing with inorganic, organic, and biological substances. Of particular interest to biochemists, biophysicists, microbiologists, chemists, physicists, and chemical engineers.

CONTENTS:

**Volume 1:** Recovery of styrene with silver fluoborate solutions, *W. Featherstone.* Reverse osmosis separations with aromatic polyamide films and hollow fibers, *R. McKinney, Jr.* Foam separation methods, *P. Somasundaran.* Hydrocarbon separation using ligand exchange reactions, *G. Davis and E. Makin, Jr.* Vapor phase extraction – a new purification method, *S. Gibbins.* Multiple extraction by Slurry systems, *E. Fuller.* Studies in melt crystallization, *G. Atwood.* Clathrates in separation processes, *E. Makin.* The principles and practice of gas-solid chromatography with salt-modified adsorbents, *J. Okamura and D. Sawyer.*

## PICKERING *Modern Analytical Chemistry*

by WILLIAM F. PICKERING, *The University of New Castle, New South Wales, Australia*

640 pages, illustrated. 1971

"This is a powerful new approach adopted by Pickering—one which demands attention. The material is very well organized and the writing is clear and easy to read. . . . I heartily recommend Professor Pickering's new approach to all educators in analytical chemistry."—Ralph N. Adams, *Journal of Chemical Education.*

A concise text and reference book suitable for a one- or two-semester undergraduate course in modern methods of chemical analysis at junior colleges, four-year colleges, or universities. Creates a new method of teaching analytical chemistry. Treatment of subject matter is comprehensive and places a distinct emphasis on fundamental principles.

CONTENTS: Introduction • Modern methods of chemical analysis • Selection of chemical methods • Factors influencing the accuracy of results • Thermal transformations • Radiation emission and radioactivity • The absorption of radiation • Interactions with magnetic fields • Reflection, refraction, and rotation of light • Structure • Basic solution chemistry • Ionic reactions – titrimetric analysis • Ionic reactions – selective procedures and kinetics • Electrical transformations • Adsorption, diffusion, and ion exchange • Heterogeneous equilibria • The challenge of automation.

## ROBINSON *Atomic Absorption Spectroscopy*

by JAMES W. ROBINSON, *Louisiana State University, Baton Rouge*

216 pages, illustrated. 1966

CONTENTS: Introduction • Equipment • Analytical parameters • Analytical applications • Topics related to atomic absorption: oscillator strength, vacuum ultraviolet, and atomic fluorescence.

**ROBINSON** ***Undergraduate Instrumental Analysis***

by JAMES W. ROBINSON, *Louisiana State University, Baton Rouge*

400 pages, illustrated.

*(Second edition)*

A textbook designed for the student who is a non-chemistry major. Extremely useful for the engineering, physics, biology, nursing, or pre-med student who is taking instrumental analysis.

CONTENTS: Concepts of analytical chemistry • Introduction to spectroscopy • Concepts of spectroscopy • Nuclear magnetic resonance • Infrared absorption • Ultraviolet molecular absorption spectroscopy • Atomic absorption spectroscopy • Colorimetry (Spectrophotometry) and polarimetry • Flame photometry • Emission spectrography • X-ray spectroscopy • Chromatography • Thermal analysis • Mass spectrometry • Electrochemistry.

**RYAN** ***Contemporary Activation Analysis***

edited by VICTOR A. RYAN, *The University of Wyoming, Laramie*

in preparation. 1974

CONTENTS (partial): Introduction, *V. Ryan.* Prompt activation analysis in neutron fluxes, *V. Ryan.* Time-of-flight neutron spectroscopy in prompt activation analysis, *M. Peisach.* Lunar and planetary activation analysis, *J. Reed.* Particle identification and particle spectroscopy in prompt activation analysis, *R. Chanda.* Gamma ray spectroscopy: detection and computation. A review, *R. Perkins. In vivo* activation analysis, *H. Palmer.*

**SACKS and MARK** ***Simplified Circuit Analysis:*** *Digital-Analog Logic*

by RICHARD D. SACKS, *University of Michigan, Ann Arbor, and* HARRY B. MARK, JR., *University of Cincinnati, Ohio*

176 pages, illustrated. 1972

A text designed to introduce the reader to the basic electronic principles incorporated in the analog-digital computing system. Acquaints him with the salient features, scope, basic methodology, capabilities, and limitations of experiment-oriented analog-digital computing systems for high speed data acquisition and experiment control. Extremely useful for senior undergraduate and graduate chemistry majors and also benefits students and researchers in any nonelectrical engineering field who use electronic instrumentation in their work.

CONTENTS: **Basic Principles of the Electronic Analog Computer:** Introduction • The Laplace transform method of analysis of passive networks • The operational amplifier and its application as a computing element • The analog computer. **Basic Principles of Digital Circuitry:** Introduction • Number systems, codes, and binary arithmetic • Computer logic • Solid-state devices and circuits • Memory devices and applications.

**SLADE and JENKINS** ***Thermal Analysis***

*(Techniques and Methods of Polymer Evaluation Series, Volume 1)*

edited by PHILIP E. SLADE, JR., *Monsanto Company, Pensacola, Florida,* and LLOYD T. JENKINS, *Chemstrand Research Center, Durham, North Carolina*

264 pages, illustrated. 1966

CONTENTS: Instrumentation, techniques, and applications of differential thermal analysis, *E. M. Barrall, II and J. F. Johnson.* Transition temperatures by differential thermal analysis, *D. David.* Instrumentation, techniques, and applications of thermogravimetry, *H. C. Anderson.* Quantitative calculations in thermogravimetric analysis, *C. D. Doyle.* Effluent-gas analysis, *A. S. Kenyon.*

**SLADE and JENKINS** ***Thermal Characterization Techniques***

*(Techniques and Methods of Polymer Evaluation Series, Volume 2)*

edited by PHILIP E. SLADE, JR., *Monsanto Company, Pensacola, Florida,* and LLOYD T. JENKINS, *Chemstrand Research Center, Durham, North Carolina*

384 pages, illustrated. 1970

Most comprehensive work on the subject and contains a complete and up-to-date review of the literature. Of interest to the technician in polymer chemistry research and industrial production.

CONTENTS: Differential scanning calorimetry theory and applications, *E. M. Barrall, II and J. F. Johnson.* Pyrolysis–gas chromatographic techniques for polymer identification, *G. M. Brauer.* Stress-strain temperature relations in high polymers, *J. A. Sauer and A. E. Woodward.* Torsional braid analysis: A semimicro thermomechanical approach to polymer characterization, *J. K. Gillham.* Thermal conductivity of polymers, *D. E. Kline and D. Hansen.* Electrothermal analysis of polymers, *D. Seanor.*

**SNYDER** ***Principles of Adsorption Chromatography:*** *The Separation of Nonionic Organic Compounds*

*(Chromatographic Science Series, Volume 3)*

by LLOYD R. SNYDER, *Union Oil Company of California, Brea*

432 pages, illustrated. 1968

CONTENTS: Introduction • The chromatographic process and techniques of separation • General aspects of adsorption • Importance of sample size in adsorption chromatography. Isotherm linearity • Bed efficiency. Bandwidth versus separation conditions • General role of adsorbent type and activity • Individual adsorbents • Role of the solvent • Gas-solid chromatography • Role of sample structure. Primary effects • Role of sample structure. Secondary effects • Separation temperature as a variable • Some related topics.

**STEVENS** ***Characterization and Analysis of Polymers by Gas Chromatography***

*(Techniques and Methods of Polymer Evaluation Series, Volume 3)*

by MALCOLM P. STEVENS, *American University of Beirut, Lebanon*

216 pages, illustrated. 1969

A short, up-to-date monograph applying the techniques of gas chromatography directly to polymers. Of interest to polymer, paint, and coatings chemists.

CONTENTS: Introduction • Analysis of volatile materials in polymers • Characterization by chemical degradation • Characterization by thermal degradation • Miscellaneous methods of analysis and characterization • Monomer purity.

**VEILLON and WENDLANDT**
***Handbook of Commercial Scientific Instruments***

a series edited by CLAUDE VEILLON and WESLEY W. WENDLANDT, *University of Houston, Texas*

**Vol. 1** ***Atomic Absorption***
(by CLAUDE VEILLON)
192 pages, illustrated. 1972

**Vol. 2** ***Thermoanalytical Techniques***
(by WESLEY W. WENDLANDT)
176 pages, illustrated. 1973

A series which presents detailed information on various groups of scientific instruments commercially available in the United States. Each volume describes as fully as possible the instrumentation available for a particular field. Specifications, descriptions, schematic drawings, photographs, approximate prices, features, and accessories are included for each instrument. The volumes are indispensable tools for all scientists and technicians in academic, industrial, or government institutions and laboratories.

CONTENTS:

**Volume 1** ***(Atomic Absorption):*** Introduction • Bausch & Lomb • Beckman • Bendix • Coleman • Corning • Heath • Hilger • Instrumentation Laboratory • Jarrell-Ash • Jobin-Yvon • OCLI • Optica • Perkin-Elmer • Philips • Spectrametrics • Varian • Zeiss • Summary.

**Volume 2** ***(Thermoanalytical Techniques):*** Differential thermal analysis (DTA)–Introduction; difference between DTA and DSC; Introduction to commercial DTA instruments • Thermobalances–introduction; commercial thermobalances; Non-Cahn type thermobalances.

**VIJH** ***Electrochemistry of Metals and Semiconductors***

*(Monographs in Electronanalytical Chemistry and Electrochemistry Series)*

by ASHOK K. VIJH, *Hydro-Quebec Institute of Research, Varennes, Quebec*

336 pages, illustrated. 1973

Surveys those areas of electrochemistry on which solid state materials science has an important bearing, and deals with a variety of electrode reactions, focusing on the role of the solid in the kinetics of charge transfers at solid-electrolyte interfaces. Directed to investigators and technicians in the areas of electrochemistry, electroanalytical chemistry, physical chemistry, corrosion science, energy science, and materials science.

CONTENTS: Introduction – the nature of solid materials • Electrode reactions on semiconductors: General principles • Electrode reactions on elemental semiconductors • Electrode reactions on bulk compound semiconductors • Anodic oxidation of metals and non-metals • Features of electrode reactions on metals covered with solid surface films • Relationship between solid state properties and electrochemical behavior of metals • Electrochemistry in the solid state • A synopsis of the industrial impact of electrochemistry – solid state interdisciplinary area.

**WAGNER and HULL** ***Inorganic Titrimetric Analysis:*** *Contemporary Methods*

*(Treatise on Titrimetry Series, Volume 1)*

by WALTER WAGNER, *University of Detroit, Michigan,* and CLARENCE J. HULL, *Indiana State University, Terre Haute*

240 pages, illustrated. 1971

The first conveniently sized single volume to be published in the last decade which provides the practicing chemist with a rapid, handy, up-to-date guide to modern titrimetric methods. Affords sufficient practical information for the reader to determine whether a given method of titrimetric analysis suits his needs in terms of applicability, specificity, precision, and accuracy, as well as the required reagents and instruments. An indispensable aid for graduate students, laboratory directors, and practicing chemists engaged in research, development, analysis, manufacture, process control, and other areas.

CONTENTS: Synoptic survey and outline of recommended contemporary methods of titrimetric analysis for the elements of groups 0 through VIII.

**WALTON and REYES** ***Modern Chemical Analysis and Instrumentation***

by HAROLD F. WALTON, *Department of Chemistry, University of Colorado, Boulder,* and JORGE REYES, *University of Trujillo, Peru*

440 pages, illustrated. 1973

A combined textbook and laboratory manual designed to provide the beginning chemistry student with practical experience in the use of analytical instruments.

CONTENTS: Potentiometry and the pH meter • Coulometric and conductimetric titration • Electrogravimetry • Polarography • The absorption of radiation • Visible and ultraviolet spectrometry • Infrared spectroscopy • Atomic absorption spectroscopy • Chromatography • Paper and thin-layer chromatography • Gas chromatography.

**WHITE and ARGAUER**
***Fluorescence Analysis:*** *A Practical Approach*

by CHARLES E. WHITE, *Department of Chemistry, University of Maryland, College Park,* and ROBERT J. ARGAUER, *United States Department of Agriculture, Beltsville, Maryland*

400 pages, illustrated. 1970

Provides a readily available source for the basic theory and practice of fluorescence and illustrates the various disciplines where fluorescence trace analysis is used in the fields of chemistry, biology, and agriculture. Of use to teachers; graduate and undergraduate students in chemistry; research scientists in catalyst development, air and water pollution, pesticides, and drug analysis; and research workers in pharmaceutical, medical, and clinical chemistry.

CONTENTS: Introduction and nature of fluoresence • Apparatus • Correction of excitation and emission spectra • Fluorescent metal chelates • Quantitative fluorescence methods for metals and nonmetals • Fluorescent indicators • Fluorescence in qualitative inorganic analysis • Spectrofluorometry with column, gas, paper and thin-layer chromatography • Application of fluorescence in agriculture • Aflatoxins • Proteins and amino acids • Polynuclear aromatic hydrocarbons • Organic compounds • Vitamins • Steroids • Chemiluminescence • X-ray fluorescence and atomic resonance fluorescence • Clinical fluorometric procedures.

## OTHER BOOKS OF INTEREST

**ALTGELT and SEGAL**
***Gel Permeation Chromatography***

edited by KLAUS H. ALTGELT, *Chevron Research Company, Richmond, California,* and LEON SEGAL, *South Regional Research Laboratory, U.S.D.A., New Orleans, Louisiana*

672 pages, illustrated. 1971

**ANDERSON** ***The Raman Effect***

**In 2 Volumes**

edited by ANTHONY ANDERSON, *University of Waterloo, Ontario*

**Vol. 1** ***Principles***
416 pages, illustrated. 1971

**Vol. 2** ***Applications***
640 pages, illustrated. 1973

**CIACCIO** ***Water and Water Pollution Handbook***

edited by LEONARD L. CIACCIO, *School of Theoretical and Applied Sciences, Ramapo College of New Jersey, Mahwah*

**Vol. 1** 480 pages, illustrated. 1971

**Vol. 2** 368 pages, illustrated. 1971

**Vol. 3** 528 pages, illustrated. 1972

**Vol. 4** 512 pages, illustrated. 1973

**HELFFERICH and KLEIN**
***Multicomponent Chromatography:***
*Theory of Interference*

*(Chromatographic Science Series, Volume 4)*

by FRIEDRICH HELFFERICH, *Shell Development Co., Emeryville, California,* and GERHARD KLEIN, *University of California, Richmond*

432 pages, illustrated. 1970

**LEVINE and DEMARIA** ***Lasers:***
*A Series of Advances*

edited by ALBERT K. LEVINE, *Division of Science and Engineering, Richmond College, Staten Island, New York,* and ANTHONY J. DEMARIA, *Quantum Physics Laboratory, United Aircraft Corporation, East Hartford, Connecticut*

**Vol. 1** 384 pages, illustrated. 1966

**Vol. 2** 456 pages, illustrated. 1968

**Vol. 3** 384 pages, illustrated. 1971

**price to be announced*

**MARTIN** ***Marine Chemistry***

**In 2 Volumes**

by DEAN F. MARTIN, *Department of Chemistry and Marine Science Institute, University of South Florida, Tampa*

**Vol. 1** ***Analytical Methods***
(*Second Edition, revised and expanded*)
408 pages, illustrated. 1972

**Vol. 2** ***Theory and Applications***
464 pages, illustrated. 1970

**PESCE, ROSÉN, and PASBY**
***Fluorescence Spectroscopy:***
*An Introduction for Biology and Medicine*

by AMADEO J. PESCE, *Michael Reese Hospital and Medical Center, Chicago,* CARL-GUSTAF ROSÉN, *Royal University of Stockholm,* TERRY PASBY, *University of Illinois, Urbana*

264 pages, illustrated. 1971

**SACKS** ***Measurements and Instrumentation in the Chemical Laboratory***

by RICHARD D. SACKS, *Department of Chemistry, University of Michigan, Ann Arbor*

in preparation. 1973

## ENCYCLOPEDIAS OF INTEREST

### ENCYCLOPEDIA OF THE ELECTROCHEMISTRY OF THE ELEMENTS

editor: ALLEN J. BARD, *University of Texas at Austin*

This new multi-volume *Encyclopedia* presents a critical review and a compilation of the descriptive electrochemistry of the elements and their compounds. Each chapter includes tables of available thermodynamic and kinetic data, a listing of known electrochemical reactions, and a discussion of the mechanism of these reactions, when known, for the element in question and its compounds. A notable feature of the *Encyclopedia* is a section on the applied electrochemistry of the elements and their compounds, and their use in electrochemical devices. Also included are references to the original literature on all data and reactions. This compilation provides a starting point for new electrochemical investigations and suggests the areas in which further research is necessary. Periodic supplements are planned in order to make available to the reader the most current information in the field.

*Encyclopedia of the Electrochemistry of the Elements* will be of utmost interest to graduate students and researchers in electrochemistry, analytical, physical, inorganic, and organic chemistry, chemical engineering, and corrosion science.